HIGH-THROUGHPUT LEAD OPTIMIZATION IN DRUG DISCOVERY

CRITICAL REVIEWS IN COMBINATORIAL CHEMISTRY

Series Editors

BING YAN
School of Pharmaceutical Sciences
Shandong University, China

ANTHONY W. CZARNIK
Department of Chemistry
University of Nevada–Reno, U.S.A.

A series of monographs in molecular diversity and combinatorial chemistry, high-throughput discovery, and associated technologies.

Combinatorial and High-Throughput Discovery and Optimization of Catalysts and Materials
Edited by Radislav A. Potyrailo and Wilhelm F. Maier

Combinatorial Synthesis of Natural Product-Based Libraries
Edited by Armen M. Boldi

High-Throughput Lead Optimization in Drug Discovery
Edited by Tushar Kshirsagar

HIGH-THROUGHPUT LEAD OPTIMIZATION IN DRUG DISCOVERY

Edited by

Tushar Kshirsagar

CRC Press
Taylor & Francis Group
Boca Raton London New York

CRC Press is an imprint of the
Taylor & Francis Group, an **informa** business

CRC Press
Taylor & Francis Group
6000 Broken Sound Parkway NW, Suite 300
Boca Raton, FL 33487-2742

First issued in paperback 2019

ISBN-13: 978-0-8493-7268-1 (hbk)
ISBN-13: 978-0-367-38771-6 (pbk)

Library of Congress Cataloging-in-Publication Data

High-throughput lead optimization in drug discovery / [edited by] Tushar Kshirsagar.
    p. ; cm. -- (Critical reviews in combinatorial chemistry)
  "CRC title."
  Includes bibliographical references and index.
  ISBN 978-0-8493-7268-1 (hardback : alk. paper)
  1. High throughput screening (Drug development) 2. Combinatorial chemistry. 3. Drug development. I. Kshirsagar, Tushar. II. Series.
  [DNLM: 1. Combinatorial Chemistry Techniques. 2. Drug Design. 3. Databases, Genetic. 4. Gene Library. QV 744 H6367 2007]

RS419.5.H537 2007
615'.19--dc22   2007035804

Visit the Taylor & Francis Web site at
http://www.taylorandfrancis.com

and the CRC Press Web site at
http://www.crcpress.com

Contents

Preface

The end of the previous millennium saw an explosion in the application of parallel synthesis techniques for making compounds for high-throughput screening (HTS). The ability to screen thousands of compounds in a short period of time, along with the ability to manage the voluminous data, led to the need for large collections of compounds. An enormous amount of effort was spent in increasing the size of screening collections. While both solid- and solution-phase techniques were employed, previously gained knowledge from peptide synthesis, the ability to use split and pool techniques, and the opportunity to use a large excess of reagents to drive reactions to completion tipped the balance in favor of solid-phase chemistry. Over time, it became clear that more thought in the design phase of library development was necessary to generate high-quality hits. Better-designed "drug-like" compounds were deemed better for the overall compound deck — even at the cost of numbers. In addition to the design component, the ability to monitor and manage the quality of the library by controlling the purity and quantity of final compounds was instrumental in generating better compound collections for HTS. The application of high-throughput purification, coupled with a better understanding of storage methods, has resulted in a shift to screening high-purity libraries against a broadening range of biological systems.

More recently, we have seen a marked shift in the use of parallel synthesis techniques for applications beyond screening collections. Our internal strategy, and that of others, has been to obtain sets of screening compounds through specialized suppliers and concentrate our internal synthetic efforts toward ongoing medicinal chemistry programs. In particular, we have employed both solid- and solution-phase parallel synthesis techniques (primarily the latter) toward three primary goals. The first has been in the area of hit-to-lead optimization for established medicinal chemistry programs. While this effort was instrumental in moving compounds forward through our development process, it has also been extremely useful in rapidly expanding our intellectual property space. The second application of parallel synthesis techniques in our internal programs has been in the design and synthesis of targeted libraries toward new early discovery projects. The goal of this effort was to enrich our compound collection with pharmacophores biased toward the biological targets of interest. The third area in which parallel synthesis has been employed is in collaboration with computation chemists to support virtual screening programs with the synthesis of small targeted sets of compounds for hypothesis evaluation.

This book attempts to compile a series of optimization projects giving the reader real-world examples of both successes as well as challenges in the application of parallel synthesis for lead optimization. In Chapter 1, Morales assembles examples from the first half of this decade and categorizes them under different classes of biological targets. It is clear from these examples that every major target class has examples of the use of small targeted libraries to aid medicinal chemistry efforts. In Chapter 2, Jennings has described parallel synthesis approaches in combination with structure-based design to afford molecules that inhibit protein–protein interactions. In their program, the data derived from x-ray co-crystal structures directed the design of targeted libraries. Chapter 3, by Ernst and Obrecht, gives examples of three targets that were explored using targeted libraries, including efforts toward the design of new insecticides and fungicides that expands the role of parallel synthesis beyond traditional drug targets. In Chapter 4, Ranise, Spallarossa, and Cesarini expand the use of medicinal chemistry design concepts in the synthesis of HIV reverse transcriptase inhibitors. They employed one-pot reaction methodology to synthesize a range of analogs that could be isolated with minimal purification. In Chapter 5, Li describes two case studies: one related to the development of novel ALK inhibitors and the other related to the development of melanocortin receptor agonists. In the first case, Li's team used homology modeling while in the second case they employed information from the natural ligand to aid the design of their libraries. Pei, Moos, and Ghosh took information from a highly flexible natural product inhibitor of adenine nucleotide translocase to design their libraries in Chapter 6. In Chapter 7, Reid and Fairlie present three case studies where they employed L- and D-amino acids for the synthesis of anticancer, antiviral, and anti-inflammatory compounds. Kundu has described three case studies for the optimization of peptides, natural products, and glycoconjugates for a range of targets in Chapter 8. These compounds impact targets that could be useful in the development of antifungals, antimalarials, and antifilarials. In Chapter 9, Sun and Lee describe the synthesis of nucleoside-based libraries as potential antibiotics. They have employed solid-phase chemistry using IRORI technology.

The book draws attention to several case studies that encompass a range of different biological targets for application in different therapeutic areas. In addition, both solid- and solution-phase techniques are described for the synthesis of directed libraries. The authors also highlight design principles used to direct the choice of templates and diversity elements. I hope that this book will be useful to those in the field, as well as to students learning about the application of parallel synthesis techniques to medicinal chemistry.

Tushar Kshirsagar

Contributors

Sara Cesarini
Dipartimento di Scienze Farmaceutiche
Università di Genova
Genova, Italy

Alexander Ernst
Polyphor AG
Allschwil, Switzerland

David P. Fairlie
Institute for Molecular Bioscience
Centre for Drug Design and Development
University of Queensland
Brisbane, Australia

Soumitra Ghosh
MitoKor, Inc.
San Diego, California

Lee D. Jennings
Wyeth Research
Pearl River, New York

Bijoy Kundu
Medicinal Chemistry Division
Central Drug Research Institute
Lucknow, India

Richard E. Lee
Department of Pharmaceutical Sciences
The University of Tennessee Health Science
  Center
Memphis, Tennessee

Rongshi Li
High-Throughput Medicinal Chemistry
ChemBridge Research Laboratories
San Diego, California

Walter H. Moos
MitoKor, Inc.
San Diego, California

Guillermo A. Morales
Semafore Pharmaceuticals
Indianapolis, Indiana

Daniel Obrecht
Polyphor AG
Allschwil, Switzerland

Yazhong Pei
MitoKor, Inc.
San Diego, California

Angelo Ranise
Dipartimento di Scienze Farmaceutiche
Università di Genova
Genova, Italy

Robert C. Reid
Institute for Molecular Bioscience
Centre for Drug Design and Development
University of Queensland
Brisbane, Australia

Andrea Spallarossa
Dipartimento di Scienze Farmaceutiche
Università di Genova
Genova, Italy

Dianqing Sun
Department of Pharmaceutical Sciences
The University of Tennessee Health Science
  Center
Memphis, Tennessee

1 Introduction and Review

Guillermo A. Morales

CONTENTS

1.1   INTRODUCTION

Combinatorial chemistry has evolved greatly since its conception by Merrifield in 1963 where the first solid-phase synthesis of peptides was reported.[1] For the next 30 years, the primary use of solid-phase chemistry remained dedicated to peptide synthesis, although there were some efforts dedicated to the development of functionalized resins for their use in solution-phase synthesis. However, it was not until the early 1990s that combinatorial chemistry moved into the area of small molecule synthesis when the first publication of a solid-phase combinatorial synthesis of benzodiazepines appeared.[2] Industry immediately noticed this breakthrough, embraced it, and began investing more resources into the development of combinatorial chemistry to expedite the drug discovery process. As a result, from the mid-1990s to the end of the last decade, an explosion of reports appeared in the literature where traditional solution-phase methodologies were ported and/or adapted to solid-phase combinatorial chemistry.

During this time, more novel functionalized polymer supports or highly fluorinated tags were developed and used as either catalysts or scavengers to expedite the synthesis or purification of reactions in solution phase. Also, the wide adaptation and use of microwave irradiation synthesis began to emerge, entering the toolbox used by chemists.

With the power to synthesize a large number of compounds in a relatively short period of time, combinatorial chemistry was sought and viewed as the solution to the drug discovery bottleneck of filling up industry's drug candidate pipelines. Unfortunately, these high expectations did not materialize, and industry found itself with large collections of compounds and still no new drug candidates.

One of the main problems during the fast evolution of combinatorial chemistry was that efforts usually focused on what compounds could be made (quantity) instead of which ones should be made (quality). In 1997, Lipinski and colleagues[3] brought this important observation to the attention of the scientific community with the publication of a seminal report on the analysis of small-molecule drugs. In their report, experimental and computational studies were conducted to estimate the ideal physicochemical profile of drug candidates based on absorption and permeability observations.

This work set the foundation for what now is known as "the rule of 5" (a.k.a. RO5, R-o-5), which predicts that for a compound to exhibit good absorption or permeation, it should have (1) no more than five H-bond donors (sum of OHs and NHs), (2) no more than ten H-bond acceptors (sum of Os and Ns), (3) a molecular weight less than 500 Daltons (Da), and (4) a calculated Log P (cLogP) value of less than 5 (or MlogP < 4.15).[3–5]

Then in 2002 Veber and colleagues[6] reported a study refining the rules that a compound must meet to achieve good oral bioavailability. Analysis of rat oral bioavailability data indicated that a molecular weight cut-off of 500 and a cLogP < 5 were not enough to distinguish compounds with good oral bioavailability. Close observation of the molecular structures of the compound set revealed that both molecular flexibility and polar exposure (polar surface area) play a significant role in the ability of a compound to exhibit oral bioavailability. Their work led to two new rules for the prediction of good oral bioavailability: A compound must have (1) ten or fewer rotatable bonds, and (2) a polar surface area (PSA) equal to or less than 140 $Å^2$ (or twelve or fewer H-bond donors and acceptors).

These empirical rules are now widely accepted for the design of drug-like compounds. However, before any molecule reaches drug-like criteria, its roots must originate from a smaller, simpler, and optimizable chemical structure (hit) with lead-like characteristics. In the majority of cases, a biologically active chemical entity undergoes scrutiny to determine the effects that different substitution patterns have on the biological profile or activity of the initial hit (SAR elucidation). After each optimization cycle, the initial hit structure typically gains key additional units (e.g., substituents, fragments, etc.), which leads to a growth in molecular mass, size, rotatable bonds, etc. Consequently, to have "room for optimization" it is highly desirable to have a small, simple, low-molecular-weight active compound. Teague and colleagues[7] analyzed low-affinity leads that eventually graduated into drugs and concluded that molecules with a molecular weight = 100 to 350 Da and cLogP = 1 to 3 offer a good starting point for the hit-to-lead-to-drug optimization phase.

With these guidelines in hand, combinatorial chemistry began a paradigm transformation where it was no longer viewed as the drug-discovery magic bullet. The synthesis of large compound libraries and mixtures was no longer pursued at the level that it had been in the previous decade. Instead, combinatorial chemistry gained wider acceptance as a valuable research tool to be applied not as an all-compounds-or-nothing synthesis technique but as a use-it-when-appropriately-needed technique. At this point, R&D efforts centered on smaller but focused target-driven small-molecule libraries where all compounds were synthesized as singles, undergoing purification and thorough characterization prior to their biological screen and storage. Nowadays, the term "combinatorial chemistry" is commonly used as a synonym for solid- and solution-phase parallel synthesis, when functionalized polymers or highly fluorinated tags are used to catalyze reactions or trap impurities or desired products, and when microwave chemistry is used.

By the turn of the new millennium, a plethora of combinatorial chemistry synthetic methodologies had been developed for a broad range of chemical transformations. With a robust combinatorial chemistry methodology arsenal, established lead-like and drug-like guidelines, and the new combinatorial chemistry paradigm (quality over quantity), medicinal chemistry projects began adapting and using combinatorial chemistry not only for the hit-discovery process to produce high-quality hits, but also for the lead optimization process.

This chapter presents a review of examples reported in the first five years of the new millennium (2000 to 2005) where combinatorial chemistry was successfully applied primarily for the optimization of small-molecule, non-peptidic, biologically active compounds (i.e., SAR elucidation, target potency, target selectivity, pharmacokinetics profile, toxicity profile, etc.), although a few examples of peptide-like leads have been included to cover particular optimization strategies.

Because the focus of this review is the successful application of combinatorial chemistry in medicinal chemistry, the articles covered herein are organized based on biological targets.

1.2   ALDOLASE INHIBITORS

1.2.1   CASE 1: 7,8-DIHYDRONEOPTERIN ALDOLASE (DHNA)

Sanders et al.[8] at Abbott Laboratories investigated a lead compound (compound **1**) that exhibited single-digit low micromolar inhibition of 7,8-dihydroneopterin aldolase (DHNA; EC 4.1.2.25) for the development of antibacterial compounds. Bacteria cells, as opposed to human cells, cannot absorb folate from their surroundings for which they are forced to synthesize their own. Tetrahydrofolate is a key enzyme involved in several important pathways, such as purine and pyrimidine synthesis as well as nucleic and amino acid synthesis. Enzymes involved in the biosynthesis of tetrahydrofolate include dihydropteroate synthase (DHPS), dihydrofolate reductase (DHFR), and dihydroneopterin aldolase (DHNA; EC 4.1.2.25). DHNA catalyzes the conversion of 7,8-dihydroneopterin into glycoaldehyde and 6-hydroxymethyl-7,8-dihydropterin.

Using Abbott's CrystaLEAD x-ray crystallographic high-throughput screening technology, a method that relies on the ability of diffusing ligands into the active site of an enzyme in its crystal form,[9] a handful of hits were obtained. Crystal structure analysis of these complexes indicate that they all share a 2-aminopyrimidin-4(3H)-one scaffold (**1**) with a key three-point hydrogen bond motif. Using this motif, a 1,000-compound library was made, affording hits in the single micromolar range. From these hits it was noted that compound **2** formed a complex where its benzoic acid unit was oriented toward an accessible groove in the enzyme (Scheme 1.1). Taking advantage of this feature, the benzoic unit was used to analog this hit to access and probe the grove. For this, an amide library was synthesized via solution-phase chemistry using a LabTech Platform IV parallel synthesizer from

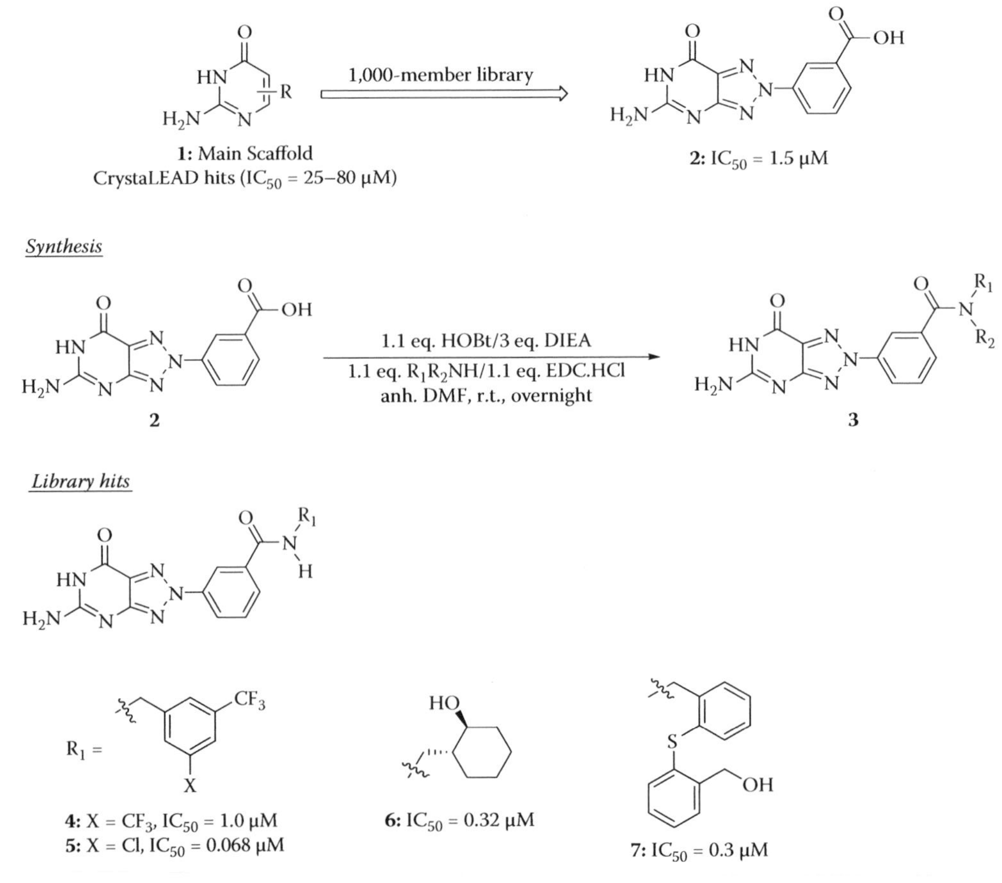

SCHEME 1.1   DHNA inhibitors.

Advanced ChemTech. A diverse set of hydrophobic (i.e., benzyl and biphenyl amines) and hydrophilic amines (i.e., amino alcohols) were used to afford several amides with sub-micromolar potency, particularly amide **5** with an $IC_{50} = 68$ nM. It is worth noting that the authors decided to use "odd-balls," in this case a benzyl amine containing a phenyl thioether (**7**) that was not expected to fit in the pocket, yet it turned out to have an $IC_{50} = 0.3$ μM. This is a good example of how using a few "odd-balls" in the synthesis of a library can afford unexpected results providing novel hits and more information on the target.

1.3 ANTIFUNGALS

1.3.1 CASE 1: HEXAPEPTIDES

Kundu et al.[10] at the Central Drug Research Institute in India published a classic study of using a positional scanning combinatorial chemistry peptide approach for the optimization of hexapeptides as antifungal agents.

Screening a peptide collection against *Candida albicans* (*Ca*) and *Cryptococcus neoformansa* (*Cn*), the hexapeptide His-D-Trp-D-Phe-Phe-D-Phe-Lys-NH$_2$ (compound **8**) was identified as a hit with $IC_{50} = 29.6$ μM and MIC $= 6.81$ μM, respectively. For this study it was decided to keep the D-amino acids at their original positions (2, 3, and 5); vary only the remaining 1, 4, and 6 positions; pool the libraries; and screen them as mixtures (**9**).

Using Rink Amide AM resin, the first iteration afforded a sublibrary where Arg at position 1 (compound **11**) exhibited good inhibition against *Ca* and *Cn*, with $IC_{50} = 84$ μM and MIC $= 28$ μM, respectively. For the second iteration, two sublibraries containing Arg and His in the first position were made. The resulting libraries were screened and the one with Arg at the first position with Ile (**12**: *Ca*: $IC_{50} = 15.8$ μM; *Cn*: MIC $= 14.1$ μM) and Leu (**13**: *Ca*: $IC_{50} = 18.0$ μM; *Cn*: MIC $= 14.1$ μM) were the most active. Based on these results, His at position 1 was dropped and a third sublibrary was made keeping Arg at position 1, and Ile and Leu at position 4.

The new sublibrary exploring position-6 was made. Although it was found that His, Lys, Leu, and Arg at position-6 (compound **14**) exhibited an MIC $= 6.7$ to 7.0 μM against *Cn*, His exhibited an $IC_{50} = 6.8$ μM against *Ca*, four times more potent than the initial lead hexapeptide **8**.

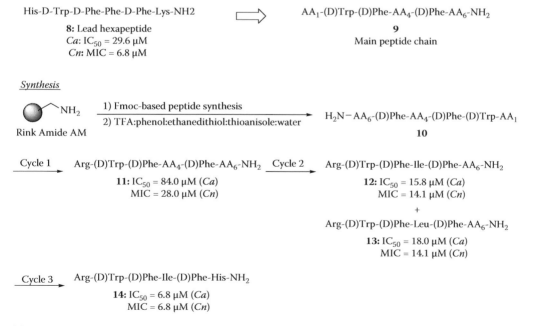

SCHEME 1.2 Antifungal hexapeptides.

1.3.2 CASE 2: 4-SUBSTITUTED, N-UNSUSBSTITUTED IMIDAZOLES

Saha et al.[11,12] at the Janssen Research Foundation recently reported the optimization of 4-substituted N-unsusbstituted imidazoles as antifungal agents. These compounds inhibit Cyt-P450 dependent sterol 14-α-demethylase leading toward ergosterol depletion, which in turn blocks the membrane synthesis process in fungal cells.

A large number of reported antifungals are triazole analogs. To move away from this compound series, the authors decided to explore an imidazole scaffold (**17**) with the goal of finding compounds that inhibit the synthesis of ergosterol with a pLADy inhibition value greater than 6.5.

Exploratory work was done immobilizing 1H-imidazole-4-carbaldehyde (**18**) on 2-chlorotrityl chloride polystyrene resin. A collection of 88 amines were anchored via reductive amination, and the resulting amines (**20**) treated with 12-15 sulfonyl (**21**) and acid chlorides to afford sulfonyl amides (**22**) and amides, respectively. From this exercise two lead compounds were found — sufonamides **26** and **27**. Molecular modeling studies of these compounds with Cyt-P450$_{cam}$ indicated that the 4-position of the benzene sulfonamide should be further explored.

A second generation of analogs was made using 4-bromo-benzenesulfonyl chloride and 4-(bromomethyl)benzene-1-sulfonyl chloride as building blocks. Using the bromo unit as a point of diversity, a series of amines, biaryls, and ethers was made affording compounds **28**, **29**, and **30** with an improved MIC profile against a panel of reference compounds Itraconazole (**15**) and Voriconazole (**16**) as well as the initial leads **26** and **27**.

1.3.3 CASE 3: TRIAZINES AND THIAZOLES

Altoper et al.[13] at Polyphor AG worked on the optimization of triazine leads **31** and **32**, which were found to have antifugal activity against *Phytophthora infestans*, *Pyrenophora teres*, and *Septoria nodum*.

A solution-phase approach was pursued where the reaction conditions were carried out and optimized in parallel focusing primarily on improving both yields and selective isomer formation. The triazine heterocycle was initially formed by condensation of (S)-methyl-isothiosemicarbazide (**36**) with glycoxals (**37**). This approach provided a route that allowed the introduction of four different functionalities at position 4 of the triazine scaffold: thiomethyl, methylsulfoxide, methylsulfone, and methoxide.

The synthetic route employed by the authors resulted in the production of triazine mixtures that required separation by flash chromatography. Selective oxidation of the thiomethyl group gave two sublibraries (sulfoxides and sulfones) that, in turn, allowed the introduction of the methoxy group.

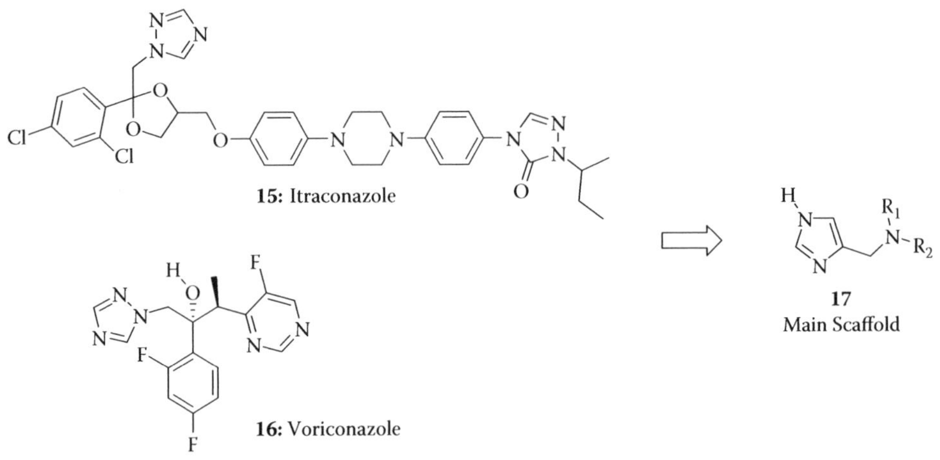

15: Itraconazole

16: Voriconazole

17
Main Scaffold

SCHEME 1.3 Antifungal imidazoles.

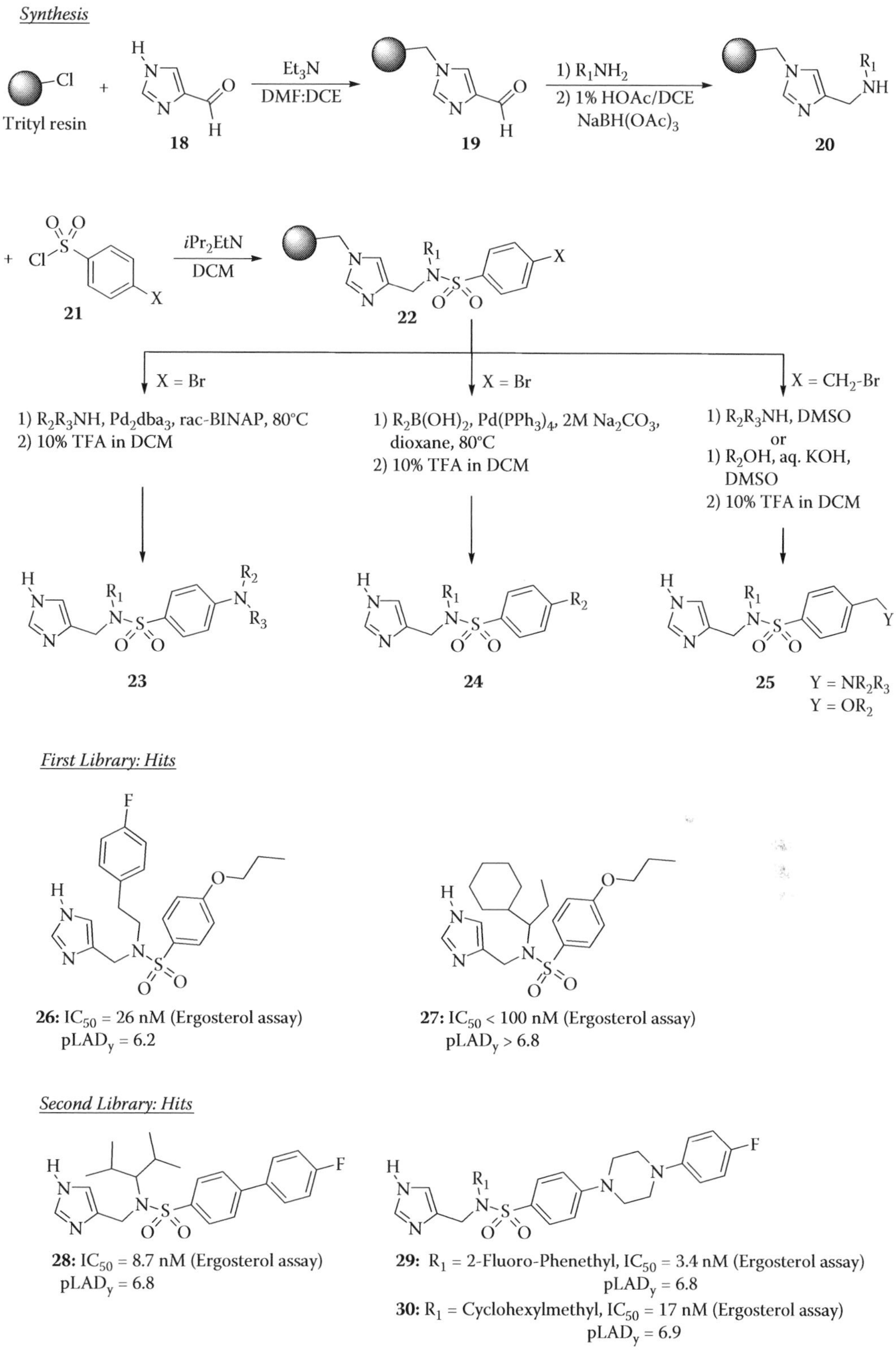
Synthesis
Trityl resin
Cl
Et3N
DMF:DCE
18
19
1) R1NH2
2) 1% HOAc/DCE
NaBH(OAc)3
20
iPr2EtN
DCM
21
22
X = Br
1) R2R3NH, Pd2dba3, rac-BINAP, 80°C
2) 10% TFA in DCM
X = Br
1) R2B(OH)2, Pd(PPh3)4, 2M Na2CO3,
dioxane, 80°C
2) 10% TFA in DCM
X = CH2-Br
1) R2R3NH, DMSO
or
1) R2OH, aq. KOH,
DMSO
2) 10% TFA in DCM
23
24
25
Y = NR2R3
Y = OR2
First Library: Hits
26: IC50 = 26 nM (Ergosterol assay)
pLADy = 6.2
27: IC50 < 100 nM (Ergosterol assay)
pLADy > 6.8
Second Library: Hits
28: IC50 = 8.7 nM (Ergosterol assay)
pLADy = 6.8
29: R1 = 2-Fluoro-Phenethyl, IC50 = 3.4 nM (Ergosterol assay)
pLADy = 6.8
30: R1 = Cyclohexylmethyl, IC50 = 17 nM (Ergosterol assay)
pLADy = 6.9

SCHEME 1.3 Antifungal imidazoles (continued).

Eventually, the authors refined the synthetic methodology to selectively make a triazine isomer. This work provides a good example of using parallel synthesis to optimize reaction conditions as well as to quickly provide sublibraries in a nonlinear approach with a reduced number of steps.

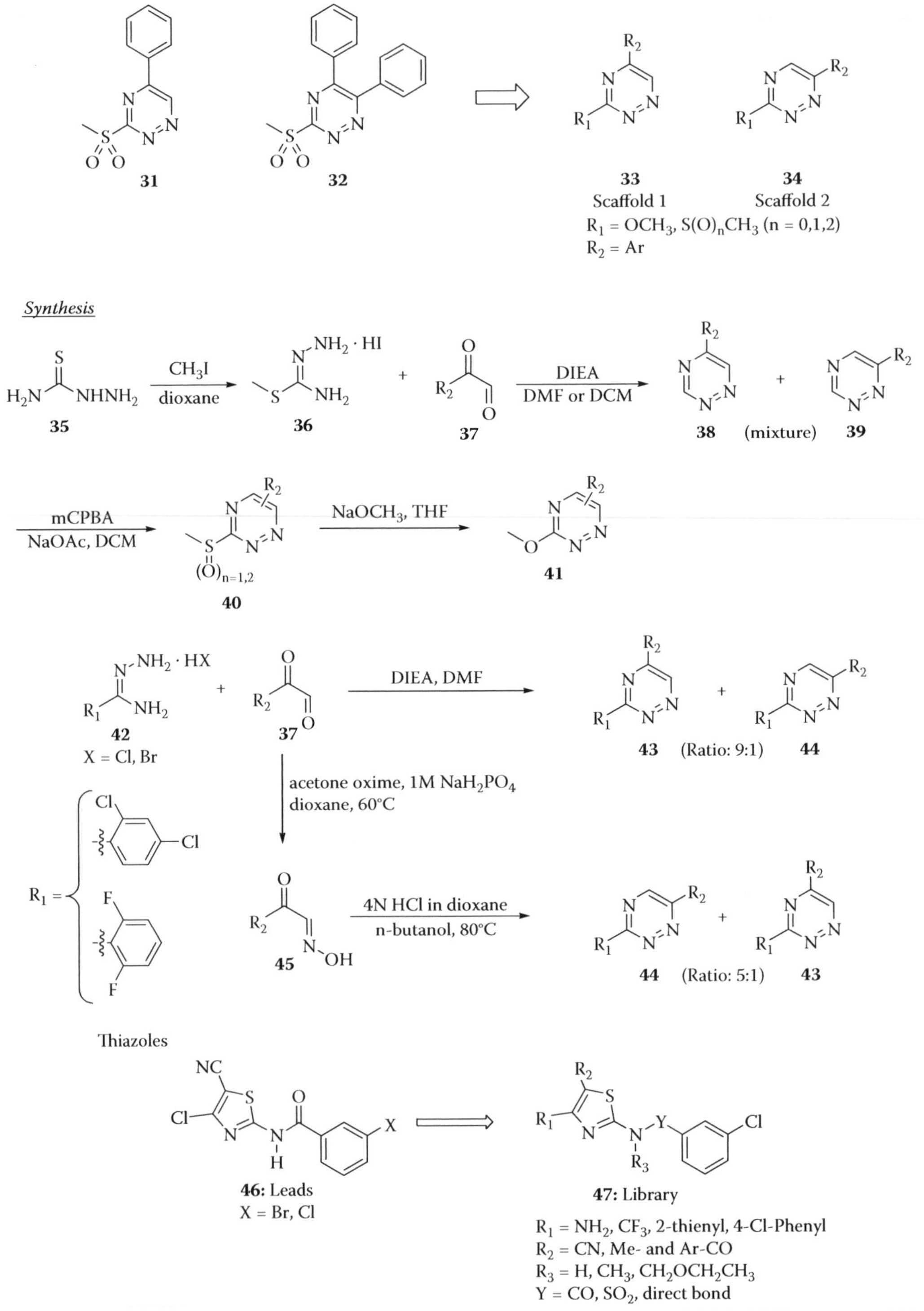

33
Scaffold 1
34
Scaffold 2
$R_1 = OCH_3, S(O)_nCH_3$ (n = 0,1,2)
$R_2 = Ar$

Synthesis

SCHEME 1.4   Antifungal triazines and thiazoles.

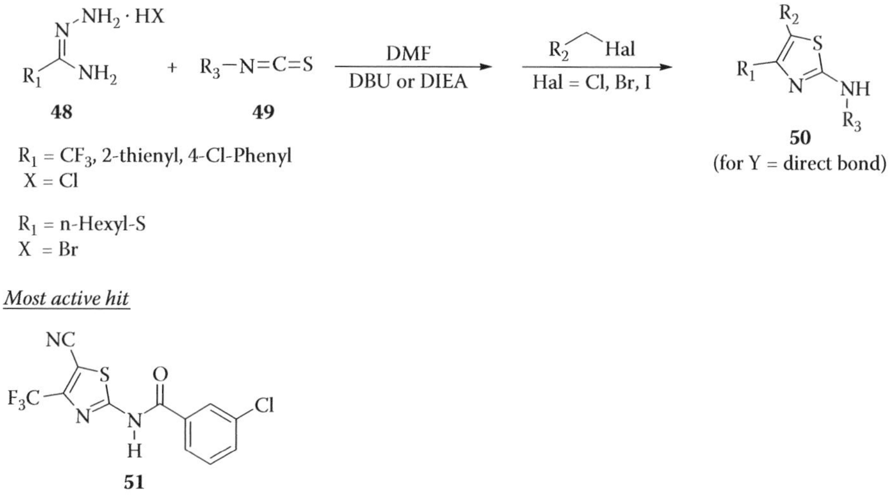

R$_1$ = CF$_3$, 2-thienyl, 4-Cl-Phenyl
X = Cl

R$_1$ = n-Hexyl-S
X = Br

Most active hit

SCHEME 1.4 Antifungal triazines and thiazoles (continued).

1.4 ANTIPARASITES

1.4.1 CASE 1: TRYPANOTHIONE REDUCTASE (TR)

Chitkul et al.[14] at the University of Southampton published the synthesis and optimization of inhibitors of trypanothione reductase (TR), a key enzyme involved in the antioxidant protective cell mechanism in parasites targeting disulfide substrates. Although glutathione reductase (GR) also targets disulfide-containing substrates in mammalian cells (e.g., glutathione disulfide), TR differs from GR in that TR's active site is wider and more hydrophobic in nature. The combination of these two factors offers an explanation of why TR can accommodate the polyamine component of trypanothione disulfide in TR's active site.

Using the natural product kukoamine A, a polyamine spermine derivative reported to selectively inhibit TR ($K_i = 1.39 \pm 0.41$ µM) over GR ($K_i > 10$ mM), the authors pursued a solid-phase synthetic approach for the identification and optimization of more potent TR inhibitors using the polyamines norspermine, spermidine, and spermine (**52, 53,** and **54,** respectively) as main scaffolds.

The primary amines in these three scaffolds were protected with the hydrazine-labile phthalimido-N-ethoxycarbonyl group (**55, 56, 57**), and subsequently reacted with a *p*-nitrophenyl carbonate allyl ester (**58**). The allyl group of the resulting carbamates was removed using Pd(PPh$_3$)$_4$, and the carboxylic acid intermediate immobilized on aminomethyl polystyrene resin using standard peptide coupling conditions (**59**). The hydroxymethyl-phenoxy-acetamido Wang-based resins were treated with hydrazine to remove the phthalimido-N-ethoxycarbonyl groups, and the free primary amines were coupled to 24 different carboxylic acids. The final polyamides were cleaved by treating the resin with TFA/H$_2$O (95:5) for 4 hr (**60, 61, 62**).

From the above 64-member library, three potent inhibitors all with K_i values in the submicromolar range were identified. The most potent inhibitor **63** was an indole-based spermine with a K_i value of 0.39 ± 0.06 µM, which is 3 times more potent than the initial lead compound kukoamine A. To optimize **63,** the authors investigated the influence of the sidechain between the indole unit and the polyamine, the linkage site on the indole, and different substituents on the indole. A follow-up library was made using 13 different indole-based carboxylic acids, where compound **64**, a 5-bromo-3-indole acetic acid spermine derivative, was identified with a K_i value of 76 ± 10 nM, which is 18 times more potent than kukoamine A and it did not inhibit yeast glutathione reductase in concentrations up to 50 µM. Interestingly, the authors pointed to the observation that **64** was a noncompetitive inhibitor for which follow-up studies will steer toward the identification of the mode of action of this potent TR inhibitor.

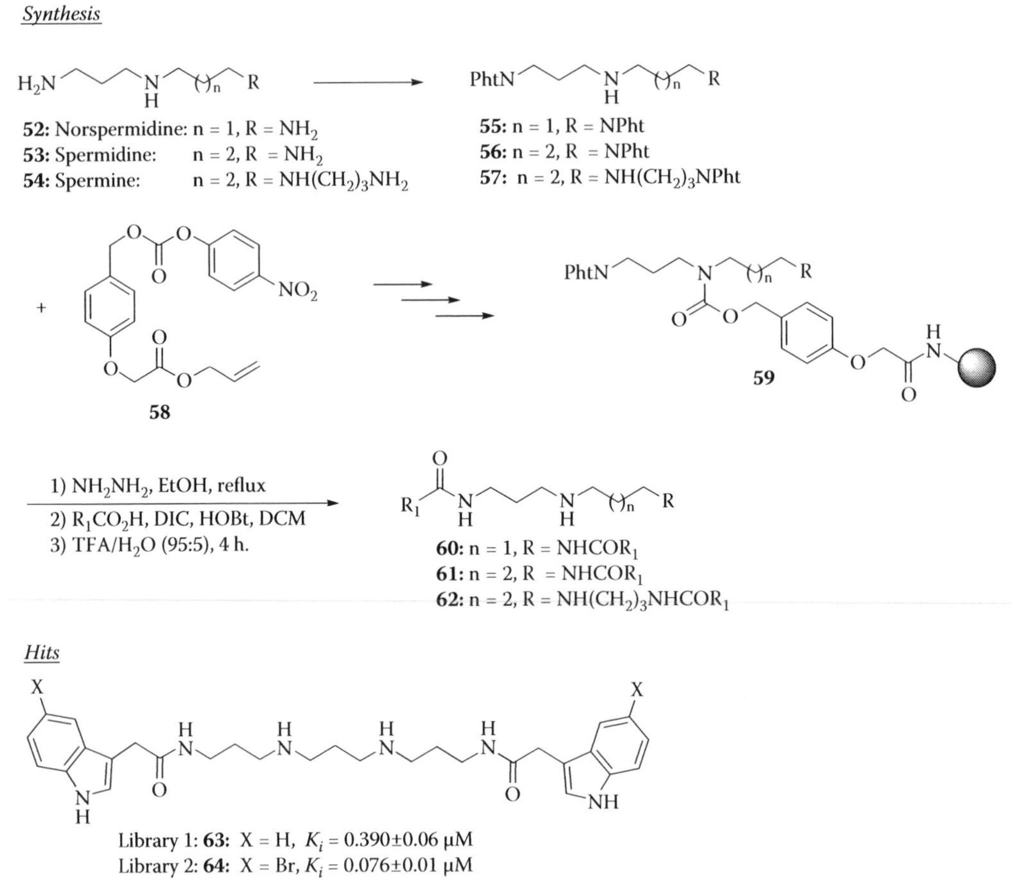

Synthesis

52: Norspermidine: n = 1, R = NH$_2$
53: Spermidine: n = 2, R = NH$_2$
54: Spermine: n = 2, R = NH(CH$_2$)$_3$NH$_2$

55: n = 1, R = NPht
56: n = 2, R = NPht
57: n = 2, R = NH(CH$_2$)$_3$NPht

58

59

1) NH$_2$NH$_2$, EtOH, reflux
2) R$_1$CO$_2$H, DIC, HOBt, DCM
3) TFA/H$_2$O (95:5), 4 h.

60: n = 1, R = NHCOR$_1$
61: n = 2, R = NHCOR$_1$
62: n = 2, R = NH(CH$_2$)$_3$NHCOR$_1$

Hits

Library 1: **63:** X = H, K_i = 0.390±0.06 µM
Library 2: **64:** X = Br, K_i = 0.076±0.01 µM

SCHEME 1.5 Trypanothione reductase (TR) inhibitors.

1.5 G PROTEIN-COUPLED RECEPTORS (GPCRS)

1.5.1 CASE 1: CCR3 ANTAGONISTS

Dhanak et al.[15] at SmithKline Beecham Pharmaceuticals reported the optimization of N-acytaled dipeptides as CCR3 antagonists. CCR3 is a chemokine receptor involved in inflammatory diseases such as asthma. The CCR3 chemokine is activated by proteins such as eotaxin, eotaxin-2, and MCP-4.

The authors used as starting point compound **65**, an N-acylated phenylalanyl methyl ester analog with an IC$_{50}$ of 5 nM against CCR3. Due to the liability of the ester group *in vivo* by esterases and the toxicity associated with methanol, the authors focused their efforts on the identification of a more suitable replacement of the methyl ester group.

The initial analogs were made via classic organic synthesis where a series of heterocyclic ester isosteres were introduced. The initial results indicated that all analogs lost potency against CCR3; but in the cases where the heterocyclic isostere contained an ethyl ester group, potency against CCR3 was regained. This observation suggested that the initial methyl ester required a higher degree of structural freedom to interact effectively with the target. The next step was directed toward the synthesis of less constrained ester and amide analogs. This new set of compounds revealed that replacing the methyl ester group with an alanine phenyl amide unit provided dipeptide **66** with an acceptable potency (IC_{50} = 190 nM) to use it for further optimization.

A polymer-assisted parallel synthesis using polymer-bound EDC as a coupling agent was used to rapidly create a 12-member compound library of **66** analogs. Different building blocks were used to probe both the phenyl unit of the phenylalanine group and the phenyl unit of the phenylbenz-amide group. This small library afforded compound **72** with an IC_{50} of 5 nM, and an IC_{50} of 15 nM in a functional assay inhibiting eotaxin-induced human eosinophil chemotaxis. Although the size of the final library was small, there is clearly more room for the further optimization of this class of compounds — not only for potency and selectivity, but also to fine-tune the physicochemical properties and profile of potential CCR3 inhibitors.

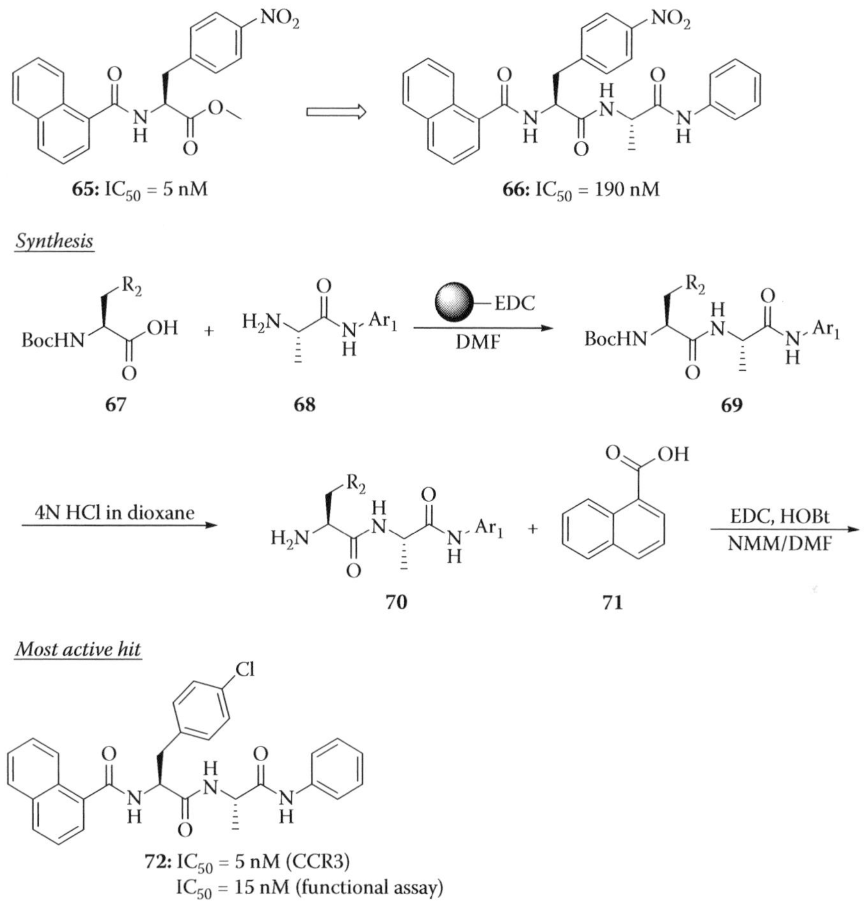

65: IC_{50} = 5 nM

66: IC_{50} = 190 nM

Synthesis

67 + **68** → **69**

70 + **71**

Most active hit

72: IC_{50} = 5 nM (CCR3)
IC_{50} = 15 nM (functional assay)

SCHEME 1.6 CCR3 inhibitors.

1.5.2 CASE 2: FOLLICLE STIMULATING HORMONE (FSH) RECEPTOR AGONISTS

Guo et al.[16] at Pharmacopeia Inc. reported the optimization of biphenyl N-alkylated diketopipera-zines as follicle stimulating hormone (FSH) receptor agonists. FSH is involved in human fertility as it promotes ovarian follicle maturation in women and spermatogenesis in men. Current therapies for low fertility treatment depend on the use of proteins that can only be administrated by injection, resulting in low patient compliance, so the authors sought small-molecule FSH receptor agonists that could be administrated either orally or transdermally.[17]

The current efforts centered on the optimization of **73** and **74**, two single-digit micromolar FSH agonists (EC_{50} = 3.0 and 1.7 μM, respectively) discovered from a combinatorial chemistry library using a cell-based CHO-hFSHR-luciferase assay. The optimization strategy involved the removal of the common amide group, substituent effects in the trimethoxyphenyl unit in **73**, incorporation of a methyl unit on C-5 of the diketopiperazine group, and changes to the urea group in **74**. To carry out this optimization campaign, a solid-phase synthesis was developed using TentaGel™-S-OH resin as solid support where the diketopiperzine core (**75**) was formed by simultaneous intramolecular cyclization/solid-phase release.

The authors constructed three libraries beginning with the solid-phase synthesis of N-substituted dipeptides (**80**) from mobilized bromo acetic acid via an ester linkage (**76**). The bromide group was used to immobilize different amines (**77**), which then were coupled with Boc-(*L* or *D*)-N-Me-4-iodophenylalanine (**79**). In library 1, the 4-iodophenylalanine unit was alkylated with several arylboronic acids under Suzuki coupling conditions. Removal of the Boc protecting group, followed by treatment with triethylamine, promoted an intramolecular cyclization forming the desired diketopiperazine core (**83**). It was determined that the removal of the sidechain amide unit was tolerated, and that N-4-alkylation of the diketopiperazine core with an n-octyl sidechain and the 3,4,5-trimethoxy groups in the biphenyl unit improved activity (**84**, EC_{50} = 400 nM). A few analogs were made by replacing the original N-1-Me group with H, ethyl, and propyl groups, finding that in all cases potency suffered, so the methyl group was subsequently held constant.

Keeping the 3,4,5-trimethoxy-biphenyl group constant, a second library was made where L- and D-amino acids were used to explore the stereochemical influence on C-5 of the diketopiperazine heterocycle. Also, to simultaneously explore different sidechains on the N-4 of the diketopiperazine,

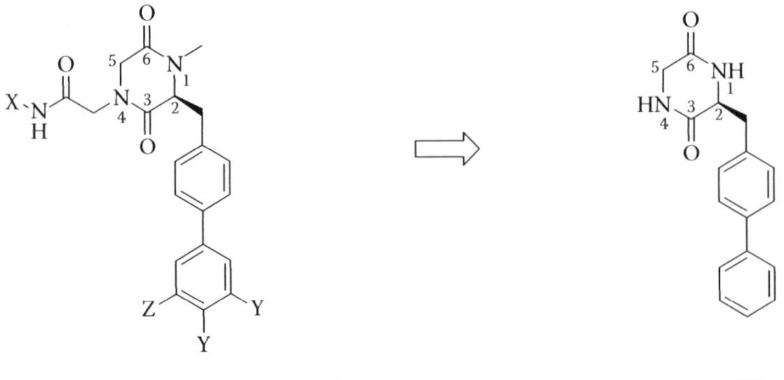

73: X = n-C_6H_{13}, Y = Z = OMe, (EC_{50} = 3.0 μM)
74: X = n-C_5H_{11}, Y = H
    Z = MeNHCON(Bu)CH_2, (EC_{50} = 1.7 μM)

75
Core Scaffold

Synthesis: Library 1

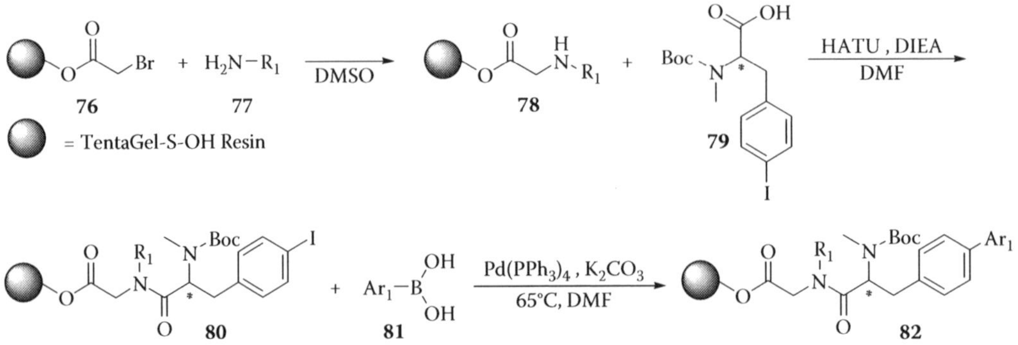

SCHEME 1.7 FHS targeted library 1.

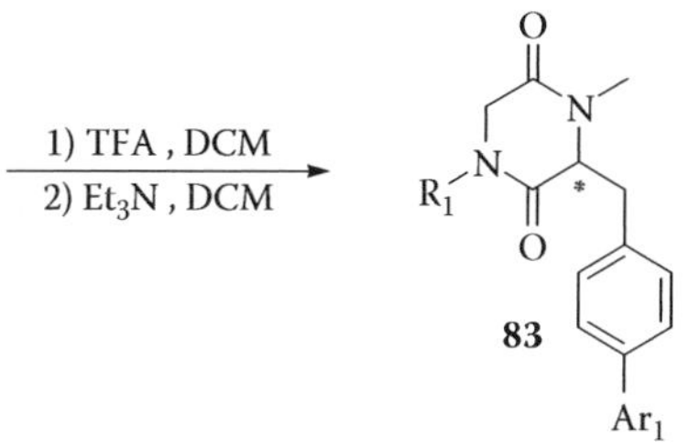

1) TFA , DCM
2) Et$_3$N , DCM

83

Library 1: Hit

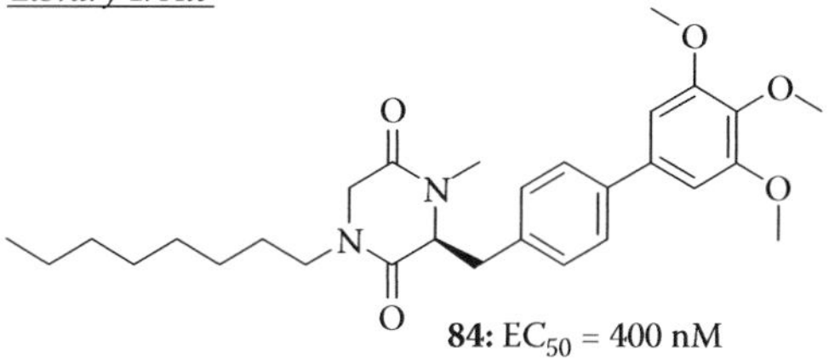

84: EC$_{50}$ = 400 nM

SCHEME 1.7 FHS targeted library 1 (continued).

Synthesis: Library 2

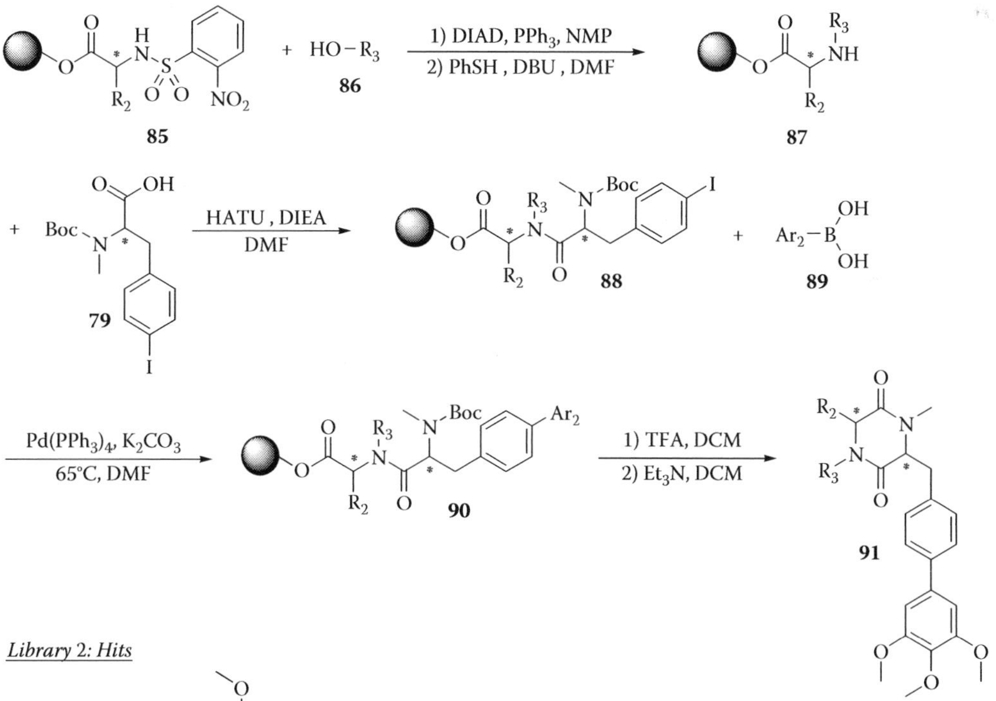

+ HO–R$_3$
86

1) DIAD, PPh$_3$, NMP
2) PhSH , DBU , DMF

85

87

+ Boc

79

HATU , DIEA
DMF

88

+ Ar$_2$–B

89

Pd(PPh$_3$)$_4$, K$_2$CO$_3$
65°C, DMF

90

1) TFA, DCM
2) Et$_3$N, DCM

91

Library 2: Hits

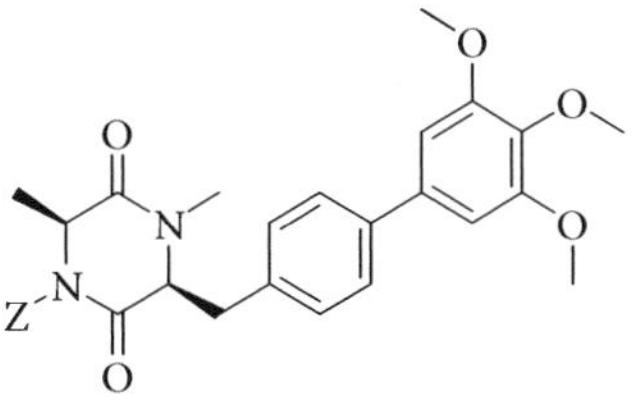

92: Z = n-C$_8$H$_{17}$, EC$_{50}$ = 13 nM
93: Z = n-C$_7$H$_{15}$, EC$_{50}$ = 25 nM

SCHEME 1.8 FHS targeted library 2.

the amino groups of immobilized amino acids were reacted with 2-nitrobenzene-sulfonyl chloride. The resulting sulfonamides (**85**) were alkylated with different alkyl alcohols (**86**) under Mitsunobu conditions. The 2-nitrobenzene-sulfonyl chloride unit was removed by treatment with thiophenol and DBU in DFM (**87**). This point allowed the authors to explore two aspects of the main scaffold: (1) the stereochemistry at C-2, and (2) the urea group found in compound **74**.

For the optimization at C-2, the authors made a second library where the amino intermediates (**87**) were acylated with Boc-(L or D)-N-Me-4-iodophenylalanine (**79**). Following the same synthetic procedure as reported above, the desired diketopiperazines were made (**91**). This second library revealed two key factors on this series: (1) the stereochemistry on C-5 and C-2 favored potency in any *S* or *R* chiral combination except for *R,R*; and (2) linear non-branched sidechains of six to eight carbons were tolerated at N-4. Screening of this second library afforded a couple of active compounds with low nanomolar cellular potency (**92, 93**: $EC_{50} = 13$ and 25 nM, respectively) (Scheme 1.8).

Based on these results, the third library directed at the exploration of the urea group found in compound **74** was made using only naturally occurring amino acids to keep the stereochemistry at C-5/C-2 S/S on the diketopiperazine core. The amino intermediates used for the second library were acylated with Boc-*L*-N-Me-4-iodophenylalanine (**79**) and then alkylated with 3-formylphenyl boronic acid. The aldehyde intermediates (**94**) were treated with several amines under reductive amination conditions and the resulting amines reacted with acid chlorides, chloroformates, and isocyanates to afford amides, carbamates, and ureas, respectively (Scheme 1.9).

This third library afforded several single-digit nanomolar inhibitors where the most potent ones had an n-C_8H_{17} alkyl group at N-4, a butyl group at the benzyl amine of the biphenyl, and a methyl group on the acylating agent affording compounds such as **97** and **98** with $EC_{50} = 2.7$ nM and 1.2 nM, respectively.

Synthesis: Library 3

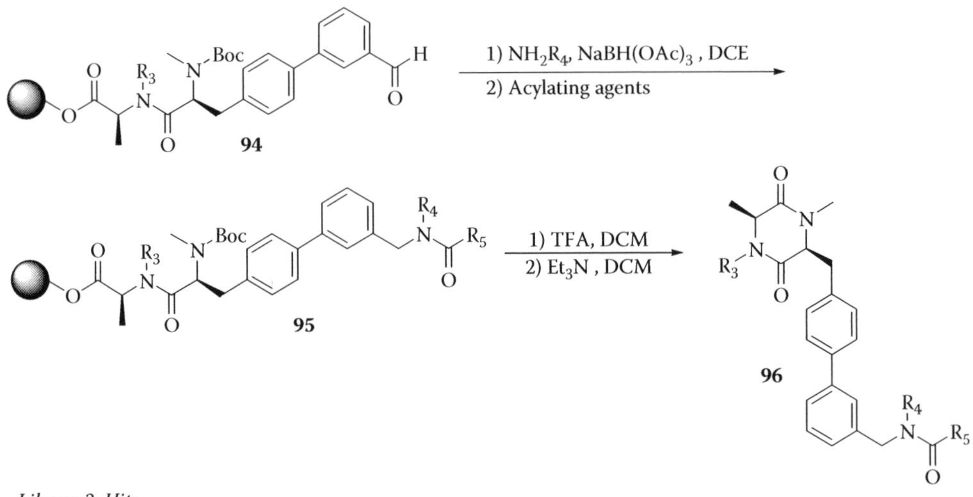

Library 3: Hits

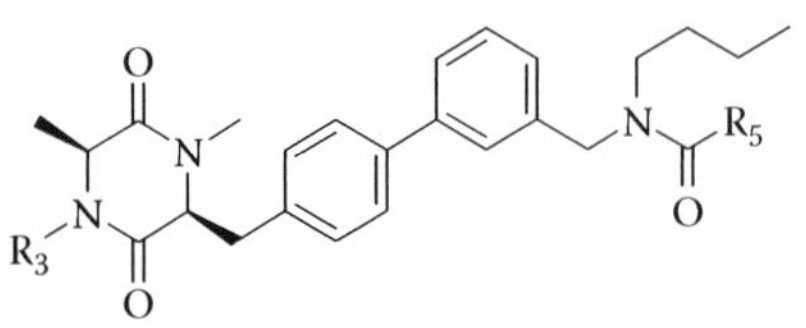

97: R_3 = n-C_8H_{17}, R_5 = OCH_3, (EC_{50} = 2.7 nM)
98: R_3 = n-C_8H_{17}, R_5 = $NHCH_3$, (EC_{50} = 1.2 nM)

SCHEME 1.9   FHS targeted library 3.

This work involved the development of three libraries and the synthesis of approximately 300 compounds, where each library was used to systematically explore a key aspect of the main scaffold. This example shows how solid-phase parallel synthesis can be used to optimize a lead with single-digit micromolar cell activity to single-digit nanomolar, a 1000-fold improvement in potency.

1.5.3   CASE 3: NOCICEPTIN/ORPHANIN FQ (N/OFQ) RECEPTOR AGONISTS AND ANTAGONISTS

Chen et al.[18] at Purdue Pharma reported the design and synthesis of novel potent and selective phenyl piperidines targeting the nociceptin/orphanin FQ (N/OFQ) receptor (NOP, also known as ORL-1), a receptor involved in metabolic and CNS disorders.

The authors began their work by developing a pharmacophore based on three reported NOP receptor ligands (**99**: $pK_i = 7.3$, **100**: $pK_i = 8.4$, and **101**). Close analysis of these three compounds lead to the following point-point, two-dimensional pharmacophore profile: (1) a large hydrophobic group connected via a 1-3 carbon bond linker to a endocyclic heteroatom, (2) the endocyclic heteroatom is a basic nitrogen, (3) the heterocyclic system is either a six- or eight-membered ring, (4) at the opposite side of the heterocyclic system there is a medium-sized hydrophobic group, and (5) at the opposite side of the heterocyclic system there is also a hydrophilic group linked to the heterocyclic system. A similarity search in the ACD (MDL® Available Chemicals Directory) database indicated that there were several piperazine-based compounds that met part of the pharmacophore criteria (pharmacophore points 4 and 5) that could be used as building blocks.

To explore the pharmacophore points 1 and 2 above, the authors used alkylating and acylating agents (**103**), using polymer-supported DIEA as a proton sponge and DMF as solvent. The synthetic approach was carried out at a 0.2 mM scale using 1-dram vials and vortexing at either room temperature or 80°C with a purity cut-off of 70% by LC/MS. Reactions meeting the purity threshold were filtered and the solvent removed using a Genevac HT8 high-throughput evaporation system.

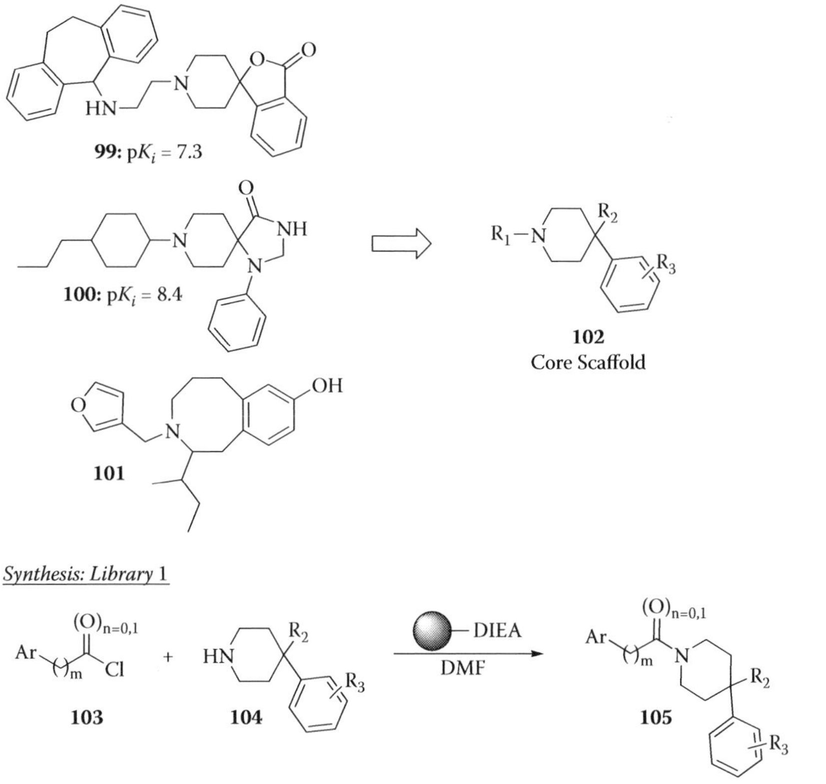

99: $pK_i = 7.3$

100: $pK_i = 8.4$

101

102
Core Scaffold

Synthesis: Library 1

103 + **104** $\xrightarrow[\text{DMF}]{\text{DIEA}}$ **105**

SCHEME 1.10   NOP targeted library 1.

Those reactions that did not pass the 70% cut-off purity criteria were submitted to an aqueous acid-and-base washing process where more than 90% of the products then passed the purity criteria.

The exploratory libraries afforded 20 hits, all exhibiting greater than 50% inhibition in the primary screen at a single concentration of 10 µM. Analysis of these hits revealed that (1) they centered around a few piperidines, (2) all but one derived from an alkylation reaction, and (3) the alkylating agents varied in both size and substituent patterns. These observations indicate the importance of the basic nitrogen in the piperazine unit and the large size of the hydrophobic pocket in the receptor.

For hit confirmation, the above 20 hits were resynthesized and their K_i values determined against the human NOP receptor expressed in recombinant HEK-293 cells (K_i ranges from 6.5 µM to 153 nM). Confirmation data indicates that 4-piperazine-based building blocks confer good affinity toward the NOP receptor. For the optimization phase, the authors used the active 4-piperazine-based building blocks (**106** through **109**) and linked hydrophobic groups (i.e., cycloalkyl groups) to the basic amine. This second library (**112**) was made using traditional synthesis (reductive alkylation and N-alkylation).

Synthesis: Library 2

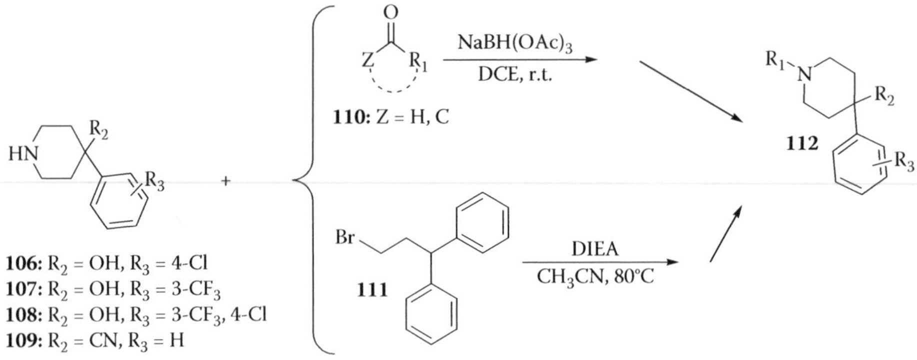

106: R_2 = OH, R_3 = 4-Cl
107: R_2 = OH, R_3 = 3-CF$_3$
108: R_2 = OH, R_3 = 3-CF$_3$, 4-Cl
109: R_2 = CN, R_3 = H

Library 2: Hits

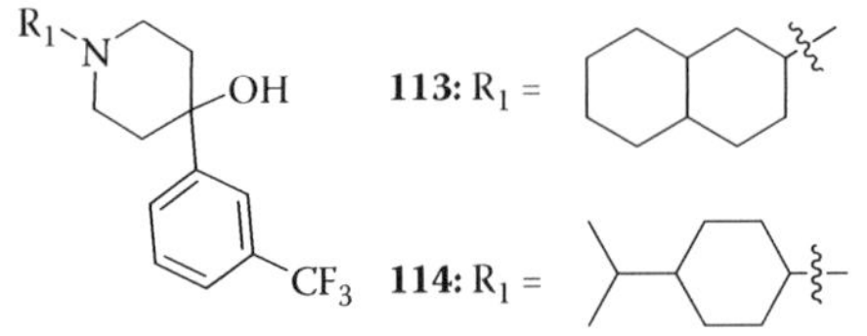

	ORL1 K_i(nM)	µ K_i(nM)	δ K_i(nM)	κ K_i(nM)	Function Agonist
113: R_1 =	12	21	10,000	NA	Agonist
114: R_1 =	12	36	5,674	4,153	Agonist

SCHEME 1.11   NOP targeted library 2.

The optimization library afforded several low-nanomolar NOP receptor agonists and antagonists exhibiting selectivity between the µ, δ, and κ receptors, in some cases up to 9000-fold (**113, 114**).

1.5.4 CASE 4: OREXIN RECEPTOR 1 (OX1) AND OREXIN RECEPTOR 2 (OX2) ANTAGONISTS

Koberstein et al.[19] at Actelion Pharmaceuticals Ltd. published a 3-year work where a combination of automated solution-phase synthesis and automated purification system was used for the synthesis and optimization of tetrahydroisoquinolines as orexin receptor 1 (OX1) and orexin receptor 2 (OX2) antagonists. These receptors are believed to modulate appetite and food intake. Two endogenous neuropetides, oxerin A and oxerin B, have recently been discovered that bind to both OX1 and OX2 (oxerin A: IC$_{50}$ (OX1) = 20 nM, IC$_{50}$ (OX2) = 36 nM; oxerin B: IC$_{50}$ (OX1) = 420 nM, IC$_{50}$ (OX2) = 38 nM).

For the library synthesis, the authors used capped glass reactors in a 4 × 6 array rack format using a Büchi Syncore system for heating and/or shaking, and a Chemspeed ASW 2000 synthesizer for reactions under inert conditions in a temperature range of −40 to +140°C. For compound purification (10- to 100-mg range), an automated HPLC (autosampler-HPLC) UV/ELSD fraction-collection triggered system was used with a purification sample time of 6 min/compound. Compound confirmation was carried out via HPLC-MS (2.5 min/sample).

Following up a hit (**115**) identified in an HTS campaign using Chinese hamster ovary (CHO) cells over-expressing human OX1 and OX2 receptors, the authors divided the main core of the hit compound into four sections for optimization. Focusing the initial efforts on the optimization of the benzyl amide site (R_1 and R_2), the authors used tetrahydropapaverine as the starting material (**117**). In a one-pot reaction, more than 100 amides were made (**118**), where tetrahydropapaverine was treated with bromoacetyl bromide and an amine in the presence of DIEA in THF. From this library, indane amide **119** was identified with 3 times higher potency toward OX1 (**119**: IC_{50} (OX1) = 49 nM, IC_{50} (OX2) = 679 nM). A follow-up synthesis of 20 additional analogs afforded 6-methoxy-indane amide **120** with affinities comparable to oxerin A and oxerin B (**120**: IC_{50} (OX1) = 19 nM, IC_{50} (OX2) = 101 nM).

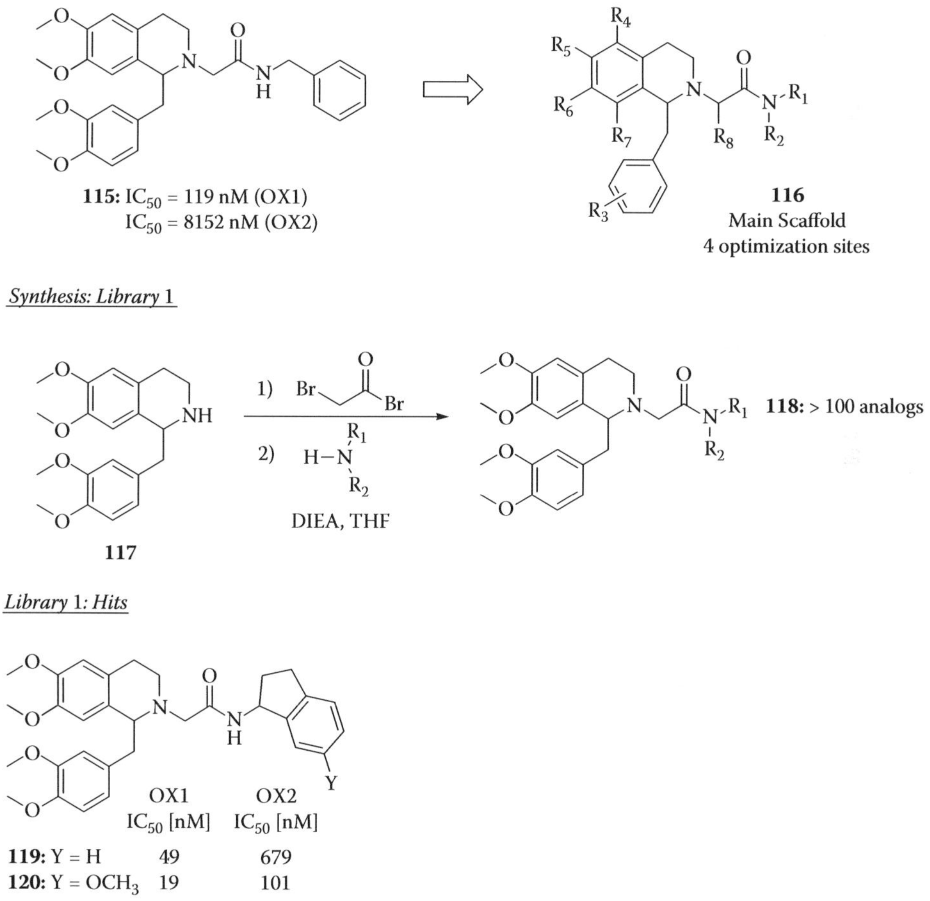

115: IC_{50} = 119 nM (OX1)
      IC_{50} = 8152 nM (OX2)

116
Main Scaffold
4 optimization sites

Synthesis: Library 1

117

118: > 100 analogs

Library 1: Hits

	OX1 IC_{50} [nM]	OX2 IC_{50} [nM]
119: Y = H	49	679
120: Y = OCH_3	19	101

SCHEME 1.12 OX1/OX2 targeted library 1.

The next step focused on the optimization of the 3,4-dimethoxybenzyl unit of the main core (R_3). Starting from homoveratrylamine (**122**), the authors synthesized nine amide intermediates that were converted into their respective tetrahydroisoquinolines (**123**) via a Bischler-Napieralski reaction. These nine tetrahydroisoquinolines were then reacted with bromoacetyl bromide and two amines (benzylamine and 1-aminoindane). It was found that removal of either methoxy group, linkage of the methoxy units via a methylene group (piperonyl analog), or replacement of both methoxy groups with fluoride groups resulted in decreased OX1 affinity. However, replacing both methoxy groups with chloride atoms brought back affinity similar to A against OX1 and with higher affinity toward OX2. Interestingly, for benzylamide analogs, it was found that replacing both methoxide groups with methyl groups (**124**), or the methyl group of the 4-methoxide with a pyrimidine group (**125**), the resulting analogs were up to 4 times more potent against OX1 and up to 44 times more potent against OX2 when compared to compound A.

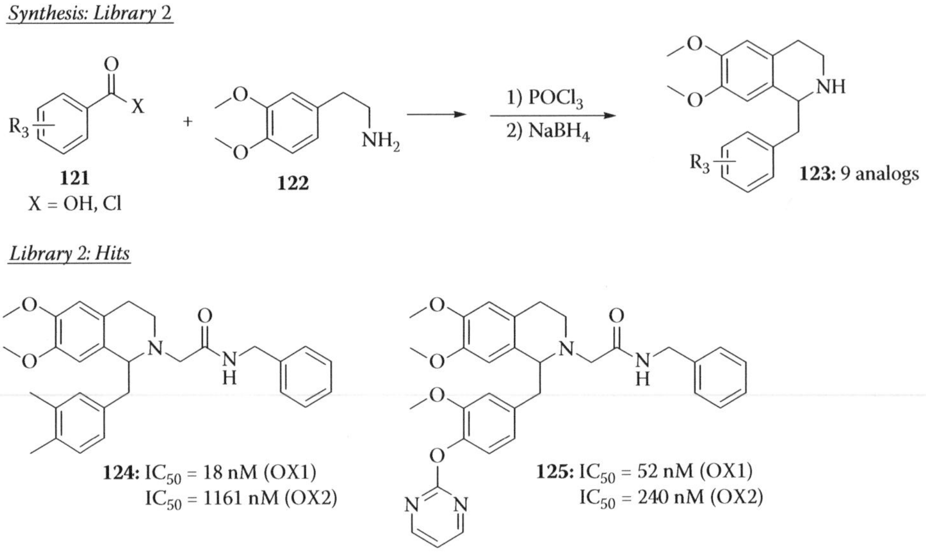

Synthesis: Library 2

Library 2: Hits

124: IC_{50} = 18 nM (OX1)
        IC_{50} = 1161 nM (OX2)

125: IC_{50} = 52 nM (OX1)
        IC_{50} = 240 nM (OX2)

SCHEME 1.13   OX1/OX2 targeted library 2.

The next phase focused on the optimization of the tetrahydroisoquinoline unit of the main scaffold. Keeping the rest of the lead compound **115** constant, the SAR exploration focused on the 6- and 7-substitution pattern (R_5 and R_6, respectively). Building a five-member library, the authors found that even small modifications on this part of the scaffold resulted in dramatic changes in activity/affinity, such as removing one of the methoxy groups. Even linking the methoxy units via a methylene group (piperonyl analog) was detrimental to activity/affinity. However, it was discovered that replacing a methoxy group for a slightly bigger alkoxy substituent on the 7-position (R_6) (ethoxy: **126**, and iso-propyloxy: **127**) afforded highly and selective OX1 antagonists (**126:** IC_{50} (OX1) = 33 nM; **127:** IC_{50} (OX1) = 17 nM; **126** and **127:** IC_{50} (OX2) > 10,000 nM).

Although this was a remarkable achievement for this project, the authors also wanted to obtain compounds with high affinity towards OX2. The breakthrough to achieve OX2 selective antagonists came when the benzyl amide group was replaced with a picolyl amide group in conjunction with the shifting of the 6- and 7-methoxy groups to the 5- and 8-positions (R_4 and R_7) (**128:** IC_{50} (OX1) > 10,000 nM, IC_{50} (OX2) = 117 nM). Further replacement of the methoxy group at the 8-position with an ethoxy group afforded **129** that was 7 times more potent and 200 times more selective towards OX2 (**129:** IC_{50} (OX1) = 5102 nM, IC_{50} (OX2) = 26 nM).

Finally, the authors investigated the influence of different substitution patterns on the glycine linker between the tetrahydroisoquinoline unit and the benzyl amide unit (R_8). It was found that phenylglycine afforded analogs that could be selective to either OX1 or OX2. A follow-up library of 200 phenylglycine-containing analogs was made where group variations were carried out at R_1 and R_3 (R_2 = H), affording compounds not only highly selective to one of the two OX receptors (**130** and **131**), but also exhibiting the desired physicochemical properties (e.g., absorption, blood-brain-barrier penetration).

Library 3: Hits

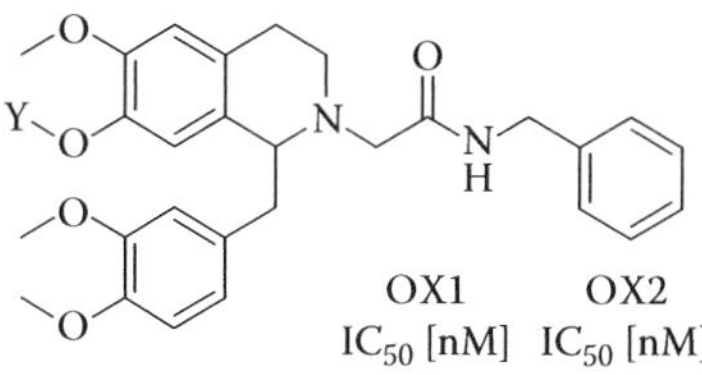

	OX1 IC$_{50}$ [nM]	OX2 IC$_{50}$ [nM]
126: Y = Ethyl	49	>10000
127: Y = Isopropyl	17	>10000

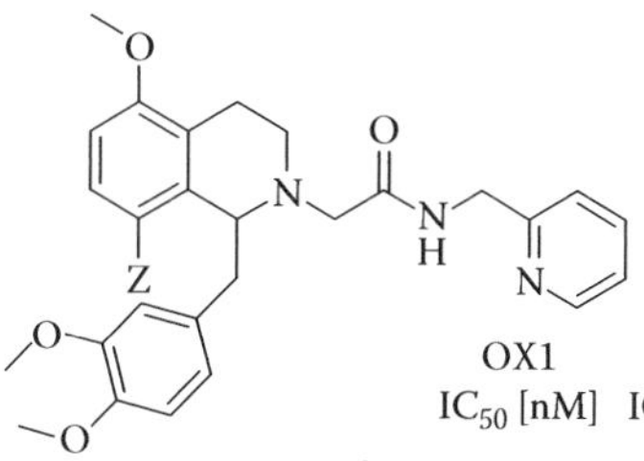

	OX1 IC$_{50}$ [nM]	OX2 IC$_{50}$ [nM]
128: Z = Methoxy	>10000	117
129: Z = Ethoxy	5102	26

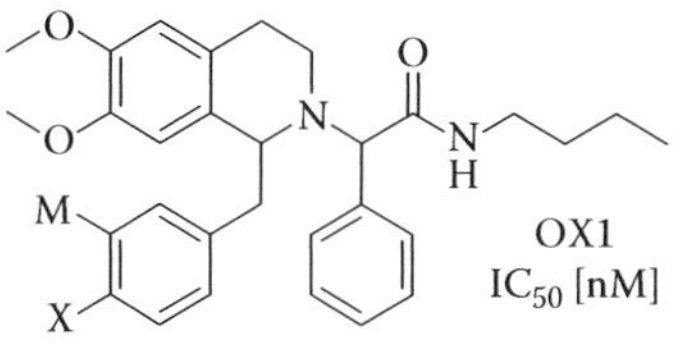

	OX1 IC$_{50}$ [nM]	OX2 IC$_{50}$ [nM]
130: M = H, X = NMe$_2$	45	1536
131: M = X = F	1906	19

SCHEME 1.14   OX1/OX2 targeted library 3.

Using a solution-phase parallel synthesis approach in combination with an automated purification system, the authors made 50 libraries, synthesizing more than 800 tetrahydroisoquinoline derivatives during the first year alone. By the end of the second year, more than 1250 discrete analogs had been made where optimization efforts were simultaneously directed toward potency/selectivity as well as physicochemical and pharmacological compound properties.

1.6   IMMUNOGLOBULINS

1.6.1   CASE 1: HUMAN IMMUNOGLOBULIN E (IgE) INHIBITORS

Berger et al.[20] at Novartis reported optimization work on phenylethylamino quinazolines as inhibitors of human immunoglobulin E (IgE) synthesis. Based on the observation that high levels of IgE have been found in allergic diseases such as asthma and rhinitis, the authors focused on the development of compounds set to block the endogenous synthesis of IgE. This work began with **132**, a hit produced during a high-throughput screening campaign.

The strategy used for a parallel synthesis approach involved the synthesis of phenylethylamino quinazolines (**133**) using 4-chloroquinazoline (**134**) as starting material, a Quest 210 parallel synthesizer (Argonaut Technologies), and polymer-based scavengers such as polymer-supported TEA and aldehyde-Wang resin. 4-Chloroquinazoline (**134**) was treated with 2 equivalents of an amine (**135**) and 3 equivalents of polymer-supported TEA to trap the generated HCl. The excess primary amine was then selectively removed over the newly made secondary amine product by treatment with 3 equivalents of aldehyde-Wang resin in combination with 1% HOAc. The authors noted that addition of 1% AcOH was crucial to the complete formation of the Schiff base of the excess primary amine over the secondary amine on the Wang resin. Filtration and solvent removal then afforded the desired products (**133**).

Beginning with **132** as the starting point, the first focus of attention was the determination of relevance of the chiral center (*S*-configuration for **132**). The opposite enantiomer of **132** was synthesized and it was found that it lacked activity against the synthesis of IgE. This fact was confirmed during the first optimization round when for some of the active *S*-compounds, their *R*-enantiomeric counterparts were synthesized and in all cases the *R*-compounds were inactive. This observation confirmed that the *S*-configuration was of extreme importance in this compound series.

The first optimization step focused on the phenylethylamino part of the main core. Substitution of the phenyl group with cyclohexyl, 2-naphthyl, and 3- and 4-substituted phenyl groups was very well tolerated, affording potent inhibitors with $IC_{50} < 100$ nM (**136:** $IC_{50} = 81 \pm 9$ nM, **137:** $IC_{50} = 14 \pm 3$ nM, **138:** $IC_{50} = 23 \pm 3$ nM). Interestingly, it was also discovered that further N-methylation of **133** or exchange of the benzyl group with a phenyl group in all cases was detrimental for activity.

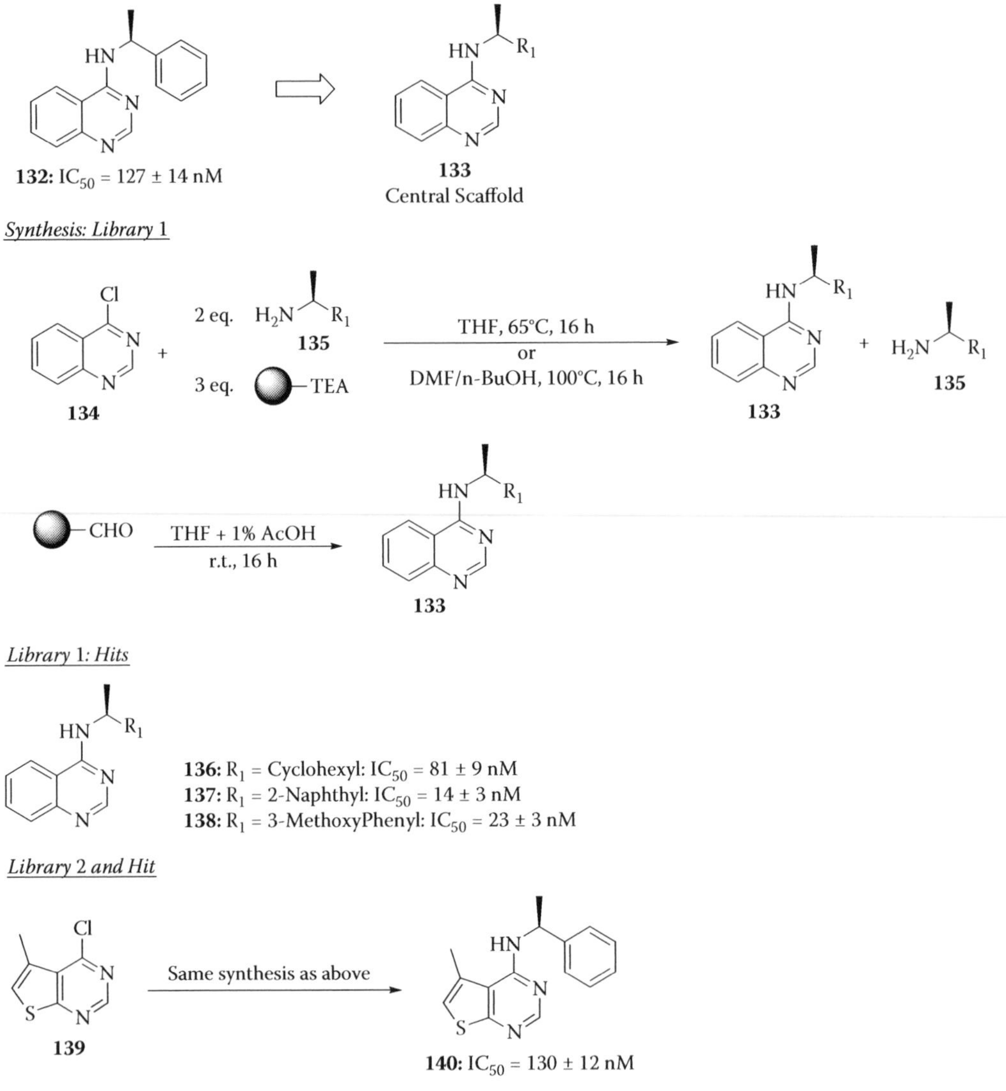

132: $IC_{50} = 127 \pm 14$ nM

133
Central Scaffold

Synthesis: Library 1

Library 1: Hits

136: R_1 = Cyclohexyl: $IC_{50} = 81 \pm 9$ nM
137: R_1 = 2-Naphthyl: $IC_{50} = 14 \pm 3$ nM
138: R_1 = 3-MethoxyPhenyl: $IC_{50} = 23 \pm 3$ nM

Library 2 and Hit

139

Same synthesis as above

140: $IC_{50} = 130 \pm 12$ nM

SCHEME 1.15 IgE targeted library.

The next efforts focused on the exploration of the heterocyclic core of the pyrimidine unit. The authors replaced the pyrimidine group with a quinoline group, resulting in complete loss of activity. They then proceeded to replace the fused group to the pyrimidine group with pyrrole, imidazole, and thiophene, finding that only the latter was tolerated, in particular the thieno[2,3-d]pyrimidin-4-yl isomer. However, only one of these thiophene-based analogs exhibited an activity barely compatible with hit **132** (**140:** $IC_{50} = 130 \pm 12$ nM).

To gain a better understanding of the observed SAR, the authors performed functional density quantum calculations to calculate the charge at the N-1 atom of the pyrimidine unit (Q_N). Plotting the N charge vs. $\log(IC_{50})$, it was found that these two parameters correlated linearly ($R = 0.94$).

These computational results indicated that increased electron-deficiency at the N-1 atom of the pyrimidine group resulted in compounds with greater activity.

This quick optimization exercise using polymer-supported parallel synthesis provided great insight into this compound series: the fact that the *S*-configuration is of extreme importance for activity as well as the electronic substitution effects on the N-1 atom of the pyrimidine group.

1.7 IMMUNOPHILINS

1.7.1 CASE 1: IMMUNOPHILIN FKBP12 INHIBITORS

Choi et al.[21] at Guilford Pharmaceuticals reported a solution-phase parallel synthesis approach for the synthesis of immunophilin FKBP12 inhibitors. Immunophilin FKBP12 is highly expressed in the brain and is known to bind to FK506 (tacrolimus), a polyketide that is shown to play a key role in nerve growth, so inhibition of FKBP12 could lead to the treatment of CNS diseases and injuries.

Using compound **141** (GPI 1046) as starting point, the authors began work on the SAR development around this compound class (**142**). Here, the strategy used involved the use of a Reacti-Therm stirring-heating module for parallel synthesis and polymer-based supports for product purification.

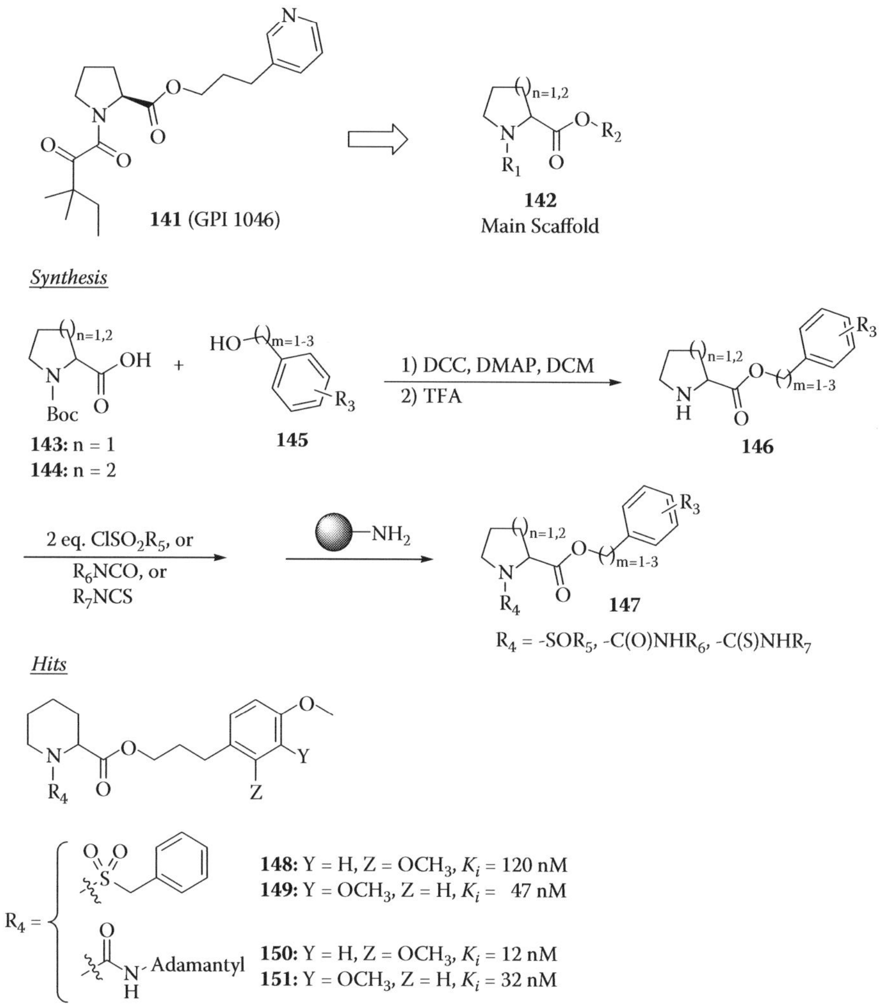

SCHEME 1.16   Immunophilin targeted library.

Using N-(*tert*-butyloxycarbonyl) proline (**143**) or N-(*tert*-butyloxycarbonyl) pipecolic acid (**144**) as starting materials, the authors first made a set of several esters (**146**) using standard DCC/DMAP/DCM reaction conditions. The esters (2 mL, 1.48 M) were arrayed into individual rows and treated with TFA to remove the Boc protecting group. Then sulfonyl chlorides, isocyanates, and thioisocyanates (2 mL, 1.48 M in DCM) were arrayed in columns, followed by the addition of TEA to afford the respective acylated products (sulfonyl amides, ureas, and thioureas, respectively). Excess acylating agent was removed by adding Tentagel-S-NH_2 resin to the wells. Final products (**147**) were purified quickly by passing each reaction mixture through a silica solid-phase column.

It was observed that the adamantyl, cyclohexyl, and benzyl groups increased activity as N-substituents of the proline and pipecolic acid units in the urea, thiourea, and sulfonamide series, respectively. Further exploration of the sulfonamide series revealed that for potency, a benzyl group was preferred over a phenyl and 4-methyl-phenyl group.

For the optimization of the ester group, a broad set of benzyl, phenethyl, and 3-phenyl-substituted propyl alcohols was used, keeping the adamantyl urea, cyclohexyl thiourea, and benzyl sulfonamide groups constant. During this optimization phase, it was observed that electron-donating groups improved activity, in particular the methoxy group. The 2,3- and 2,4-dimethoxy substitution patterns were observed to provide the most potent inhibitors (**148–151**: K_i < 120 nM).

1.8 INTEGRINS

1.8.1 CASE 1: αvβ3 INTEGRIN INHIBITORS

Gopalsamy et al.[22] at Wyeth-Ayerst Research reported the synthesis of αvβ3 integrin inhibitors utilizing a solid-phase approach. αvβ3 is involved in liver metastasis, smooth muscle and vascular endothelial cell migration and proliferation, and also in osteoclast cell adhesion and bone resorption.

RGD peptidomimetics have been found to be potent inhibitors of αvβ3 where many of them possess a 2,3-diaminopropionic acid core. However, the methodologies used for their synthesis only allow the variation of one functionality at a time. Gopalsamy and co-workers[22] devised a combinatorial solid-phase synthesis that allowed the chemical modification of both the N-terminus region as well as the α-amino group of the 2,3-diaminopropionic acid.

Using β-N-dde-α-N-Fmoc-L-diaminopropionic acid as the starting material, this orthogonally protected diamino acid was anchored to Wang resin using peptide coupling conditions (**152**). The Fmoc group was removed, and the free amine acylated with chloroformates, carboxylic acids, or isocyanates. The resulting intermediates (**153**) were treated with hydrazine to remove the dde protecting group (**154**). The N-Fmoc-protected benzoic acid (**155**) was then coupled to the β-amino group to afford (**156**), the Fmoc group removed and the free amine further functionalized into amidines, guanidines, pyrimidines, and ureas. Treatment of the resulting resins (**156**) with 50% TFA/DCM for 30 minutes afforded the desired products (**157**).

This synthetic approach allowed the authors to quickly obtain an SAR on this compound series. It was found that for the C-terminus amino group, bulky carbamates (neopentyl > ethyl, methyl) were the most potent in the series (carbamates > amides > ureas); and for the N-terminus amine group, cyclic guanidines were the best groups (cyclic guanidines > guanidine > amidines, pyridimidines) (**158, 159**: −log IC_{50} (mM) ~ 4.7).

1.8.2 CASE 2: α₄β₁/α₄β₇ INTEGRIN ANTAGONISTS

Castaneda et al.[23] at Genentech Inc. reported the solid-phase synthesis of $\alpha_4\beta_1/\alpha_4\beta_7$ integrin antagonists. Because it is believed that $\alpha_4\beta_7$ plays a role in inflammatory bowel disease, while $\alpha_4\beta_1$ plays a role in diseases such as asthma and multiple sclerosis, finding selective ligands for these integrins could serve as therapeutic agents to alleviate or control such diseases.

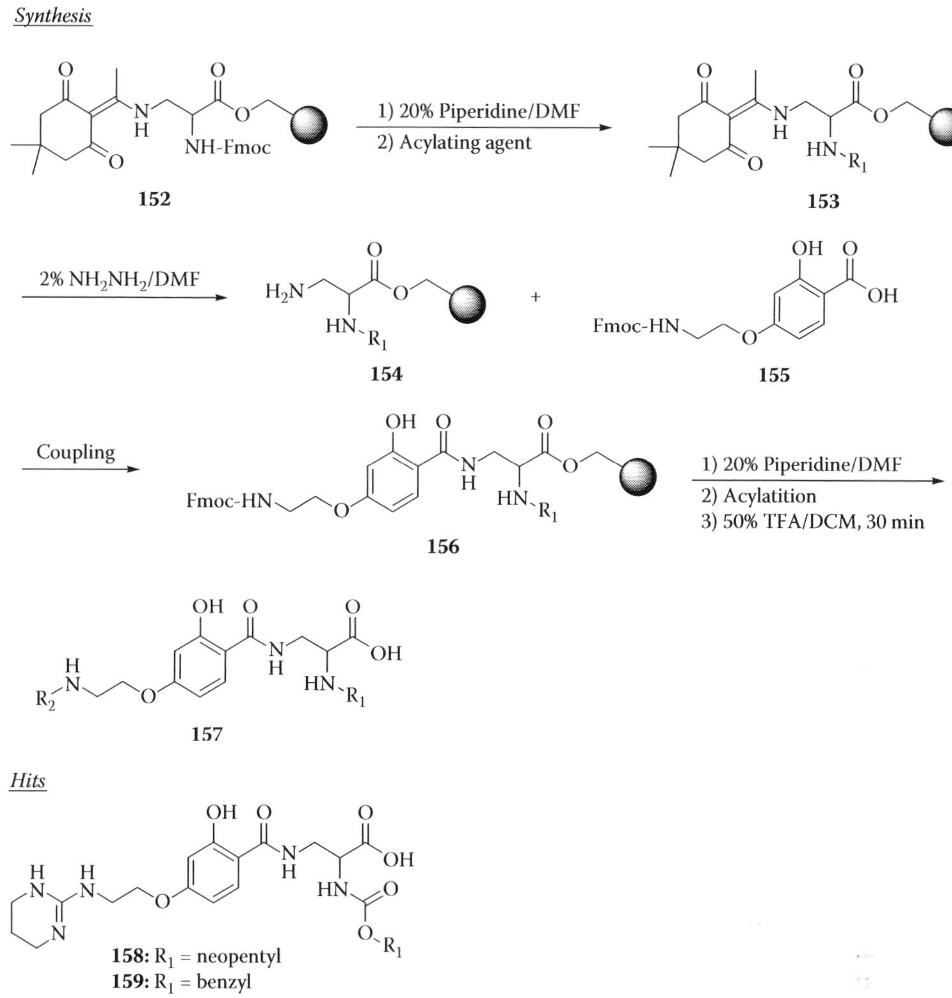

Synthesis

SCHEME 1.17 αvβ3 targeted library.

This research focused on two potent leads (**160:** IC_{50} ($\alpha_4\beta_7$) = 25 nM, IC_{50} ($\alpha_4\beta_1$) = 25.3 nM; **161:** IC_{50} ($\alpha_4\beta_7$) = 72 nM, IC_{50} ($\alpha_4\beta_1$) = 454 nM). Because these two leads share the same (S)-2--(2-chlorobenzamido)-3-phenylpropanoic acid core (**163**), the authors developed a solid-phase synthesis around this scaffold using an Argonaut Technologies' Quest 210 and a Robbins 48-well reaction block.

For the SAR exploration around lead **160,** the authors built a 100-member library anchoring 4-iodo-N-Fmoc-phenylalanine on Wang resin (**164**), removed the Fmoc group and acylated the amine with 2-chlorobenzoic acid, performed Suzuki couplings with formylarylboronic acids (**168**) in the presence of $PdCl_2(dppf)_2$, and then derivatized the aldehyde group with primary and secondary amines under reductive amination conditions. Cleavage with TFA/DCM afforded the desired aminomethylbiphenylalanines (**169**). This library indicated that the 4-hydroxylphenethylamino group was preferred at the *ortho*-position for activity (*ortho* > *meta* > *para*) and that the phenol group was essential for activity against $\alpha_4\beta_7$ (**171:** IC_{50} ($\alpha_4\beta_7$) = 736 nM, IC_{50} ($\alpha_4\beta_1$) = 51.1 nM). Interestingly, replacing the 4-hydroxylphenethylamino group with secondary amines afforded trisubstituted amines with more potent, lower-molecular-weight inhibitors (**172** and **173:** IC_{50} ($\alpha_4\beta_7$) ~ 10 nM, IC_{50} ($\alpha_4\beta_1$) ~ 20 nM).

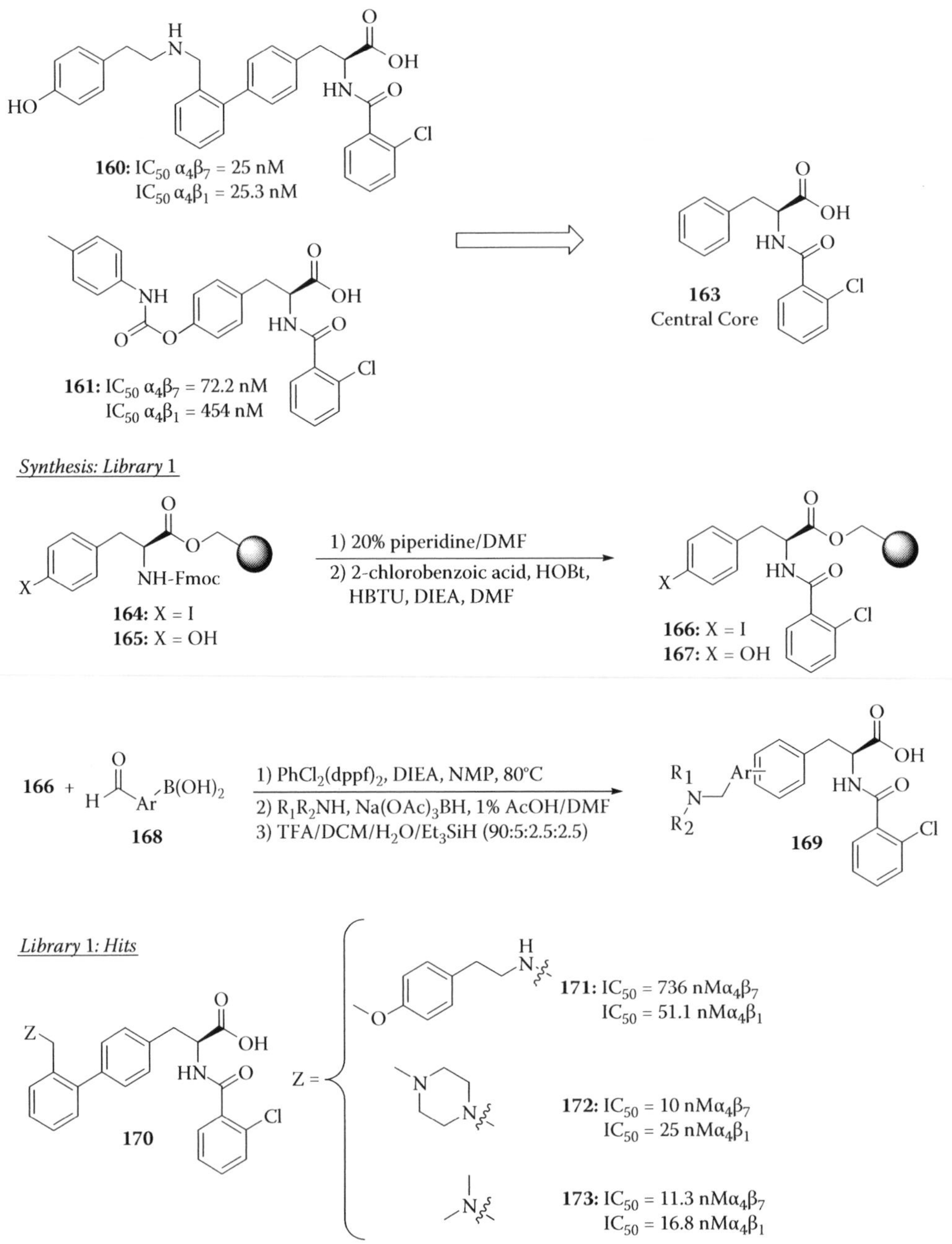

160: IC$_{50}$ $\alpha_4\beta_7$ = 25 nM
IC$_{50}$ $\alpha_4\beta_1$ = 25.3 nM

161: IC$_{50}$ $\alpha_4\beta_7$ = 72.2 nM
IC$_{50}$ $\alpha_4\beta_1$ = 454 nM

163
Central Core

Synthesis: Library 1

164: X = I
165: X = OH

1) 20% piperidine/DMF
2) 2-chlorobenzoic acid, HOBt, HBTU, DIEA, DMF

166: X = I
167: X = OH

166 + **168**

1) PhCl$_2$(dppf)$_2$, DIEA, NMP, 80°C
2) R$_1$R$_2$NH, Na(OAc)$_3$BH, 1% AcOH/DMF
3) TFA/DCM/H$_2$O/Et$_3$SiH (90:5:2.5:2.5)

169

Library 1: Hits

170

171: IC$_{50}$ = 736 nM$\alpha_4\beta_7$
IC$_{50}$ = 51.1 nM$\alpha_4\beta_1$

172: IC$_{50}$ = 10 nM$\alpha_4\beta_7$
IC$_{50}$ = 25 nM$\alpha_4\beta_1$

173: IC$_{50}$ = 11.3 nM$\alpha_4\beta_7$
IC$_{50}$ = 16.8 nM$\alpha_4\beta_1$

SCHEME 1.18 $\alpha_4\beta_1/\alpha_4\beta_7$ targeted library 1$_x$ and hits.

For the SAR exploration around lead **161,** the authors built an 80-member library anchoring N-Fmoc-tyrosine on Wang resin (**165**), removed the Fmoc group, and carried out the acylation with 2-chlorobenzoic acid as previously described (**167**). Then the phenol group was treated with 4-nitrophenyl chloroformate, and the 4-nitrophenyl carbonate intermediate was then treated with primary and secondary amines that, after TFA cleavage, produced the desired tryrosine carbamates (**174**). The SAR indicated that substituting aliphatic amines for the aniline was favorable for activity against $\alpha_4\beta_7$ (4-fold) and particularly against $\alpha_4\beta_1$ (>400-fold) (**175** and **176:** IC$_{50}$ ($\alpha_4\beta_7$) = 12–22 nM, and IC$_{50}$ ($\alpha_4\beta_1$) = 1–10 nM respectively).

Synthesis: Library 2

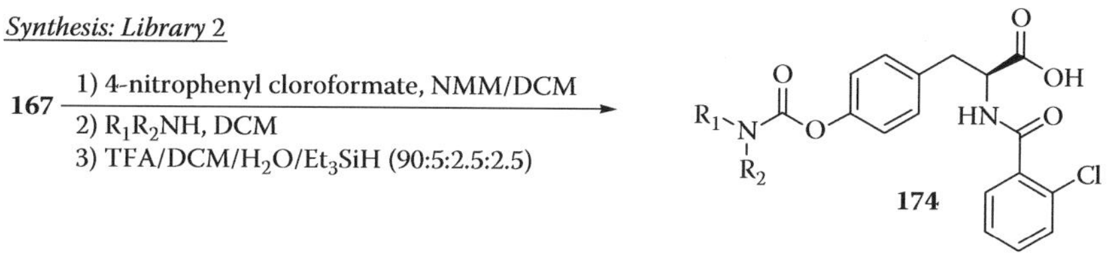

Library 2: Hits

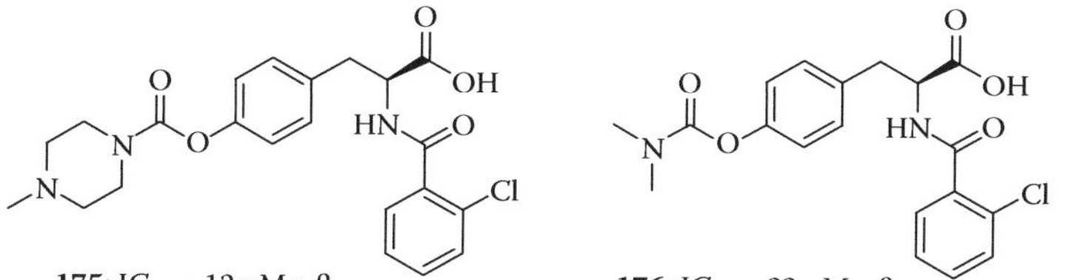

175: $IC_{50} = 12$ nM $\alpha_4\beta_7$
$IC_{50} = 1.1$ nM $\alpha_4\beta_1$

176: $IC_{50} = 22$ nM $\alpha_4\beta_7$
$IC_{50} = 9.3$ nM $\alpha_4\beta_1$

SCHEME 1.19 $\alpha_4\beta_1/\alpha_4\beta_7$ targeted library 2 and hits.

1.9 KINASE INHIBITORS

1.9.1 CASE 1: ADENOSINE KINASE

Bauser et al.[24] at Bayer AG reported the solution-phase synthesis in special glass microtiter plates of 2-aryl 7-amino oxazolo-pyrimidines (**178**) as adenosine kinase inhibitors. Adenosine (**177**) is a purine-based nucleoside with a 10- to 30-second half-life that functions as an extracellular signaling molecule mediating several physiological responses. Cells under stress conditions (e.g., ischemia, pain, inflammation) release adenosine. Adenosine then activates specific extracellular purinergic receptors (P1 receptors). inducing protective pharmacological responses such as analgesic and anti-convulsive effects. Because adenosine is taken into the cell and converted into AMP by adenosine kinase (AK), the extracellular adenosine concentration is reduced. Consequently, inhibition of AK would allow adenosine to be transported back to the extracellular space by adenosine transporters, resulting in an increase of extracellular adenosine concentration.

Using reported AK inhibitors and the AK x-ray structure, the authors proposed the 2-aryl-7-amino-oxazolo[5,4-*d*]pyrimidine (**178**) as a potential scaffold for AK inhibitors. For the parallel synthesis of this compound class, a three-step route was devised using commercially available 5-amino-4,6-dihydroxy pyrimidine (**179**) as the starting material. Acylation with benzoyl chlorides followed by treatment with polyphosphoric acid (PPA) afforded the desired 7-hydroxy oxazolo pyrimidines while condensation with benzoic anhydride produced directly the desired product (**180**). Treatment of these 7-hydroxy intermediates with $POCl_3$ provided their corresponding chlorides (**181**). The chloride was then successfully displaced with a broad range of amines (**182**), including primary, secondary, hindered amines, and anilines to produce the desired 2-aryl-7-amino-oxazolo[5,4-*d*]pyrimidines (**183**). Intermediates bearing a bromophenyl oxazolo unit (**184**) were further reacted with boronic acids (**185**) under Suzuki coupling conditions to afford biaryl compounds (**186**). Intermediates bearing a nitrophenyl oxazolo unit (**187**) were reduced to their corresponding anilines and then acylated with a set of carboxylic acids with TBTU to produce amides (**188**).

For the first exploratory library, more than 800 compounds were made where the unsubstituted phenyl group at position-2 was kept constant, changing the amine groups at position-7. Preliminary SAR indicated that aryl amines are necessary for activity with a spacer between the aryl moiety and the amine. It was also observed that activity decreased with shorter length spacers (**189** vs. **190**). The most potent analog was derived from racemic benzyl bis-pyrrolidine (**191:** $IC_{50} = 200$ nM

± 18 nM). Both enantiomers of **191** were made and the *trans*-enantiomer was found to be the one eliciting potency against AK. For practical purposes, the authors decided to use the racemic benzyl bis-pyrrolidine unit for the optimization phase.

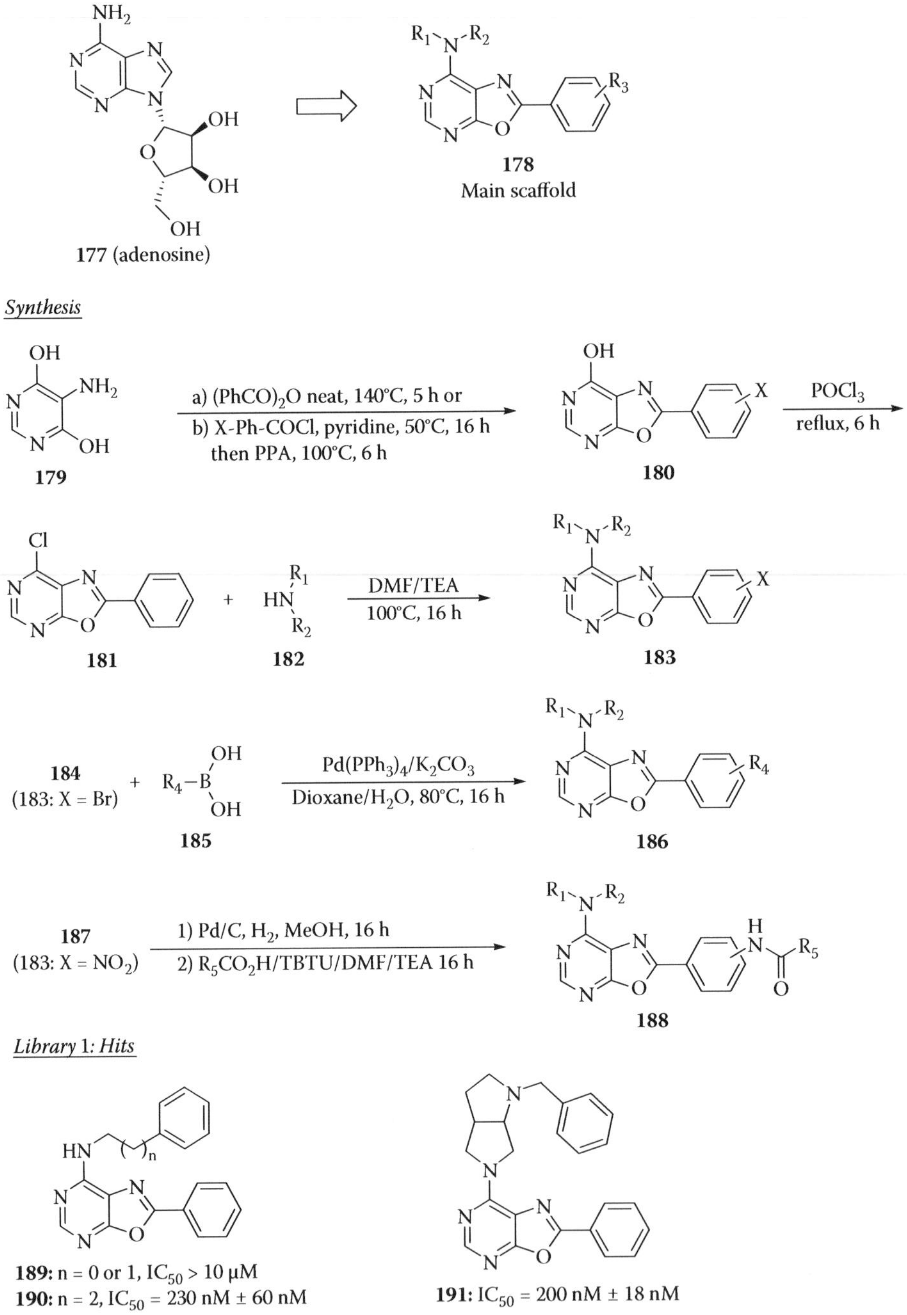

177 (adenosine)   **178**
                             Main scaffold

Synthesis

Library 1: Hits

189: n = 0 or 1, IC$_{50}$ > 10 μM
190: n = 2, IC$_{50}$ = 230 nM ± 60 nM   **191:** IC$_{50}$ = 200 nM ± 18 nM

SCHEME 1.20 AK targeted library 1.

The next efforts centered on the optimization of the aryl group at the 2-position, keeping the racemic benzyl bis-pyrrolidine unit constant. More than 60 biaryl compounds and 100 amides were made. The SAR on the biaryl compounds revealed that substitution at the *para*-position provided better activity compared to *meta*-substitution (**192:** IC$_{50}$ = 90 nM ± 40 nM).

When optimizing the amide analogs, it was also noticed that *para*-amides were more potent than the *meta*-amides (**193**: IC_{50} = 30 nM ± 80 nM). The SAR indicated that the *para*-amide analogs were more potent compared to the biaryl analogs. With the amide SAR at hand, the authors decided to explore the effects of changing the phenyl with a pyridyl group, keeping the amine group constant at the 4-position. To their delight, they found that a 3-pyridyl group with an N,N,-dimethylamino group at the 4-position exhibited activity in the single-digit nanomolar range (**194**: IC_{50} = 10 nM ± 6 nM).

Optimization library: Hits

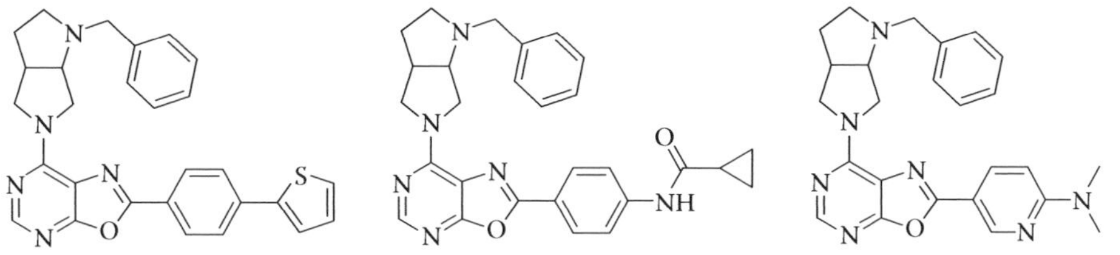

192: IC_{50} = 90 nM ± 40 nM   **193**: IC_{50} = 30 nM ± 30 nM   **194**: IC_{50} = 10 nM ± 6 nM

SCHEME 1.21   AK targeted library: optimized hits.

1.9.2   CASE 2: AKT/PKB

Zhao et al.[25] at Merck reported a microwave-based synthetic strategy approach for the synthesis of 2,3,5-trisubstituted pyridines as Akt-1 and Akt-2 dual inhibitors. Akt (a.k.a. PKB) is an important serine/threonine kinase involved in the apoptotic signaling pathway. There are three known human Akt isozymes (Akt-1/Akt-α, Akt-2/Akt-β, and Akt-3/Akt-γ) possessing an N-terminal Pleckstrin homology (PH) domain and all isozymes sharing greater than 80% homology. Extracellular stimulation of the Akt pathway induces translocation of Akt to the plasma membrane where, through its PH domain, it interacts with PIP3 forming a complex that allows Akt phosphorylation/activation.

A high-throughput screening exercise identified the 2,3-diphenylquinoxaline **195** as an Akt-1 inhibitor (IC_{50} = 2.5 μM). An ATP competition assay indicated that this hit was not an ATP-competitive kinase inhibitor. Using an additional assay with mutated Akt isozymes with no PH-domain, it was also noticed that **195** did not inhibit Akt. This led to the conclusion that **195** is an allosteric Akt kinase inhibitor. Focusing their efforts on the main scaffold (**196**), the authors developed a microwave-based synthetic methodology for the synthesis of a 2,3-diphenylquinoxaline library. This approach involved the synthesis of benzil amines (**199**) from commercially available *p*-bromomethyl benzil (**197**) and amines (**198**) in the presence of polymer-bound DIEA followed by the use of Fluorous Technonologies' fluourous-tagged thiol as a scavenger for the quick and efficient removal of excess **197**. Benzil amines (**199**) were then cyclized with 1,2-diaminobenzenes (**200**) under microwave irradiation conditions to produce the desired quinoxalines (**196**).

The first library consisted of 200 compounds, from which **201** was identified as a potent Akt-1 inhibitor (IC_{50} = 290 nM) with high selectivity over Akt-2, Akt-3, PAK, and PKC. Unfortunately, **201** lacked solubility and was found to be inactive in cell-based assays. To improve the physicochemical properties of future analogs, three additional heterocyclic compound classes were explored. Using benzil amine **202** and 24 different α-aminocarboxamides (**203**), a focused library of two 5,6-diphenylpyrazin-2(1H)-one regioisomers was made (**204**, **205**). Each regioisomer exhibited a unique inhibition profile where the nature of a substituent at position-3 (R_4) dictated the degree of Akt isozyme selectivity (**206** vs. **207**: Akt-1 > Akt-2, and **209** vs. **208**: Akt-2 > Akt-1).

Cell-based assays using A2780 human ovarian tumor cells pretreated with either **206** or **209** followed by incubation with doxorubicin resulted in a threefold caspase-3 activity vs. doxorubicin

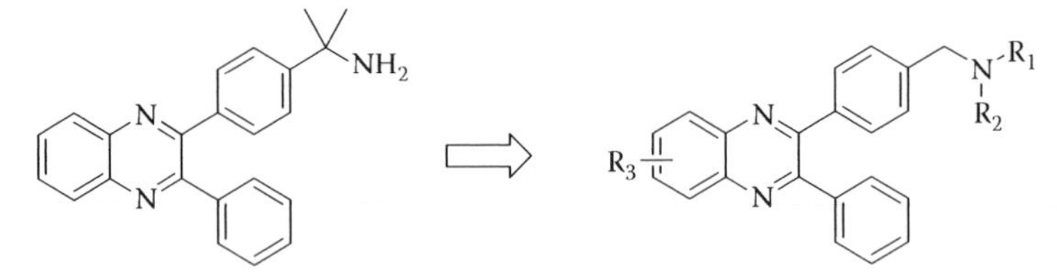
NH2
N
N
R3
R1
R2

195: IC_{50} = 2.5 μM (Akt-1)

196
Central Scaffold

Synthesis: Library 1

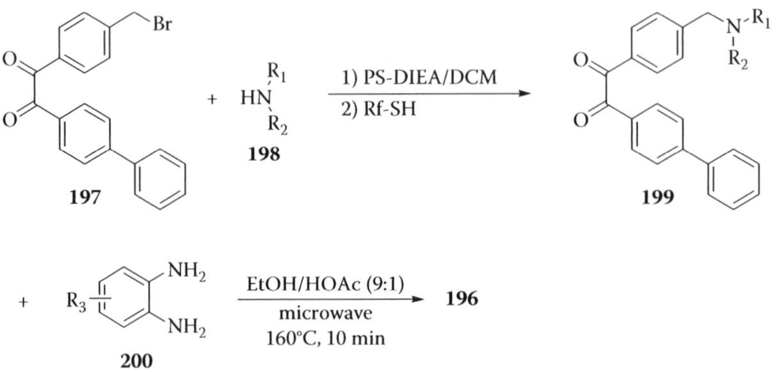
Br
O
O
+ HN
R1
R2
198
1) PS-DIEA/DCM
2) Rf-SH
197
199
+ R3
NH2
NH2
200
EtOH/HOAc (9:1)
microwave
160°C, 10 min
196

First Library: Hit

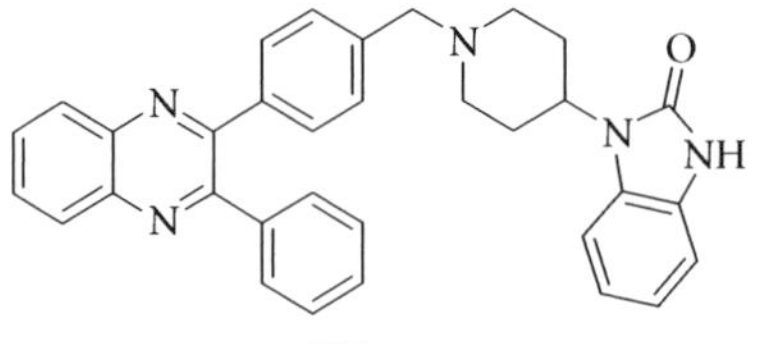
N
O
NH

201
IC_{50} = 290 nM (Akt-1)
IC_{50} = 2090 nM (Akt-2)
IC_{50} > 50,000 nM (Akt-3, PKA, PKC)

SCHEME 1.22 Akt targeted library 1.

alone. Interestingly, when A2780 cells were pretreated with a 1:1 mixture of **206** and **209**, a tenfold caspase-3 activity was detected. This led to the conclusion that a dual Akt1/Akt2 kinase inhibition approach provides a higher apoptotic response compared to targeting a single Akt isozyme.

With this new piece of information, the authors pursued the synthesis of dual Akt-1/Akt-2 inhibitors. Leveraging the synthetic methodology already developed, the authors used **202** and 50 diverse aryl-1,2-diamines (**200**) to explore a further SAR on the quinoxaline core. This focused library afforded tetrazoles (regioisomeric mixtures), 1H-pyrazole, and imidazole analogs with potent dual Akt-1/Akt-2 inhibition. Tetrazole analogs did not show activity in cells, likely due to their zwitterionic nature. On the other hand, imidazole quinoxaline analog **210** was found to be a potent dual Akt1/Akt2 inhibitor in A2780, LNCaP, HT29, and MCF7 cancer cells (Akt-1: IC_{50} = 59 nM, Akt-2: IC_{50} = 210 nM). *In vivo* testing of **210** in a mouse model (50 mpk IP administration, 3 doses, every 90 min) showed that this analog had a plasma concentration (Cp) of 1.5 to 2.0 μM, and it was efficacious in inhibiting phosphorylation of both Akt-1 and Akt-2 in the lungs of mice injected with IGF.

Library 2: Regioisomers

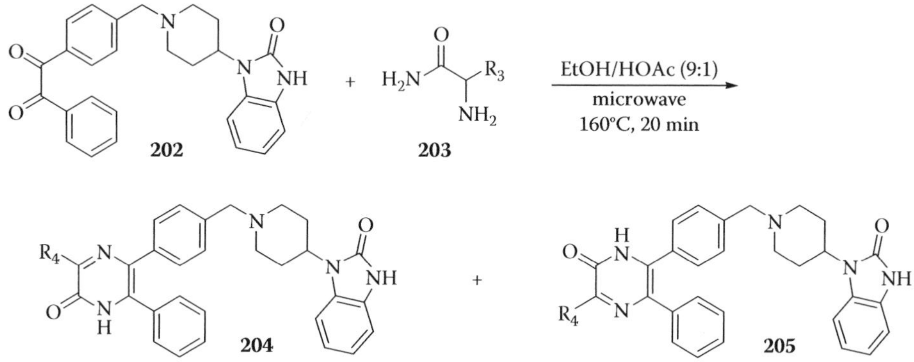

Library 2: Hits

206: R_4 = CH$_4$, IC$_{50}$ = 760 nM (Akt-1)
        IC$_{50}$ > 24,000 nM (Akt-2)
208: R_4 = isobutyl, IC$_{50}$ > 18,000 nM (Akt-1 & Akt-2)

207: R_4 = CH$_3$, IC$_{50}$ ~ 1000 nM (Akt-1 & Akt-2)
209: R_4 = isobutyl, IC$_{50}$ > 20,000 nM (Akt-1)
        IC$_{50}$ = 325 nM (Akt-2)

Library 3: Dual Akt-1/Akt-2 inhibitor

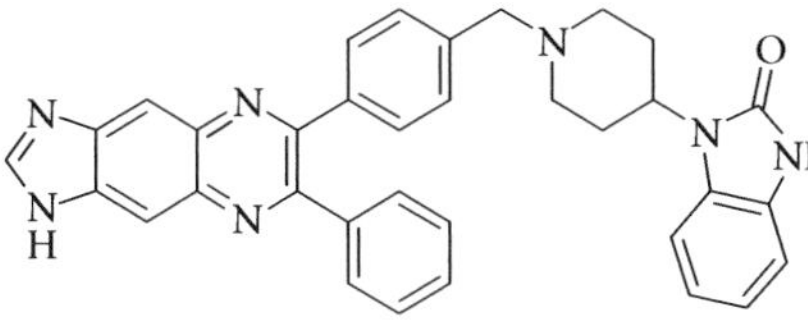

210: IC$_{50}$ = 58 nM (Akt-1)
        IC$_{50}$ = 210 nM (Akt-2)

SCHEME 1.23   Akt targeted library 2 and library 3.

1.9.3   CASE 3: AURORA-A

Fancelli et al.[26] at Nerviano Medical Sciences disclosed the solid-phase synthesis of a pyrrole-fused 3-amino-pyrazole bicyclic scaffold as potent and selective Aurora kinase inhibitors. The Aurora kinases are Ser/Thr protein kinases that play a crucial role in cell mitosis, such as bipolar spindle formation and centrosome duplication. There are three known isozymes: Aurora-A, Aurora-B, and Aurora-C. Because these kinases are found highly expressed in cancer cells, they offer a potential route for the use of Aurora inhibitors as anti-cancer therapeutic agents.

The authors designed the pyrrole-fused 3-amino-pyrazole bicyclic scaffold (**211**) based on the fact that the 3-amino-pyrazole moiety has been reported as an adenino mimetic pharmacophore that binds to the backbone of the enzyme in the kinase hinge region of the ATP pocket. For the rapid synthesis of analogs, a solid-phase route was developed using polystyrene isocyanate resin (**212**) to immobilize Boc-protected tetrahydropyrrolopyrazole (**213**) via the NH-pyrazole. The free amino group of the Boc-protected urea-linked intermediate (**214**) was then acylated with a set of 22 acyl chlorides (**215**). The resulting amides were then treated with 50% TFA/DCM to remove the Boc protecting group. Then the NH of the pyrrole unit was acylated with a 46-compound set of both iso-cyanates (**216**) and carboxylic acids (**217**) to afford the corresponding urea and amide analogs (**218** and **219**, respectively). The final products were cleaved using a combination of aqueous NaOH in MeOH at 40°C for 3 days. The crude mixtures were then neutralized with HCl and filtered through silica-packed cartridges. This methodology afforded a 1012-compound library (**220** and **221**), iso-lating about 10 to 40 mg of final product. The authors also used a nearly identical solution-phase approach where the resin was changed with COOEt.

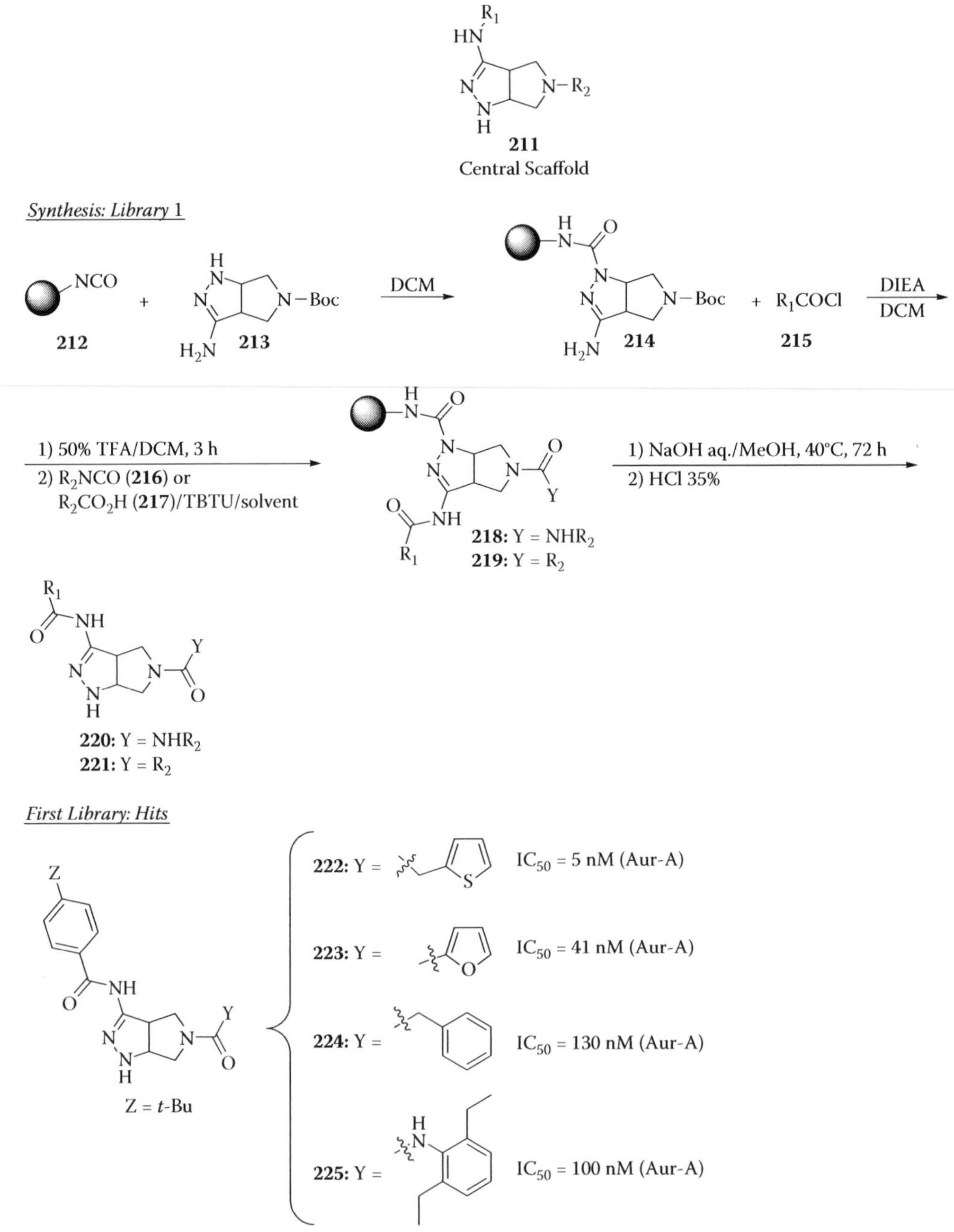

211
Central Scaffold

Synthesis: Library 1

212   **213**   **214**   **215**

1) 50% TFA/DCM, 3 h
2) R_2NCO (**216**) or
   R_2CO_2H (**217**)/TBTU/solvent

218: Y = NHR$_2$
219: Y = R$_2$

1) NaOH aq./MeOH, 40°C, 72 h
2) HCl 35%

220: Y = NHR$_2$
221: Y = R$_2$

First Library: Hits

Z = *t*-Bu

222: Y =   IC$_{50}$ = 5 nM (Aur-A)

223: Y =   IC$_{50}$ = 41 nM (Aur-A)

224: Y =   IC$_{50}$ = 130 nM (Aur-A)

225: Y =   IC$_{50}$ = 100 nM (Aur-A)

SCHEME 1.24   Aurora targeted library 1.

Screening the library against Aurora-A clearly indicated that acylation of the 3-amino group with 4′-*tert*-butyl-substituted benzamido groups was favored for potency. On the pyrrole site, four acylating agents (three acyl chlorides and one isocyanate) provided potent compounds (**223** to **225**) down to the single-digit nanomolar potency (**222**). However, all these compounds exhibit a low-solubility profile in buffer pH 7, a serious obstacle for further development.

To improve the solubility profile of these compounds, the authors changed the hydrophobic *tert*-butyl group with a more water-soluble 4-(4-N-methyl)piperazin-yl group. This strategy paid off as the corresponding analogs of **222** to **225** not only exhibited a better solubility profile, they also exhibited nanomolar potency in both *in vitro* and cell-based assays in the HCT-116 cell line (**226** to **229**). The most potent urea compound **229** proved more soluble in buffer pH 7 (>200 µM), and both enzyme and cell-based inhibition potency below 45 nM. The x-ray structure of **229** with Aurora-A was resolved and it proved that **229** interacted with the enzyme in the hinge region as initially hypothesized.

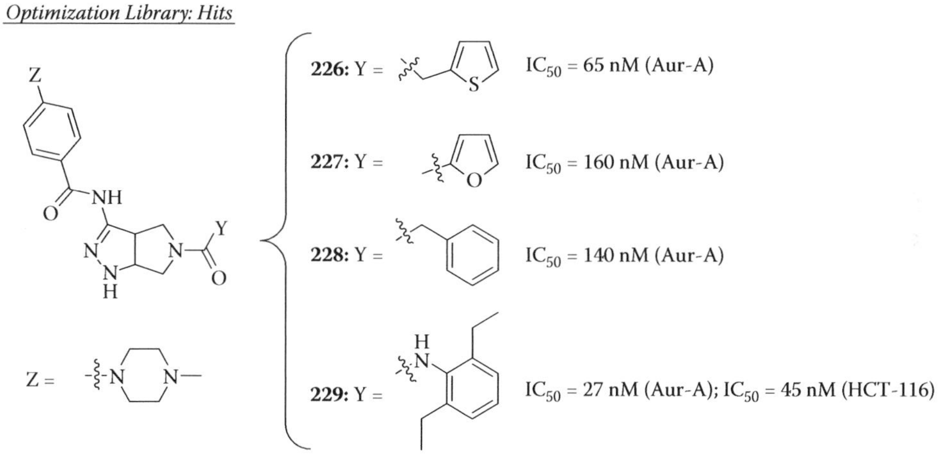

Optimization Library: Hits

226: Y = IC_{50} = 65 nM (Aur-A)

227: Y = IC_{50} = 160 nM (Aur-A)

228: Y = IC_{50} = 140 nM (Aur-A)

229: Y = IC_{50} = 27 nM (Aur-A); IC_{50} = 45 nM (HCT-116)

SCHEME 1.25 Aurora targeted optimized hits.

1.9.4 CASE 4: CDK2

Kim et al.[27] at Bristol-Myers Squibb reported the synthesis of aminothiazoles as inhibitors of cyclin-dependent kinase 2 (CDK2). Cyclin-dependent kinases (CDKs) are heterodimeric serine/threonine protein kinases that control and regulate cell cycle, transcription, cell proliferation, as well as several growth-regulatory signals. The cell cycle consists of four well-defined phases: (1) G1 (Gap1) to initiate DNA synthesis, (2) G2 (Gap2) for cell division, (3) S (Synthesis) for DNA replication, and (4) M (Mitosis) for cell division. There are 13 different human CDKs identified, many of which form complexes with cyclin proteins (cyc) that then participate in the cell cycle process. It has been determined that misregulation of CDKs is frequently found in cancer cells, particularly CDK2 and CDK4. Agents that bind to CKDs to negatively regulate cdk and cyclin/cdk complexes could have a significant impact in cancer treatment.

A high-throughput screening campaign at Bristol-Myers Squibb for CDK inhibition identified the 2-amino-5-thio-substituted thiazole **230** as a low micromolar inhibitor of CDK1/cyc B, CDK2/cyc E, and CDK4/cyc D, with particular selectivity for CDK2/cyc E (IC_{50} = 0.17 µM). Centering on scaffold **231**, the authors devised both a solution- and solid-phase approach for the rapid exploration of this chemotype.

For the solution-phase approach, the authors synthesized **232** as the key building block for the exploration of the thio ether unit. In this case, **232** was treated with an alkoxide (KOtBu or NaOEt) to make *in situ* the thiolate intermediate **233,** which then was alkylated with a diverse range of electrophiles (e.g., epoxides, primary alkyl bromides, chlorides, methyl sulfonates, etc.) to afford desired products (**234**).

For the solid-phase approach, the authors used the acid-labile SASRIN resin (235). In this approach, the benzyl alcohol of SASRIN was converted into the chloride using triphosgene and PPh$_3$ in DCM. The resulting SASRIN chloride (236) resin was then alkylated with 2,2,2-trifluoro-N-(5-thiocyanatothiazol-2-yl)acetamide (237) in the presence of DIEA in DCM. The resulting resin-bound thiocyanate (238) was then treated with dithiothreitol (DTT), affording the resin-bound thiol intermediate 239, which was then alkylated with several bromides and chlorides in the presence of DBU in DMF at 70°C (240). The following step involved the removal of the trifluoroacetamide group of the thio ether intermediates (240) with NaBH$_4$ in EtOH/THF (241). The resulting secondary amine analogs (241) were then acylated with several acyl chlorides. The final products (243) were then cleaved from the resin using 50% TFA in DCM.

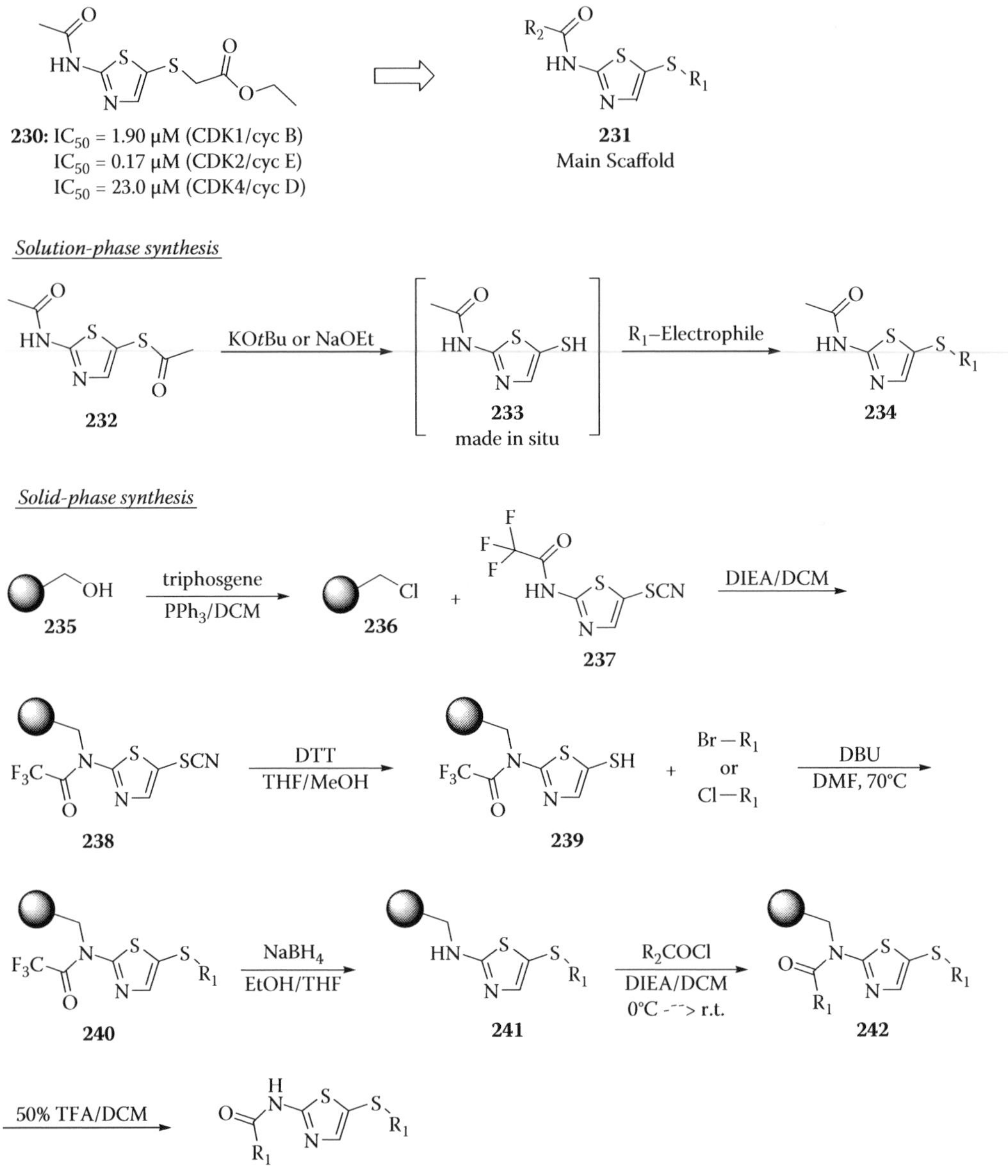

230: IC$_{50}$ = 1.90 μM (CDK1/cyc B)
IC$_{50}$ = 0.17 μM (CDK2/cyc E)
IC$_{50}$ = 23.0 μM (CDK4/cyc D)

231
Main Scaffold

Solution-phase synthesis

Solid-phase synthesis

SCHEME 1.26 CDK2 targeted libraries.

Screening of the first total library (solution-phase + solid-phase) showed that replacing the ethyl ester group of **230** with bulky substituents improved activity threefold (**244**: $IC_{50} = 0.05$ μM for CDK2/cyc E). Ketones, ethers, and alcohols proved less potent (single-digit micromolar potency), and replacing the ethyl ester moiety with a carboxylic acid proved detrimental for activity (**245**: $IC_{50} > 25$ μM for CDK2/cyc E). The authors noted that **230** did not exhibit any activity in a cell proliferation assay using the ovarian tumor cell line A2780. An explanation for this was that the ester group of **230** was metabolized and converted into its corresponding carboxylic acid **245**, which was already proved to lack activity in the enzyme-based assay. To avoid future metabolism liabilities, the authors decided to find a suitable ester isostere to improve compound stability. It was determined that 5-alkyloxazole substituents proved very potent, and that potency increased with branched sp^3 alkyl groups along with the bulkiness of the substituents (Me < Et < *i*-Pr, < *t*-Bu ≈ CH_2-c-hexyl). From all these combinations it was the *t*-Bu analog that exhibited low nanomolar potency in both enzyme-based and cell proliferation assays along with high selectivity against CDK2/cyc E (**246**: $IC_{50} = 6$ nM, $IC_{50} = 40$ nM in A2780 cells; CDK2/cyc E > CDK1/cyc B1 >> CDK4/cyc D1; ratio = 1:8:138, respectively).

Library 1: Hits

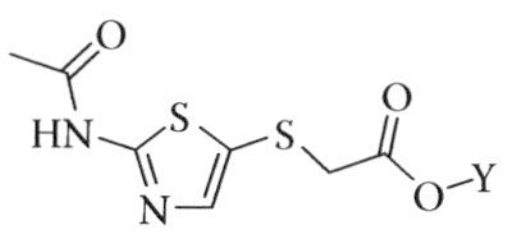

244: Y = *t*-Bu, $IC_{50} = 0.05$ μM (CDK2/cyc E)
245: Y = H, $IC_{50} > 25$ μM (CDK2/cyc E)

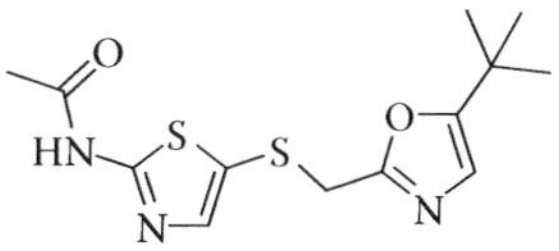

246: $IC_{50} = 6$ nM (CDK2/cyc E)
$IC_{50} = 60$ nM (CDK1/cyc B1)
$IC_{50} = 690$ nM (CDK4/cyc D1)
$IC_{50} = 50$ nM (A2780 cells)

First optimization library: Hit

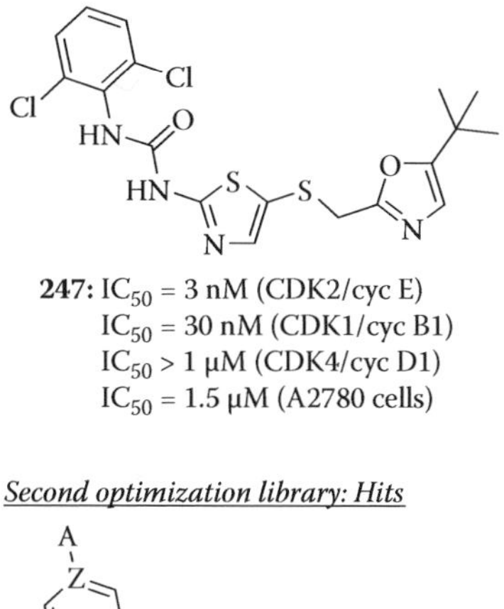

247: $IC_{50} = 3$ nM (CDK2/cyc E)
$IC_{50} = 30$ nM (CDK1/cyc B1)
$IC_{50} > 1$ μM (CDK4/cyc D1)
$IC_{50} = 1.5$ μM (A2780 cells)

Second optimization library: Hits

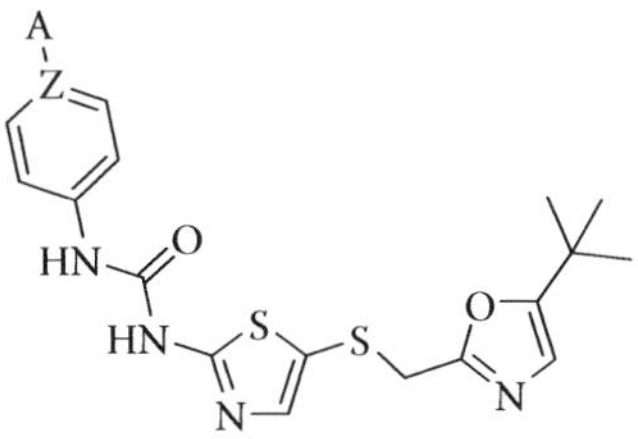

248: Z = C, A = $CH_2NHCH(CH_2OH)_2$
$IC_{50} = 3$ nM (CDK2/cyc E)
$IC_{50} = 30$ nM (CDK1/cyc B1)
$IC_{50} = 100$ nM (CDK4/cyc D1)
$IC_{50} = 30$ nM (A2780 cells)

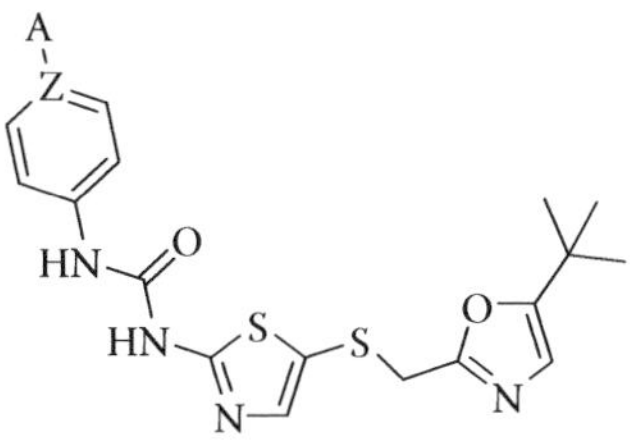

249: Z = N, A = none
$IC_{50} = 5$ nM (CDK2/cyc E)
$IC_{50} = 80$ nM (CDK1/cyc B1)
$IC_{50} = 1090$ nM (CDK4/cyc D1)
$IC_{50} = 240$ nM (A2780 cells)

SCHEME 1.27 CDK2 targeted libraries: hits.

For the next optimization library, the authors decided to maintain constant the 5-*tert*-butyl oxazole substituent and explore substituent effects on the 2-amino group of the thiazole ring. The authors probed the effects of amides, carbamates, and ureas, and found that while the carbamates were the less potent ones, some amides and ureas exhibited single-digit nanomolar potency in the enzyme inhibition assay. However, the SAR between the amides and ureas were opposite of one another. While small alkyl sp^3 substituents increased potency for the amides (Me ≈ *i*-Pr > *t*-Bu > Ph), for the ureas substituted phenyl groups were more potent than simple alkyl substituents. Urea analog **247** showed good potency and selectivity against CDK2/cyc E (**247**: CDK2/cyc E: IC$_{50}$ = 3 nM, CDK1/cyc B1: IC$_{50}$ = 30 nM, CDK4/cyc D1: IC$_{50}$ > 1 µM). Unfortunately, the pharmacokinetic profile of **247** was less favorable, resulting in a dramatic loss of activity in the cell proliferation assay (IC$_{50}$ = 1.5 µM).

To address this issue, the authors removed the two chloro atoms and included a polar amino alcohol group on the phenyl ring (**248**) as well as a pyridyl group (**249**). Both changes afforded inhibitors with a similar enzyme-inhibition and selectivity profile as **247**, but they differ in the cell-based assay where amino-alcohol-containing analog **248** exhibited more potency than **249** (**248**: IC$_{50}$ = 30 nM vs. **249**: IC$_{50}$ = 240 nM in A2780 cells). Screening **248** against a kinase panel (e.g., several PKC isozymes, Lck, EMT, Zap70, Her1, Her2, IGFR), it was determined that this compound is very selective toward CDKs because inhibition for other kinases was in the micromolar range (IC$_{50}$ > 5600 nM). Compound **248** also exhibited a favorable pharmacokinetics profile in mice (IV dose, 10 mg/kg) with a t$_{1/2}$ = 2.2 hr, MRT = 1.8 hr, %F = 24 (oral PO dose, 10 mg/kg), and % protein binding = 79.

1.9.5 CASE 5: DNA-PK

Hardcastle et al.[28] at the Northern Institute for Cancer Research (United Kingdom) developed a solution- and a solid-phase approach for the synthesis of chromen-4-one-based analogs as DNA-dependent protein kinase (DNA-PK) inhibitors. DNA-PK is a trimeric nuclear serine/threonine kinase complex that belongs to the phosphatidylinositol (PI) 3-kinase-like (PIKK) family, and it is involved in the signaling of cellular stress responses. DNA-PK binds to DNA double-strand breaks (DSBs) at which point it becomes an active kinase posphorylating itself (autophosphorylation) as well as additional downstream targets. DNA-PK plays a role in the DNA repair process. It has been observed that cells with a defective DNA-repair mechanism are both highly radiosensitive and susceptible to cytotoxic agents that induce DNA DSBs repair.

Using reported chromenone-based morpholine-containing DNA-PK inhibitors as leads (**250** to **253**), the authors produced several small focused libraries around scaffold **254** to devise an SAR for this compound class.

Both solution- and solid-phase approaches were based practically on the same strategy. Substituted 2-hydroxylpher methyl ketones (**255**) were treated with CS$_2$ to produce 4-hydroxy-2H-chromene-2-thiones (**256**). To immobilize **256**, the authors used Merrifield resin (**257**) in the presence of DBU, where the 4H-chromen-4-one thiols were immobilized as thio ethers (**259**). For the solution-phase approach, **256** were reacted with ethyl iodide (**258**) to afford 2-(ethylthio)-4H-chromen-4-ones (**259**). Here, the authors derivatized Br or OTf intermediates **259** into biaryl **261** using Suzuki coupling conditions. A solid-phase approach was used to alkylate phenol-containing **260** using alkyl bormides or alcohols (Mitsunobu alkylations). Polymer-supported thioether intermediates **261** were oxidized into their respective sulfones (**262**) with mPCBA. Solid-supported sulfones **262** were then reacted with secondary amines (**263**) to afford the desired chromenones **254**. Thioethyl ether intermediates **261** did not need to be oxidized, and direct treatment with secondary amines afforded the desired chromenones **254**.

Initial efforts centered on analogs of **251** because this regioisomer was the most potent one. Direct analogs were made where the morpholine group was changed with different amines. The SAR obtained showed that short-length linear, branched, benzylic, and cyclic (five-member rings)

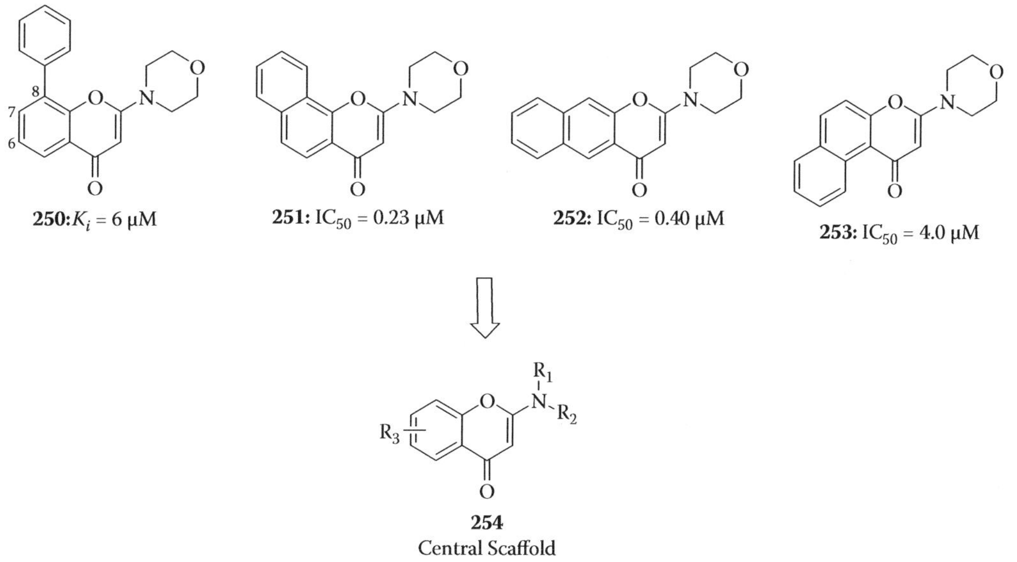

250: K_i = 6 μM   **251:** IC$_{50}$ = 0.23 μM   **252:** IC$_{50}$ = 0.40 μM   **253:** IC$_{50}$ = 4.0 μM

254
Central Scaffold

SCHEME 1.28   DNA-PK leads.

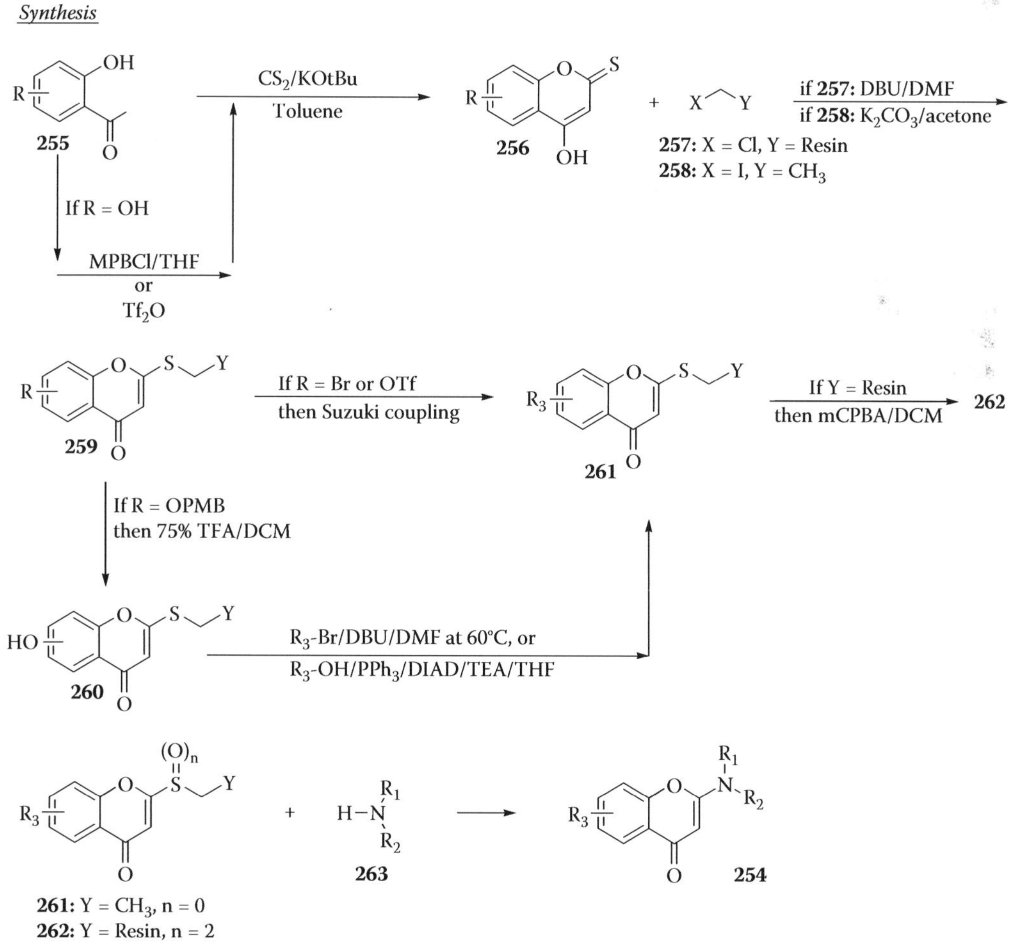

Synthesis

CS$_2$/KOtBu
Toluene

255

256

X-Y

257: X = Cl, Y = Resin
258: X = I, Y = CH$_3$

if **257:** DBU/DMF
if **258:** K$_2$CO$_3$/acetone

If R = OH

MPBCl/THF
or
Tf$_2$O

259

If R = Br or OTf
then Suzuki coupling

261

If Y = Resin
then mCPBA/DCM   **262**

If R = OPMB
then 75% TFA/DCM

260

R$_3$-Br/DBU/DMF at 60°C, or
R$_3$-OH/PPh$_3$/DIAD/TEA/THF

261: Y = CH$_3$, n = 0
262: Y = Resin, n = 2

H–NR_1R_2

263

254

SCHEME 1.29   DNA-PK targeted library synthesis.

did not improve activity at all. In fact, only two six-member ring secondary amines (**264:** piperidine, **265:** thiomorpholine) exhibited inhibition in the single-digit micromolar range; however, neither one was more potent than **251** (morpholine group).

The authors then turned their efforts to the optimization of **250**. Keeping the morpholine group constant, the authors used phenol analogs probing systematically both substitution effects as well as phenol isomer effects (**260:** 6- and 7-hydroxy-chromenones). The SAR obtained indicated that 7-alkoxy-chromenones were, in general, more potent than the corresponding 6-alkoxy-chromenones. Although no clear SAR was observed on the 7-alkoxy-chromenone series, the observed trend was that benzyl analogs were more potent than phenetyl and alkyl analogs. All three monochloro-substituted benzyl analogs, in particular, reached sub-micromolar activities (**266** to **268:** *para-* ~ *meta-* > *ortho-*).

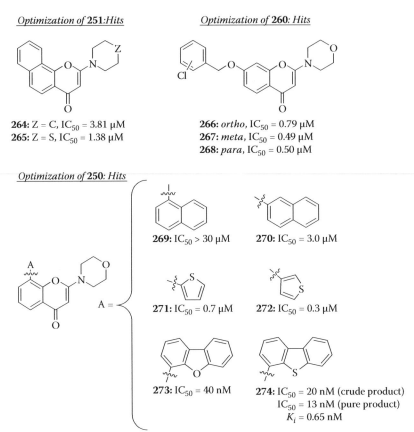

Optimization of **251**:*Hits*

264: Z = C, IC$_{50}$ = 3.81 μM
265: Z = S, IC$_{50}$ = 1.38 μM

Optimization of **260**: *Hits*

266: *ortho,* IC$_{50}$ = 0.79 μM
267: *meta,* IC$_{50}$ = 0.49 μM
268: *para,* IC$_{50}$ = 0.50 μM

Optimization of **250**: *Hits*

269: IC$_{50}$ > 30 μM   **270:** IC$_{50}$ = 3.0 μM

271: IC$_{50}$ = 0.7 μM   **272:** IC$_{50}$ = 0.3 μM

273: IC$_{50}$ = 40 nM   **274:** IC$_{50}$ = 20 nM (crude product)
                                 IC$_{50}$ = 13 nM (pure product)
                                 K_i = 0.65 nM

SCHEME 1.30   DNA-PK targeted library: hits.

For direct analogs of **250**, the authors synthesized biaryl and 1,2-diphenylethene derivatives (**261**) from bromide or triflate intermediates (**259**) using a diverse array of aryl and vinyl boronic acids. The authors systematically synthesized the same analogs for the three aromatic isomers (6-, 7-, and 8-chromenones). The SAR in this series clearly indicated that only the 8-substituted chromenones exhibited activity against DNA-PK. Replacing the phenyl group in the core by 1-naphthyl was detrimental for activity, although the 2-naphthyl analog exhibited single-digit micromolar potency (**269:** IC$_{50}$ > 30 μM and **270:** IC$_{50}$ = 3.0 μM, respectively). Substitution of the phenyl group with the well-known thiophene isostere increased potency into the sub-micromolar range, the 3-thionyl analog being the most potent one and twice as potent as the 2-thionyl analog (**271:** IC$_{50}$ = 0.7 μM, **272:** IC$_{50}$ = 0.3 μM). *Ortho*-phenyl substituents showed a dramatic loss in activity, indicating that steric constraints were playing a significant role in this series.

From all the substituents used, the dibenzofuran analog **273** showed significant improvement in activity in the low nanomolar range (**273:** $IC_{50} = 40$ nM). Interestingly, when the furan oxygen was replaced with a sulfur atom (dibenzothiophene analog **274**), this new analog exhibited a twofold increase in potency (**274:** $IC_{50} = 20$ nM). Compound **274** was resynthesized, purified, and re-assayed against DNA-PK to give an IC_{50} of 13 nM, confirming its inhibitory potency against DNA-PK. Compound **274** was also confirmed to be an ATP competitor with a Ki of 0.65 nM, the most potent DNA-PK inhibitor reported thus far.

1.9.6   CASE 6: GSK-3

Naerum et al.[29] at Novo Nordisk reported a work using a combination of real and virtual high-throughput screening using topological patterns with predicted molecular descriptors for the discovery of new glycogen synthase kinase-3 (GSK-3) inhibitors via scaffold hopping. GSK-3 is a serine-threonine kinase that plays an important role in glycogen synthesis, being directly implicated in metabolic diseases such as diabetes. GSK-3 also targets proteins implicated in neurological disorders and cancer. To date, there are two documented GSK-3 isozymes: GSK-3α and GSK-3β.

To find new chemotypes for GSK-3 inhibition, a high-throughput screening exercise was carried out using the Novo Nordisk compound collection and screening them against GSK-3. Although this HTS campaign produced 47 hits, it was determined that only a few were suitable for follow-up optimization. To find better leads, the authors used a computational chemistry approach for scaffold hopping. Using CATS2, a customized version of the original CATS (Chemical Advanced Tempalte Search) methodology, the authors compared 32 virtual libraries against the 47 hits found in the HTS campaign. Here, one of the virtual libraries stood out from the rest with a high hit rate (~800 virtual compounds). Using a fragment-based, two-dimensional fingerprint diversity analysis approach, the hits were distilled down to a total of 324 hits. This approach took lead **275** into scaffold **276**.

For the initial exploratory library, the authors developed a solid-phase synthetic approach wherein a formyl-functionalized resin (**277**) was used to anchor amines (**278**) via reductive amination conditions using $NaCNBH_3$. The immobilized amines (**279**) were then acylated with halocarboxylic acids (**280**) using DIC. The resulting halo amides (**281**) were then reacted with thiols (**282**) to afford the corresponding thio ethers (**283**). The final products (**276**) were released from the support using neat TFA.

The preliminary SAR indicated that a 2-substituted 5-(1-pyridyl)-[1,3,4]-oxadiazol moiety linked to the benzyl thio ether is important for activity (**284:** $IC_{50} = 1.2$ μM). Also of particular importance is the methylene unit between the sulfur atom and the phenyl ring. With this information in hand, the authors developed a computational HQSAR model to predict activities within this compound series. A set of 96 compounds was selected and synthesized. Although none of the analogs was more potent than **284**, this follow-up library confirmed the observation that the oxadiazol-pyridyl moiety was critical for activity.

For the optimization of **284**, the authors synthesized more than 100 compounds using a solution-phase approach similar to the solid-phase method used. Because the initial solid-phase made libraries also indicated that the amide substituent was not essential for activity, the new library replaced the substituted amide fragment with smaller substituents while also probing the pyridyl group. It was found that the 4-pyridyl group was necessary for activity as the 3- and 2-pyridyl analogs lost all potency against GSK-3. Replacing a sulfur atom for the oxygen atom of the oxadiazole ring was well tolerated, keeping activity within the same range, but a nitrogen atom was detrimental for activity (**285:** $IC_{50} = 8$ μM, **286:** $IC_{50} = 11$ μM, **287:** $IC_{50} > 250$ μM; O ~ S >> N, respectively).

To examine the optimum substitution pattern on the benzyl ring, it was found that replacing the 4-acetic acid unit with halides (I > Cl > F) was favorable, particularly at the 3-position, reaching sub-micromolar activities (**288** to **291**).

275
Scaffold Hopping
276
New Scaffold
Synthesis
277
NaCNBH$_3$
NMP/cat. H$_2$O/HOAC
278
279
280
DIC/toluene
DCP/NMP
281
HS−R$_2$
282
DIEA
NMP
283
TFA
276
First library: Hit
284: IC$_{50}$ = 1.2 μM

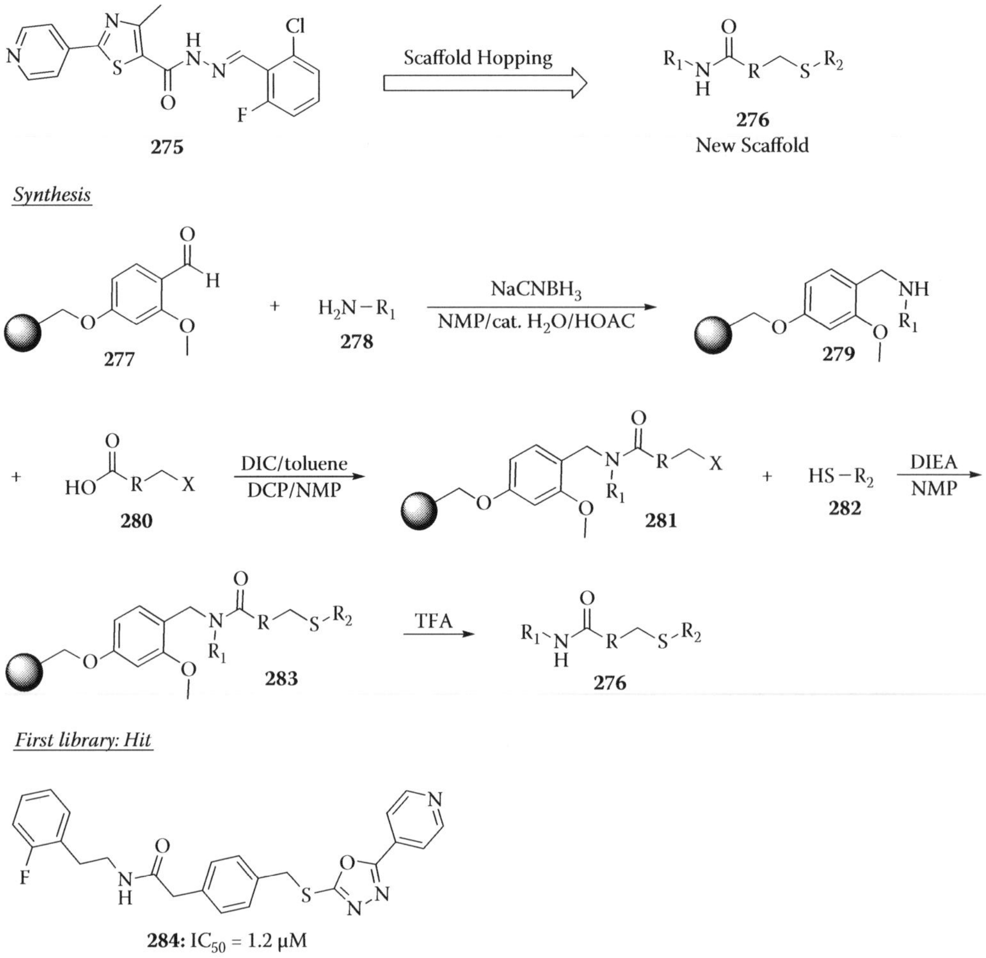

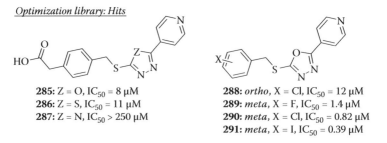

SCHEME 1.31 GSK3 targeted library and hits.

Optimization library: Hits
285: Z = O, IC$_{50}$ = 8 μM
286: Z = S, IC$_{50}$ = 11 μM
287: Z = N, IC$_{50}$ > 250 μM
288: *ortho*, X = Cl, IC$_{50}$ = 12 μM
289: *meta*, X = F, IC$_{50}$ = 1.4 μM
290: *meta*, X = Cl, IC$_{50}$ = 0.82 μM
291: *meta*, X = I, IC$_{50}$ = 0.39 μM

SCHEME 1.32 GSK3 targeted library: optimized hits.

1.9.7 Case 7: Lck

Chen et al.[30] at Bristol Myers Squibb used a parallel solution- and solid-phase synthetic approach for the discovery and optimization of imidazoquinoxalines as Lck (p56^{Lck}) inhibitors. The protein p56^{Lck} is an Src-like tyrosine kinase that contains five domains (N-terminal, SH3, SH2, kinase, and a noncatalytic C-terminal domain); it is highly expressed in T cells and natural killer (NK) cells, and it plays a critical role in T-cell-mediated immune responses. Because T cells are involved in inflammatory disorders such as rheumatoid arthritis and multiple sclerosis, as well as autoimmune responses (e.g., transplant rejection), Lck inhibition could provide treatments for T-cell-mediated autoimmune and inflammatory disorders.

A high-throughput screening exercise of the Bristol-Myers Squibb compound collection identified 1,5-imidazoquinoxaline **292** as a potent nanomolar selective Lck inhibitor. The authors validated this hit by synthesizing regioisomer **293**, where it was determined that regioisomer **292** is indeed the one exhibiting high affinity toward Lck (**292**: IC_{50} = 170 nM, **293**: IC_{50} > 50 mM).

The first library (approximately 160 compounds; Scheme 1.33) was made via a parallel solution-phase approach where efforts focused on the exploration of the bromo-phenyl ring of **292**. Treating imidazoquinoxalinone **295** with POCl3 afforded the key intermediate 4-chloroimidazo[1,5-a]quinoxaline (**296**), which was then reacted with a variety of anilines, aminopyridines, phenols, and thiophenols (**297**). The SAR from this library indicated that removal of the 2-bromo group as well as its replacement with fluoro, methoxy, and nitro groups was deleterious for activity. Removal of the aniline NH or its substitution with O or S was also detrimental for activity. Interestingly, incorporation of an NHCO linker increased potency almost threefold (**299**: IC_{50} = 60 nM). Positioning the 2-bromo at the 4-position also resulted in a dramatic loss of activity. However, exchanging the bromo group in **292** with chloro increased potency about threefold (**300**: IC_{50} = 60 nM). Surprisingly, it was also found that additional methyl or halo groups (bromide and chloride) at the 6-position increased potency further down to single-digit nanomolar potency (**301** to **304**: IC_{50} = 50–9 nM).

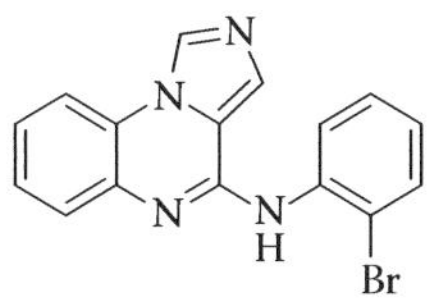

292: IC_{50} = 170 nM   393: IC_{50} > 50 μM 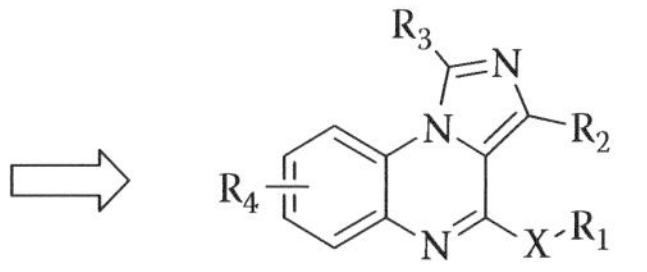   294
Main Scaffold

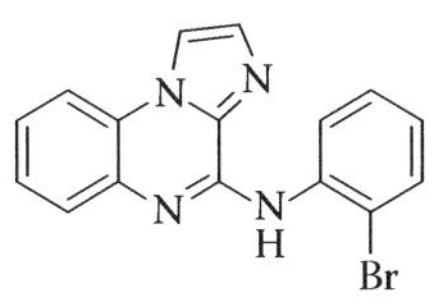

First Library (solution phase)

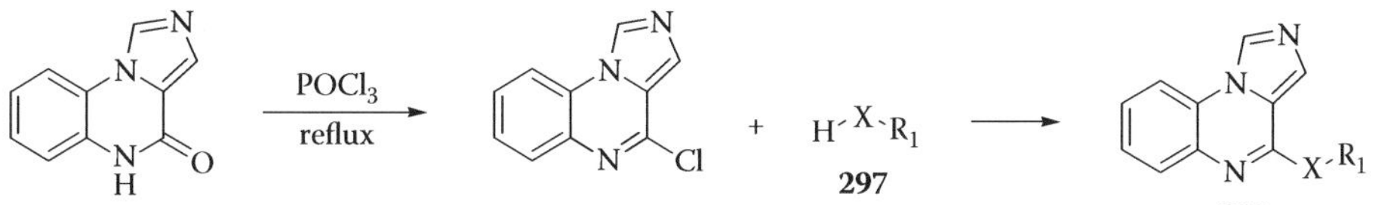

298
X = no linker, NH, O, S
R_1 = substituted Phenyl, Pyridyl

First Library: Hits

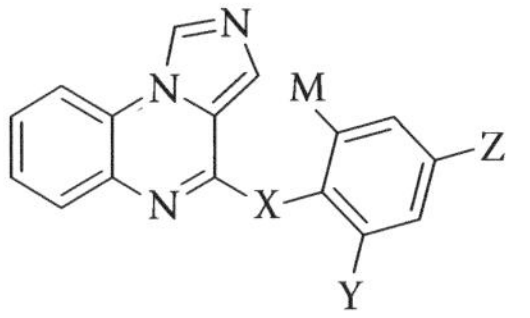

299: X = NHCO, Y = Br, Z = M = H, IC_{50} = 60 nM
300: X = NH, Y = Cl, Z = M = H, IC_{50} = 60 nM
301: X = NH, Y = M = Br, Z = H, IC_{50} = 50 nM
302: X = NH, Y = M = CH$_3$, Z = H, IC_{50} = 16 nM
303: X = NH, Y = M = Cl, Z = H, IC_{50} = 9 nM
304: X = NH, Y = Cl, M = CH$_3$, Z = H, IC_{50} = 9 nM

SCHEME 1.33 Lck targeted library: first library and hits.

For the optimization phase, the authors focused their efforts on compounds **302** and **303**. For this phase, a solid-phase approach was devised where solid-supported anilines **305** were reacted in a 3CC multicomponent reaction with 2,3-dichloroquinoxalines **306** and isonitriles **307** to give the desired polymer-supported imidazoquinoxalinones **308**. Final products **309** were obtained by cleavage from the solid support.

The SAR obtained from this library indicates that small-sized substituents (e.g., methyl) on the imidazole ring are well tolerated (**310**, **311**) while the introduction of a phenyl group led to complete loss of activity. However, the incorporation of small, electron-donating groups (e.g., methoxy) on the phenyl ring increased potency dramatically. These groups, in combination with either 2-Cl, 6-Me, or 2,6-di-Me groups on the aniline ring, resulted in further increased potency against Lck, reaching single-digit nanomolar activities (**315**, **316**: IC_{50} ~ 2 nM). Testing compound **316** in an anti-CD3/anti-CD28-induced PBL proliferation cell-based assay (**316**: PBL assay: IC_{50} = 670 nM).

Optimization via solid-phase synthesis

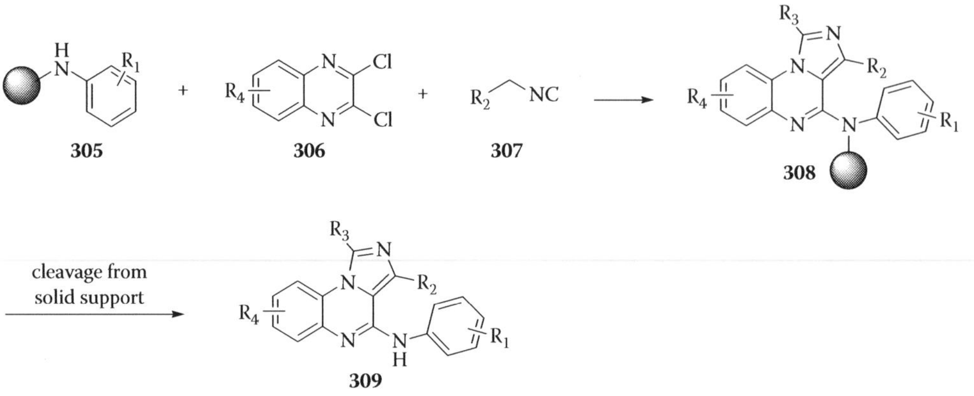

Optimization phase: Hits

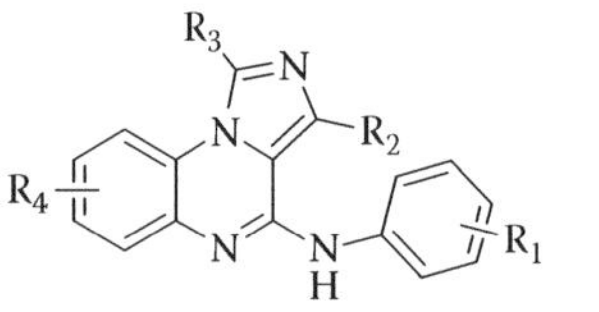

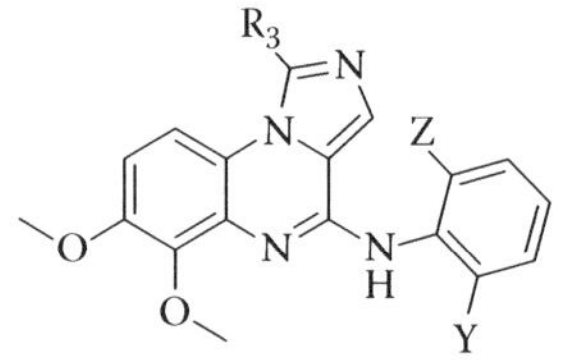

292: R_1 = 2-Br, IC_{50} = 170 nM
300: R_1 = 2-Cl, IC_{50} = 60 nM
310: R_1 = 2-Cl, R_2 = CH_3, IC_{50} = 30 nM
311: R_1 = 2-Br, R_3 = CH_3, IC_{50} = 180 nM

312: Y = Cl, Z = CH_3, R_3 = CH_3, IC_{50} = 7.4 nM
313: Y = Cl, Z = CH_3, R_3 = CH_2OH, IC_{50} = 6.2 nM
314: Y = Cl, Z = CH_3, R_3 = CHN=OH, IC_{50} = 4.3 nM
315: Y = Z = CH_3, IC_{50} = 2.4 nM
316: Y = Cl, Z = CH_3, IC_{50} = 2 nM (Lck)
                                 IC_{50} = 670 nM (cell-based assay)

SCHEME 1.34   Lck targeted library: optimization library and hits.

1.9.8   CASE 8: P38

Dumas et al.[31] at Bayer Research Center reported the automated parallel solution-phase synthesis of biaryl ureas as p38 kinase inhibitors. Mitogen-activated protein (MAP) kinases are involved in the transduction of extracellular signals received by cell-surface receptors. These signals then target a diverse set of downstream targets, resulting in a cellular response. The p38 MAP kinase is a 38-kDa enzyme that consists of four known isozymes: p38α, p38β, p38γ, and p38δ. It was been determined that p38 plays a key role in the cyctokine signaling network, which is implicated in chronic inflammatory conditions such as rheumatoid arthritis. Consequently, inhibition of p38 could offer a way for the therapeutic treatment of cytokine-regulated afflictions such as rheumatoid arthritis.

From a previous combinatorial chemistry effort, the authors discovered urea **317** as a potent p38 inhibitor (IC_{50} = 53 nM). At the time of this discovery, ureas were not known to act as kinase inhibitors, and thus efforts were invested in the optimization of this hit.

Toward this goal, the authors developed an automated parallel solution-phase synthetic approach for the synthesis of a 1000-member library. For this, a Gilson 215 robotic liquid handler was used in conjunction with a 100-vessel dry-block heater with an orbital shaker from J-KEM Scientific.

The synthesis involved the reaction of amines (**319**) with isocyanates (**320**) in DMF at 80 to 95°C for 18 hr. Unreacted isocyanates were quenched by the addition of MeOH and the solvents removed using a Savant SpeedVac evaporator.

The first efforts centered on the SAR elucidation around the pyrazole unit of the main scaffold **318**, for which the 2,3-dichlorophenyl group and the *tert*-butyl group on the 5-member ring were kept constant. It was determined that replacing the pyrazole unit with an isoxazole was well tolerated, observing that the two possible isoxazole regioisomers exhibited similar activity in the enzyme-inhibition assay (**321, 322**: IC_{50} = 58 and 36 nM, respectively) as well as in the single-point SW1353 (human chondrosarcoma) cell-based assay (**321, 322**: % Inh at 0.5 μM = 88 and 93, respectively). Replacing the pyrazole with either thiophene or thiadiazole resulted in loss of activity (greater than threefold).

The second optimization round focused on isoxazole **322**, where substitution effects on the isoxazole ring were explored. Substitution of the *tert*-butyl group by smaller alkyl (e.g., Me, *i*-Pr, *n*-Bu) and bulkier groups (e.g., cyclobutyl, adamantyl) resulted in loss of p38 inhibition potency.

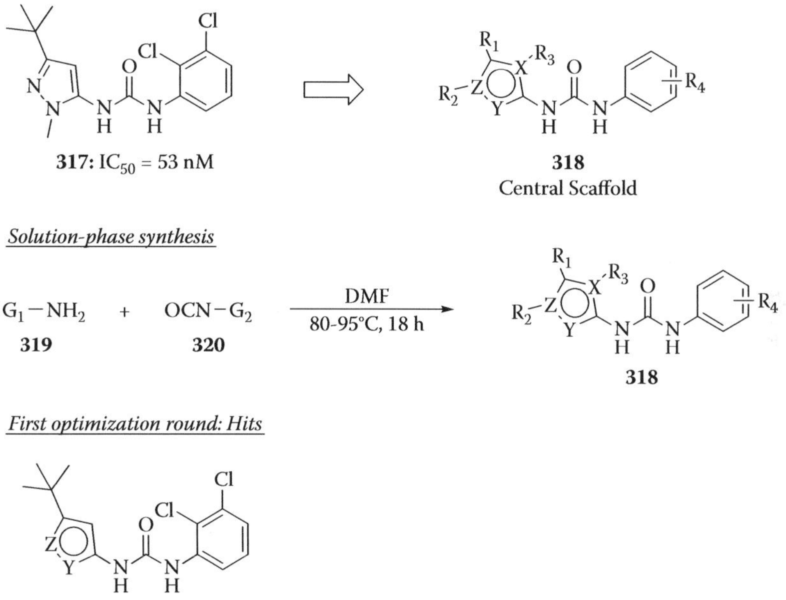

Solution-phase synthesis

$$G_1-NH_2 \quad + \quad OCN-G_2 \quad \xrightarrow[\text{80-95°C, 18 h}]{\text{DMF}}$$

319   **320**

318

First optimization round: Hits

321: Y = N, Z = O, IC_{50} = 58 nM, %Inh (0.5 μM) = 88 (SW1553 cells)
322: Y = O, Z = N, IC_{50} = 36 nM, %Inh (0.5 μM) = 93 (SW1553 cells)
IC_{50} = 0.82 μM (SW1553 cells)

SCHEME 1.35   p38 targeted library.

Having confirmed that the *tert*-butyl isoxazole ring in **322** was essential for activity, the third optimization round focused on the phenyl ring. However, noticing that both **321** and **322** exhibited similar activity against p38, both regioisomers were used during this optimization phase. In both regioisomers, it was found that both 2,3-chloro groups were crucial for activity as their partial or complete replacement only resulted in loss of activity, leaving compound **322** as the most potent inhibitor in the p38 enzyme-based assay. Further testing of **322** in the SW1553 cell-based assay showed that this compound induced a dose-response inhibition with sub-micromolar potency ($IC_{50} = 0.82$ μM), nominating this compound as a potent lead for future efforts.

1.9.9 CASE 9: RAF-1

Smith et al.[32] at Bayer Research Center published the synthesis of biaryl ureas as Raf-1 kinase inhibitors. Raf is a serine/threonine kinase that functions as a downstream effector of Ras. The Raf kinase family is comprised of three isozymes: A-Raf, B-Raf, and C-Raf (Raf-1). The Ras signaling pathway triggers a cascade of cellular events such as proliferation, apoptosis, differentiation, and transformation. The blockade of Raf results in disruption of Ras signaling, offering the means to regulate the fate of cancer cells.

A high-throughput screening assay of Bayer's corporate compound collection identified 3-thienyl urea **323** as a moderate Raf-1 inhibitor ($IC_{50} = 17$ μM). Leveraging on the synthetic methodology previously developed for the p38 kinase inhibition program (see Section 1.9.8), the same methodology was used for the exploration and optimization of **323**.

The first optimization round focused on the phenyl moiety. The SAR obtained clearly indicated that small lipophilic substituents at the *para*-position (e.g., Me, Cl) increased potency three- to tenfold (**325, 326**: $IC_{50} = 17$ and 6.8 μM, respectively) while the incorporation of bulkier steric substituents, electron-withdrawing and electron-donating groups, resulted in loss of activity. Interestingly, replacement of the sulfur atom in the thiophene ring with O and N (furan and pyrrole, respectively) was well tolerated, keeping potency in the single-digit micromolar range (**327–329**). Replacement of the phenyl group with heterocyclic systems showed that thiadiazoles bearing small lipophilic groups exhibited similar but not better potency than **325** (**330–332**: $IC_{50} = 1.2$–2.0 μM). The observed heteroaryl potency trend was thiadiazoles > pyrazoles and pyrroles > triazoles, isoxazoles, and oxazoles.

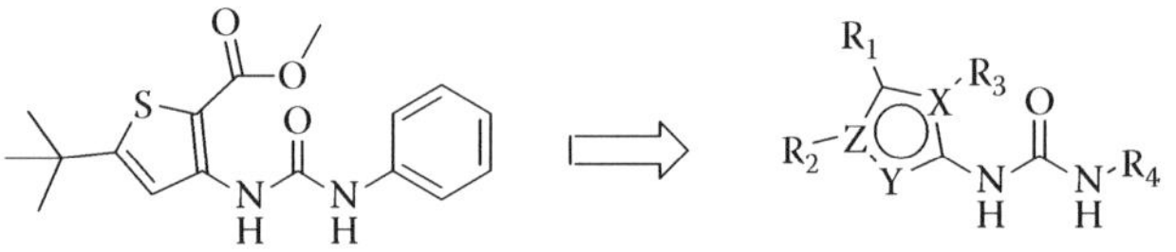

323: $IC_{50} = 17$ μM

324
Central Scaffold
R_4 = Aryl, Heteroaryl

First optimization round: Hits

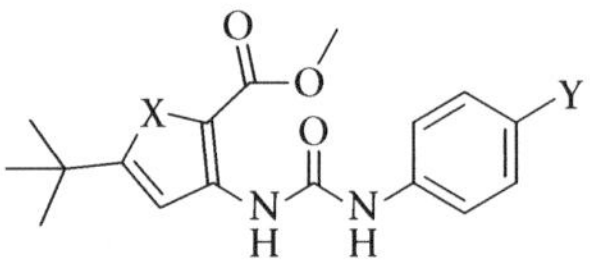

325: X = S, Y = CH$_3$, $IC_{50} = 1.7$ μM
326: X = S, Y = Cl, $IC_{50} = 6.8$ μM
327: X = O, Y = CH$_3$, $IC_{50} = 4.0$ μM
328: X = NH, Y = CH$_3$, $IC_{50} = 3.0$ μM
329: X = NCH$_3$, Y = CH$_3$, $IC_{50} = 5.0$ μM

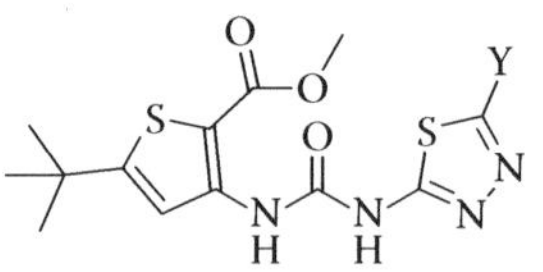

330: Y = CH$_3$, $IC_{50} = 1.2$ μM
331: Y = C$_2$H$_5$, $IC_{50} = 2.0$ μM
332: Y = cyPr, $IC_{50} = 1.9$ μM

Crossover Hit

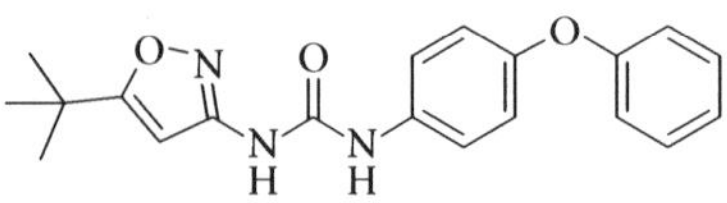

333: $IC_{50} = 540$ nM (Raf-1)
$IC_{50} = 360$ nM (p38)

SCHEME 1.36 Raf1 targeted library.

For convenience, the *para*-methyl phenyl moiety was kept constant and the next optimization phase focused on the thiophene ring. Replacement of the methyl ester with unsubstituted, mono-substituted, and di-substituted amides resulted in moderate potency, but none was as potent as the methyl ester group. Removal of the methyl ester group or substitution of it with small alkyl groups resulted in total loss of activity. In a similar fashion, replacement of the *tert*-butyl group with small polar functionalities or larger lipophilic groups was also deleterious for activity.

Having established that both the *tert*-butyl group on the heterocyclic system and the *para*-methyl phenyl moiety were essential for activity, the authors proceeded to explore alternate heterocycles for the thiophene, keeping the former two groups constant. Surprisingly, no other heterocyclic system afforded analogs with activity against Raf-1.

Interestingly, this work overlapped with the above effort on p38, and a crossover compound was found active against both Raf-1 and p38 with submicromolar potency (**333**: Raf-1: $IC_{50} = 540$ nM, p38: $IC_{50} = 360$ nM).

1.10 NUCLEAR RECEPTORS

1.10.1 CASE 1: 11β-HYDROXYSTEROID DEHYDROGENASE TYPE 1 (11β-HSD1) INHIBITORS

Copola et al.[33] at Novartis published a parallel solution-phase synthesis for the synthesis of perhydro-quinolylbenzamides as 11β-hydroxysteroid dehydrogenase type 1 (11β-HSD1) inhibitors. 11β-HSD1 is a microsomal membrane-bound enzyme that catalyzes the conversion of inactive ketohormones (e.g., cortisol) into their corresponding alcohols. Cortisol stimulates glucogenesis. Over-expression of cortisol has been linked to conditions such as hypertension, obesity, and impaired glucose tolerance (diabetes). There are two 11β-HSD isoforms: (11β-HSD1 and 11β-HSD2), the former mainly expressed in the liver whereas the latter in the kidney. Selective inhibition of 11β-HSD1 over 11β-HSD2 could lead to the treatment of metabolic conditions (e.g., diabetes, obesity).

A high-throughput screening campaign of 300,000 compounds from their compound collection against 11β-HSD1 identified benzamide **334** as a single-digit micromolar inhibitor (**334**: CGP013289: $IC_{50} = 9.7$ μM for pig-derived 11β-HSD1 and $IC_{50} = 128$ μM for human-derived 11β-HSD1). The authors noted that initially for the HTS assays, pig liver 11β-HSD1 was used due to the amounts of enzyme needed; but when human 11β-HSD1 was used, it was observed that the activities were much lower. Nevertheless, human 11β-HSD1 was used for this research.

Centering on the benzamide core (**335**), their synthetic approach consisted of a mixture of traditional solution-phase synthesis with automated solution-phase and polymer-scavenger based protocols. For the automated parallel solution-phase synthesis, the authors used the Novartis AutoChem system and Argonaut Technologies' Quest 210 synthesizer for the polymer-assisted synthesis. For the AutoChem protocol, the authors used 0.5 M stock solutions of anilines (**336**) in DMF, a 1.0 M solution of an acid chloride (**337**) in THF, and a 1.0 M solution of NMM (**338**) in THF. The reactions were performed at room temperature using a ratio of aniline:NMM:acid chloride of 1:1.5:1.5 equivalents for 18 hr. For the polymer-bound scavenger approach, the authors first mixed 0.5 mmol of an amine (**339**) and polymer-bound DIEA (1.5 equivalents) in DCM (10 mL); then 0.55 mmol of a benzoyl chloride (**340**) was added. After agitation for 18 hr, polymer-bound trisamine (0.75 equivalents) was added and the mixture further agitated for 3 hr. For the AutoChem approach, products were automatically purified by HLPC. For the scavenger-based approach, the products were obtained after solvent removal.

The first library indicated that removal of the phenolic group in **334** improved activity slightly and that replacing the 3,4,5-trimethoxy phenyl group with a 2,4-dichlorophenyl group was very well tolerated, improving potency sixfold (**341**: $IC_{50} = 19.67$ μM). Replacing the piperidine amide with 2-mono or 2,6-dialkyl piperazines increased potency by almost one order of magnitude (single-digit micromolar) (**342**: $IC_{50} = 1.06$ μM, **343**: $IC_{50} = 1.83$ μM). But it was the bicyclic piperazine analogs (**344**, **345**) that provided nanomolar potency and good selectivity for 11β-HSD1 (**345**: $IC_{50} = 0.22$ μM, ratio 11β-HSD1:11β-HSD2 = 190:1).

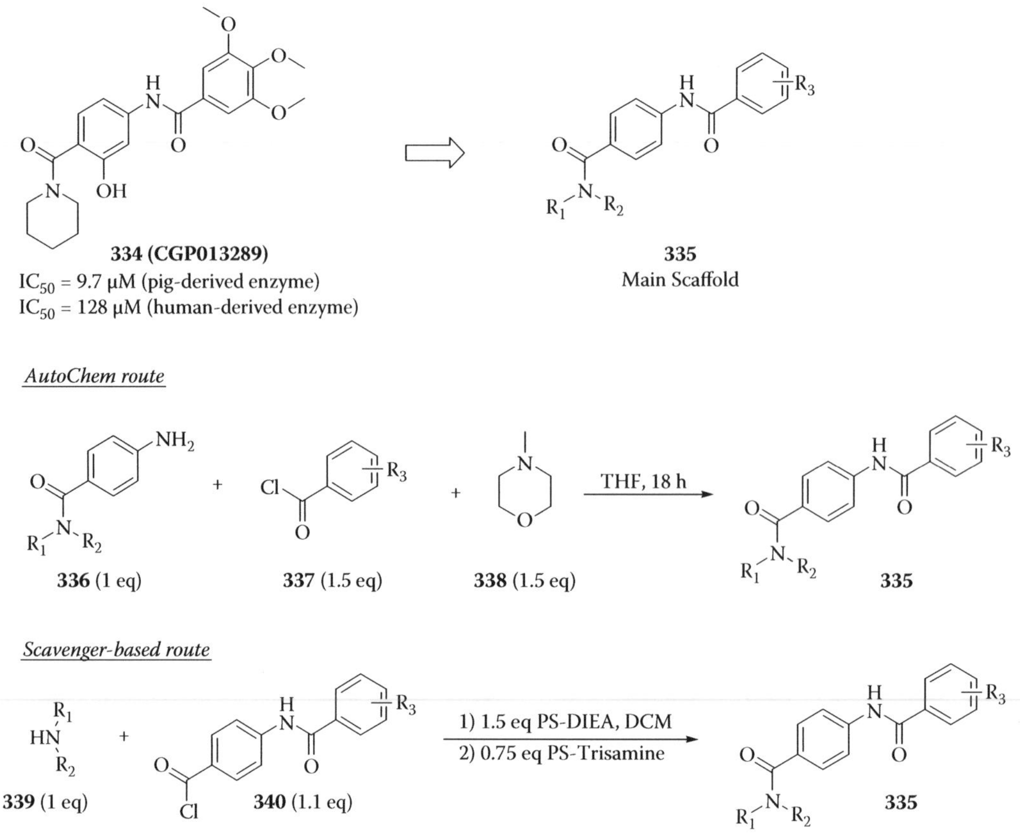

SCHEME 1.37   11β-HSD1 targeted library: synthesis.

For the optimization of **344**, the authors kept the decahydroquinoline amide portion constant and explored different substitution patterns in the central aromatic ring. Although no clear SAR pattern was observed, it was, however, determined that addition of both electron-donating groups, halides, positive charges, and benzannulation (**346**: IC$_{50}$ = 0.10 μM) improved activity to the 100-nM range, still with remarkable selectivity for 11β-HSD1. So for the next optimization step, the authors kept both the decahydroquinoline amide and middle phenyl ring constant, and probed for changes on the 2,4-dichlorobenzoyl group. Replacement of the 2,4-dichlorobenzoyl group with 4-fluorobenzyl, and replacement of the 2,4-dichlorobenzamide with N-methyl-4-benzamide improved potency twofold in both cases (**347, 348**: IC$_{50}$ = 0.05 μM). Interestingly, it was **349**, the regioisomer of analog **344**, that afforded the most potent compound in the series (**349**: IC$_{50}$ = 0.014 μM).

From the 64 compounds made in this study, 30 were tested in a cell-based 11β-HSD1 inhibition assay and 22 were found to exhibit greater than 50% inhibition at 1 μM. These 22 compounds were assayed in an *in vivo* ADX mouse model to screen for liver corticosterone levels. It was found that compounds **345** and **346**, although not the most potent compounds in the enzyme inhibition assay, were in fact the most efficacious *in vivo*, eliciting a maximum response of greater than 70% corticosterone reduction.

1.10.2   CASE 2: NONSTEROIDAL LIVER X RECEPTOR (LXR) AGONISTS

Collins et al.[34] at GlaxoSmithKline reported the synthesis and optimization of tertiary amines as nonsteroidal liver X receptor (LXR) agonists. LXRs belong to the nuclear hormone receptor superfamily of activated transcription factors. There are two known LXR subtypes: LXRα (NR1H3) and LXRβ

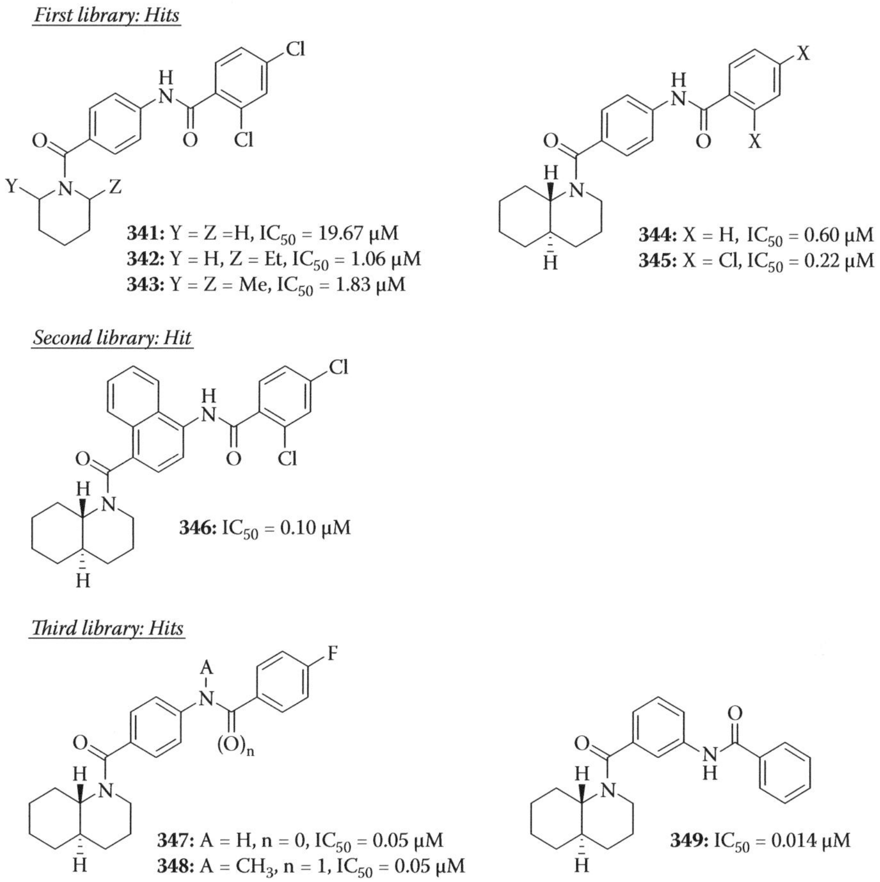

First library: Hits

341: Y = Z =H, IC$_{50}$ = 19.67 μM
342: Y = H, Z = Et, IC$_{50}$ = 1.06 μM
343: Y = Z = Me, IC$_{50}$ = 1.83 μM

344: X = H, IC$_{50}$ = 0.60 μM
345: X = Cl, IC$_{50}$ = 0.22 μM

Second library: Hit

346: IC$_{50}$ = 0.10 μM

Third library: Hits

347: A = H, n = 0, IC$_{50}$ = 0.05 μM
348: A = CH$_3$, n = 1, IC$_{50}$ = 0.05 μM

349: IC$_{50}$ = 0.014 μM

SCHEME 1.38   11β-HSD1 targeted library: optimization and hits.

(NR1H2). The LXRs form heterodimers with the 9-*cis*-retinoic acid receptor RXR (NR2B). These heterodimers, in turn, regulate the expression of the ABC transporter ABCA1, a transporter involved in reverse cholesterol transport. The high-density lipoprotein (HDL) is known to mediate reverse cholesterol transport, aiding in the transfer of cholesterol from peripheral cells back to the liver. This process is beneficial to the human body as it reduces the risk of cardiovascular disease-related conditions such as stroke, atherosclerosis, ischemic heart disease, and coronary heart disease.

Collins' team began efforts around compound **350** (EC$_{50}$ = 260 nM), a partial LXRa agonist tertiary amine molecule found from a high-throughput screening campaign using a cell-free ligand-sensing assay (LiSA) for human LXRα. This assay measures the ligand-dependent recruitment of a 24-amino acid peptide fragment of the steroid receptor coactivator 1 (SRC1) to the nuclear receptor. Using the putative LXRα–agonist 24(S),25-epoxycholesterol (**351**) as reference, it was determined that **350** was only 17% effective in recruiting the SRC1 protein when compared to **351**.

To improve the efficacy of **350**, the authors developed a solid-phase approach to explore variations of both the bezamide unit as well as the tertiary amine. Their strategy involved the immobilization of phenolic carboxylic acids (**353**) onto two different resins — Sasrin resin to make carboxylic acid analogs and Rink amide resin to make amide analogs. Since during the method development phase it was observed that Rink resin provided the highest overall yields, the authors chose this resin for the initial library synthesis.

The first library consisted of 120 compounds made by first anchoring four phenolic carboxylic acids (**353**) using HATU and 2,6-lutidine in NMP, followed by the alkylation with 3-bromopropanol (**355**) under Mitsunobu conditions. The resulting bromides (**356**) were then treated with 2,2-diphenylethanamine (**357**) in DMSO and the resulting secondary amines (**358**) were then derivatized with 30 different benzaldehydes (**359**) under reductive amination conditions. The final products (**361**) were obtained after cleavage with 10% TFA in DCM.

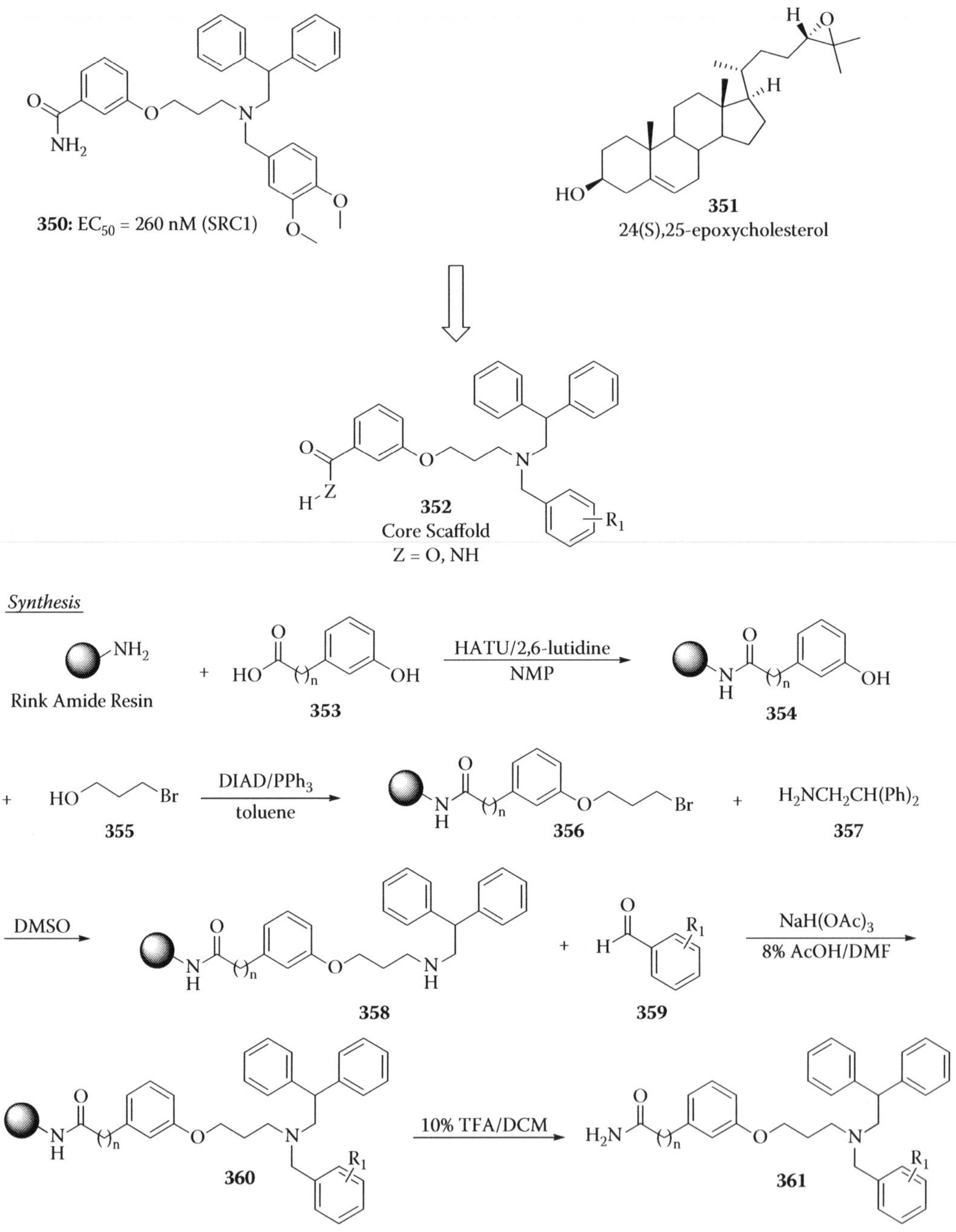

350: EC$_{50}$ = 260 nM (SRC1)

351
24(S),25-epoxycholesterol

352
Core Scaffold
Z = O, NH

Synthesis

353

HATU/2,6-lutidine
NMP

354

355

DIAD/PPh$_3$
toluene

356   +   H$_2$NCH$_2$CH(Ph)$_2$
357

DMSO

358   **359**

NaH(OAc)$_3$
8% AcOH/DMF

10% TFA/DCM

360   **361**

SCHEME 1.39   LXR targeted library: synthesis.

From this first library, compound **362** exhibited nanomolar affinity (EC$_{50}$ = 260 nM) for LXRα and 80% SRC1 recruiting. However, this compound was found inactive in the LXRα- GAL4 cell-based assay. To improve cell-based activity, its corresponding carboxylic acid **363** was made using Sasrin resin. Although **363** was less potent in the SRC1 assay, this analog exhibited single-digit cellular micromolar activity in the GAL4 assay (~8 μM). Although several amides containing the 4-methoxy-benzyl group with additional methoxy or fluoro groups on the benzyl aromatic ring were also found to exhibit sub-micromolar activity in the SRC1 assay, those containing a *meta*-trifluoromethyl group exhibited activity in the double-digit nanomolar range (**364–366**). Of the analogs made, compound **366** not only was the more potent in the SRC1 assay ($EC_{50} = 45$ nM), but it was also the most potent in the cell-based assay ($EC_{50} = 425$ nM).

Hits

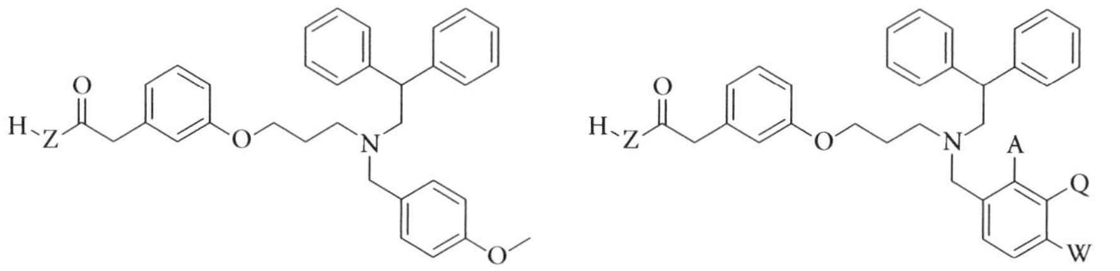

362: Z = NH, EC_{50} = 260 nM (SRC1)
363: Z = O, EC_{50} = 860 nM (SRC1)

364: Z = NH, A = W = H, Q = CF_3, EC_{50} = 85 nM (SRC1)
365: Z = NH, A = H, Q = CF_3, W = F, EC_{50} = 85 nM (SRC1)
366: Z = NH, A = Cl, Q = CF_3, W = H, EC_{50} = 45 nM (SRC1)
367: Z = O, A = Cl, Q = CF_3, W = H, EC_{50} = 125 nM (SRC1)
EC_{50} = 190 nM (GAL4)

SCHEME 1.40 LXR targeted library: optimization and hits.

Having improved cellular potency in this series, the authors then proceeded to change the amide functionality of **366** into its carboxylic acid analog **367**. Although **367** lost some activity in the SRC1 assay ($EC_{50} = 125$ nM), this carboxylic acid analog exhibited improved potency in cells ($EC_{50} = 190$ nM). Further *in vivo* testing in mice indicated that **367** possesses an acceptable pharmacokinetic profile with a $t_{1/2} = 2$ hr, $C_{max} = 12.7$ μg/mL, and F (%) = 70 when administered at 10 mg/kg. Moreover, **367** was also found to increase HDL plasma levels as well as ABCA1 expression.

1.11 OXIDOREDUCTASES

1.11.1 CASE 1: DIHYDROOROTATE DEHYDROGENASE (DHODH) INHIBITORS

Haque et al.[35] at Bristol-Myers Squibb published a solid-phase synthetic approach for the optimization of pyrazoles as *Helicobacter pylori* dihydroorotate dehydrogenase (DHODH) inhibitors. *H. pylori* is a microaerophilic bacterium responsible for several gastrointestinal disorders, playing a role in afflictions such as gastritis, gastric ulcers, and even gastric cancer. Current treatment involves multi-drug therapy which, as a side effect, tends to eradicate normal gastrointestinal flora. Dihydroorotate dehydrogenase (DHODase) is involved in the *de novo* biosynthesis of pyrimidine. Pyrimidines are crucial for bacterial survival; consequently, interrupting the pyrimidine production by DHODase could lead to the eradication of *H. pylori*, offering a viable treatment for gastrointestinal disorders.

Using lead compounds **368** and **369**, the authors devised a two-prong strategy to quickly synthesize related analogs based on core **370** to quickly provide early SAR and clues for rapid lead optimization. The synthetic strategy was based on the use of BAL linker on a polystyrene support (**371**) using Robbins Scientific's filtration/reaction blocks. In one approach, amines (**372**) were first immobilized via reductive amination conditions. The immobilized amines (**373**) were then acylated with 1-N-alkyl monoacid/monoester pyrazoles (**374**). The resulting pyrazole ester amides (**375**) were saponified (**376**) and then coupled with another set of amines (**377**) to afford the desired immobilized pyrazole carboxy diamides (**378**). Final treatment of the resin with 50% TFA in DCM afforded desired products **370**.

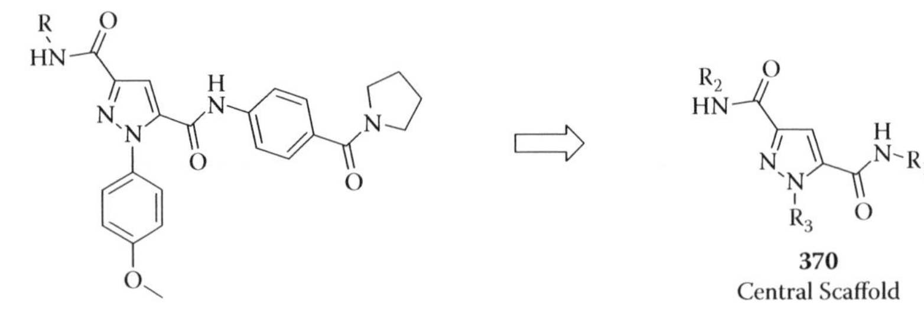

368: R = benzyl, K_i = 26 nM, MIC = 3 µg/mL
369: R = 2-OH-Ph, K_i = 50 nM

Synthesis: Route 1

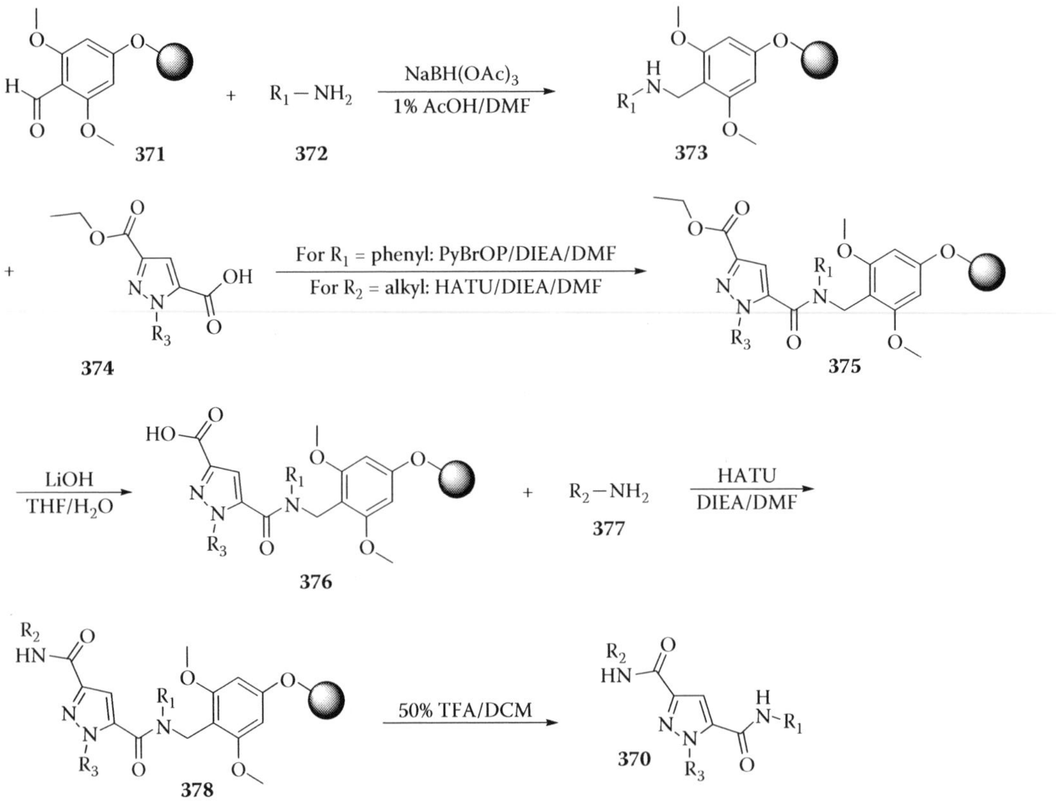

SCHEME 1.41 DHODH targeted library: synthetic route 1.

During the course of this synthesis, it was noted that coupling of anilines frequently did not work as expected. To alleviate this problem, the authors developed an alternate synthesis for the introduction of anilines. For the second approach, the second set of amines (**377**) was first immobilized on the BAL-polystyrene support (**371**) using reductive amination conditions. The immobilized amines (**379**) were then acylated with 5-(furan-2-yl)-1-alkyl-1H-pyrazole-3-carboxylic acids (**380**). Then the furan group of the immobilized intermediates (**381**) was oxidized to a carboxylic acid by treatment with $RuCl_3$ and $NaIO_4$ in $CH_3CN/H_2O/CCl_4$. Then the carboxylic acid intermediates (**382**) were coupled with anilines (**372**) using PyBrOP coupling conditions.

Synthesis: Route 2

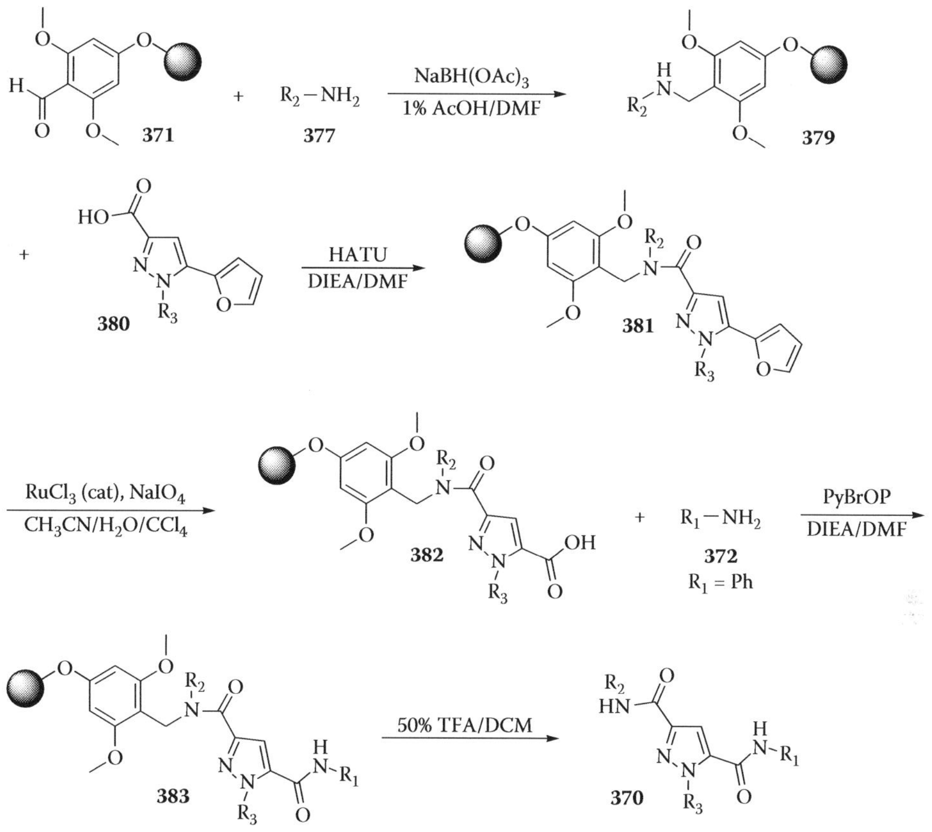

SCHEME 1.42 DHODH targeted library: synthetic route 2.

For the first library the 4-methoxy-phenyl group of hits **368** and **369** was kept constant. Although this library (77 compounds) did not produce a compound with sub-micromolar inhibition potency, two promising analogs were identified (**384**, **385**), where **384** is an isomer of **368**. These two hits share a benzylamide functionality.

For the second library, a total of 44 compounds were made where the focus was on probing changes on the benzyl group (R_2). Different substitution patterns on the aromic unit of the benzyl group were fairly well tolerated, although it was observed that loss of activity increased proportionally with the size increase of substituents at the *para*-position (cyclohexyl > methoxy > chloro > fluoro). Nonetheless, this library afforded several potent sub-micromolar compounds, where it was observed that replacing the benzyl group with small alkyl groups led to **386**: $K_i = 138$ nM, and **387**: $K_i = 331$ nM.

For the third optimization library, 160 compounds were made where the authors probed the influence of the methoxy group (R_3) of the phenyl pyrazole core as well as the second amide group (R_1). To obtain compounds with acceptable drug-like characteristics, the authors focused on the use of fragments with low molecular weight as well as groups that would improve water solubility (e.g., pyridine). It was determined that replacement of the methoxy group with a chloro group was well tolerated and that reducing the original amide group to an unsubstituted benzamide group was beneficial for both enzyme affinity and cell activity (**388**: $K_i = 4$ nM, MIC = 0.125 μg/mL; **389**: $K_i = 4$ nM, MIC = 0.250 μg/mL).

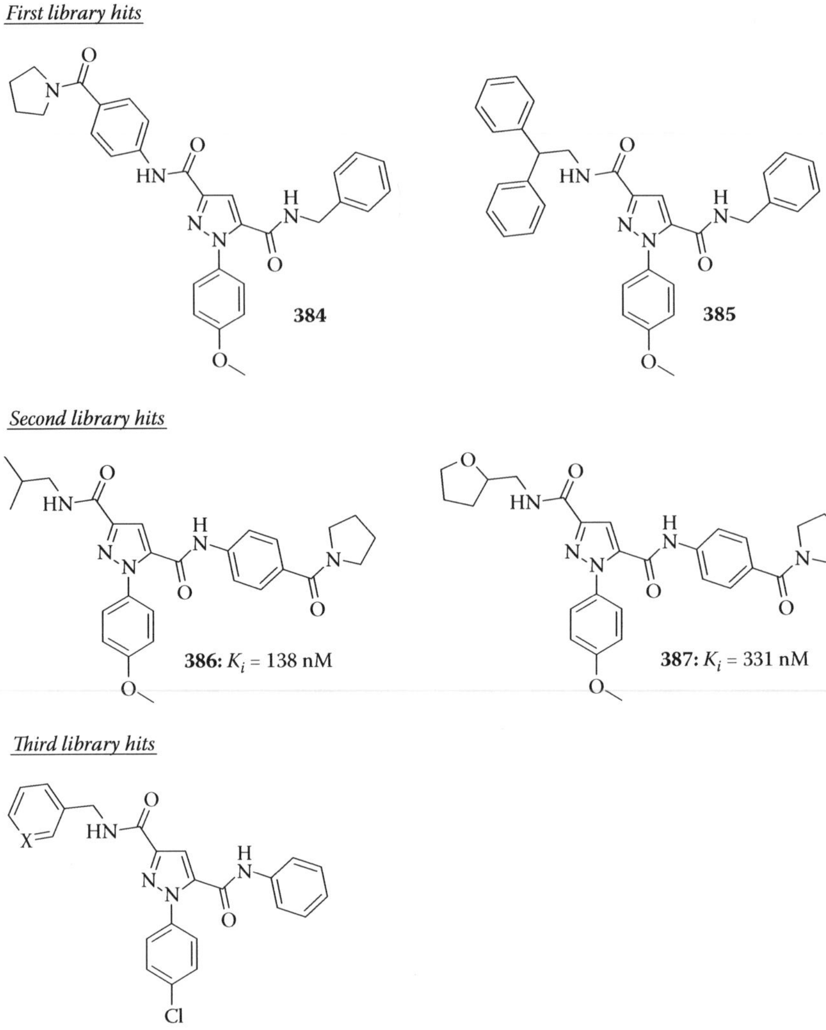

First library hits

384

385

Second library hits

386: K_i = 138 nM

387: K_i = 331 nM

Third library hits

388: X = C, K_i = 4 nM, MIC = 0.125 µg/mL
389: X = N, K_i = 4 nM, MIC = 0.250 µg/mL

SCHEME 1.43   DHODH targeted library: hits.

1.12   PHOSPHATASE INHIBITORS

1.12.1   CASE 1: CDC25B

Fritzen et al.[36] at Pharmacia & Upjohn disclosed a solid-phase synthetic approach for the synthesis of racemic tetrahydroisoquinolines as inhibitors of the protein phosphatase Cdc25B. Phosphatases catalyze the dephosphorylation of proteins. Cdc25B, in particular, acts on cyclin-dependent kinases (CDKs). CDKs are protein kinases involved in cell mitosis that, upon dephosphorylation, become active and start the mitosis cycle. Because mitosis is often deregulated in cancer cells, inhibition of Cdc25B could be a therapeutic alternative for cancer treatment.

After screening Pharmacia & Upjohn's corporate compound collection against Cdc25B, the hit compound **390** was identified (PNU-108937, IC_{50} = 70–100 µM). The authors devised a strategy in which commercially available 3,5-diiodo-L-tyrosine (**392**) could serve as the main starting material. The authors synthesized the key building block **393** by first forming the tetrahydroisoquinoline unit via Pictet-Spengler cyclization. The secondary amine was subsequently Fmoc protected and then the phenolic unit was orthogonally protected as a silyl ether to afford key intermediate **393**. To anchor this tetrahydroisoquinoline carboxylic acid **393** to Wang resin (**394**), it was necessary to use 2-chloro-1,3-dimethyl-2-imidazolinium hexafluorophosphate (CIP) because conditions such as DCC/DMAP/HOBT or PPh$_3$/DEAD (Mitsunobu alkylation conditions) gave poor yields. Although it was determined by chiral HPLC that CIP coupling conditions epimerized the enantiomerically pure starting material, it was found that activity was not affected by this as both enantiomeric products were equally active against Cdc25B.

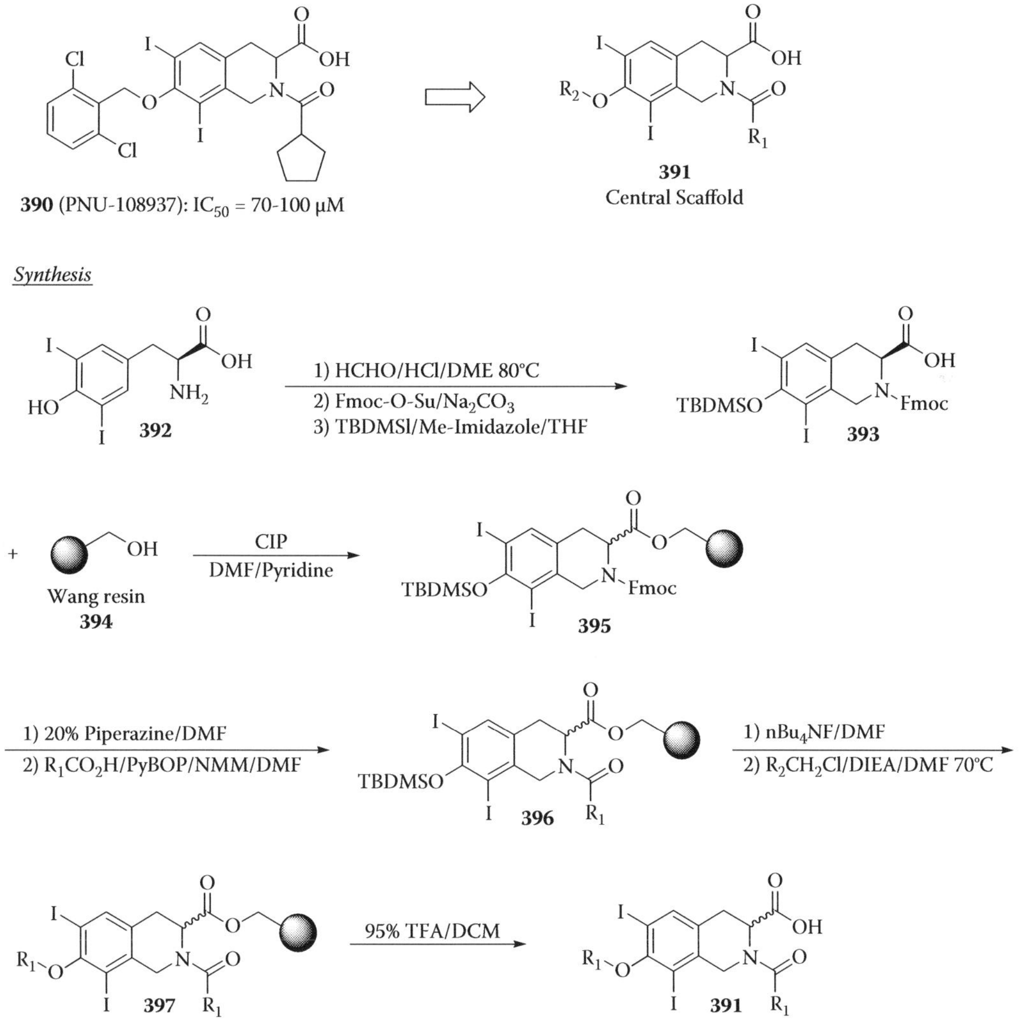

390 (PNU-108937): IC$_{50}$ = 70-100 μM

391
Central Scaffold

Synthesis

SCHEME 1.44 Cdc25B targeted library: synthesis.

Once intermediate **393** was on the solid support (**395**), the Fmoc group was removed using 20% piperazine/DMF and then acylated with four different carboxylic acids. Each amide product (**396**) was split in four portions, each portion was treated with *n*-Bu$_4$NF/DMF to remove the silyl protecting group, and the resulting phenols were then alkylated with six different primary chlorides in the presence of DIEA in DMF at 70°C (**397**). Final cleavage with 95% TFA/DCM provided a 24-member library with good purities (**391**). This initial library produced 22 compounds with IC$_{50}$ < 100 μM, identifying compound **398** as the most potent one (IC$_{50}$ = 15 μM). This compound was resynthesized via solution phase and re-assayed against Cdc25B to confirm its activity (IC$_{50}$ = 23 μM).

First library: Hit

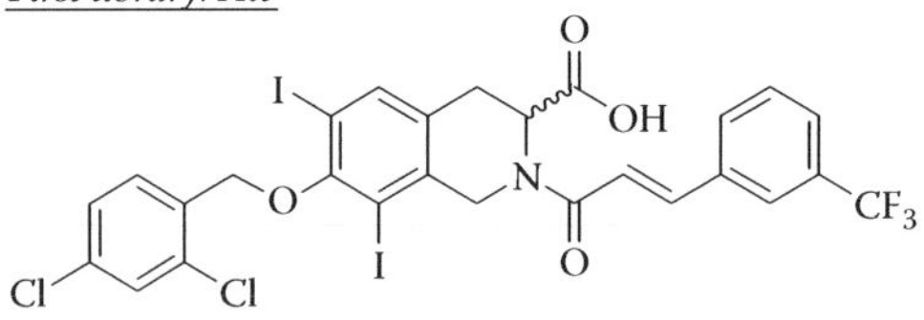

398: IC_{50} = 15-23 µM

Optimization library: Hits

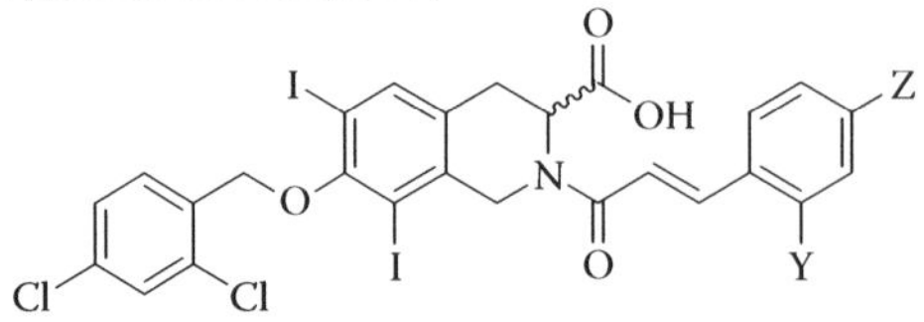

399: Y = H, Z = Br, IC_{50} = 23 µM
400: Y = Cl, Z = H, IC_{50} = 15 µM

SCHEME 1.45   Cdc25B targeted library: hits.

Based on these results, a 51-compound follow-up library was made, where different cinnamoyl acids were used at N-2 and a variety of alkylating agents at the C-7 phenol. Although no clear SAR could be established, some of the isolated products exhibited inhibitory activities against Cdc25B in the 15- to 25-µM range (**399, 400**).

1.13   PROTEASES AND METALLOPROTEASES

1.13.1   PROTEASES

1.13.1.1   Case 1: Caspase-3 Inhibitors

Linton et al.[37] at Idun Pharmaceuticals reported the optimization of acetylated dipeptides based on Ac-DEVD-H (**401**), a tetrapeptide with inhibitory activity against a panel of caspases, particularly caspase-3 with an IC_{50} of 3.5 nM. The authors' goal was to identify compounds exhibiting inhibition toward a broad spectrum of caspases (caspase-1, caspase-3, caspase-6, caspase-7, and caspase-8). With this goal in mind, the authors began their work by truncating **401**, finding that the 2-naphthyloxy acetyl-LD-H dipeptide (**402**) exhibited sub-micromolar inhibition selectively against caspase-3 (IC_{50} = 0.94 µM) in the same caspase panel.

With this lead compound in hand, a solid-phase approach was developed (**403**). The strategy began with the synthesis of Fmoc-protected 3-amino-4-oxobutanoic acid *tert*-butyl ester (**405**), followed by its coupling to an aminoethylphenoxyacetic acid semicarbazide linker (**406**) via imine formation. The resulting semicarbazone (**407**) was then immobilized onto aminomethylpolystyrene resin (**408**), the Fmoc group was then removed, and the free amino group coupled to Fmoc-Leu-OH. The Fmoc group of the resulting dipeptide was removed, the free amino group coupled to a broad range of carboxylic acids, the Boc group removed, and the products removed under acidic conditions to afford the desired N-acetylated LD-H dipeptides (**403**).

This library produced quinolinyloxy and isoquinolinyloxy analogs that display selective sub-micromolar inhibition against caspase-3. However, the most potent compound against the entire caspase panel was the 1-naphthyloxy acetyl-LD-H dipeptide (**411**) with IC_{50} values in the sub-micromolar region for four out of the five caspases in the caspase panel. Although the authors do not report additional information on the most potent and/or selective inhibitors such as physicochemical properties, cell-based profile, or why they wanted to pursue inhibitors that would simultaneously inhibit multiple caspases (a "dirty bomb" approach), this work presents a good example of

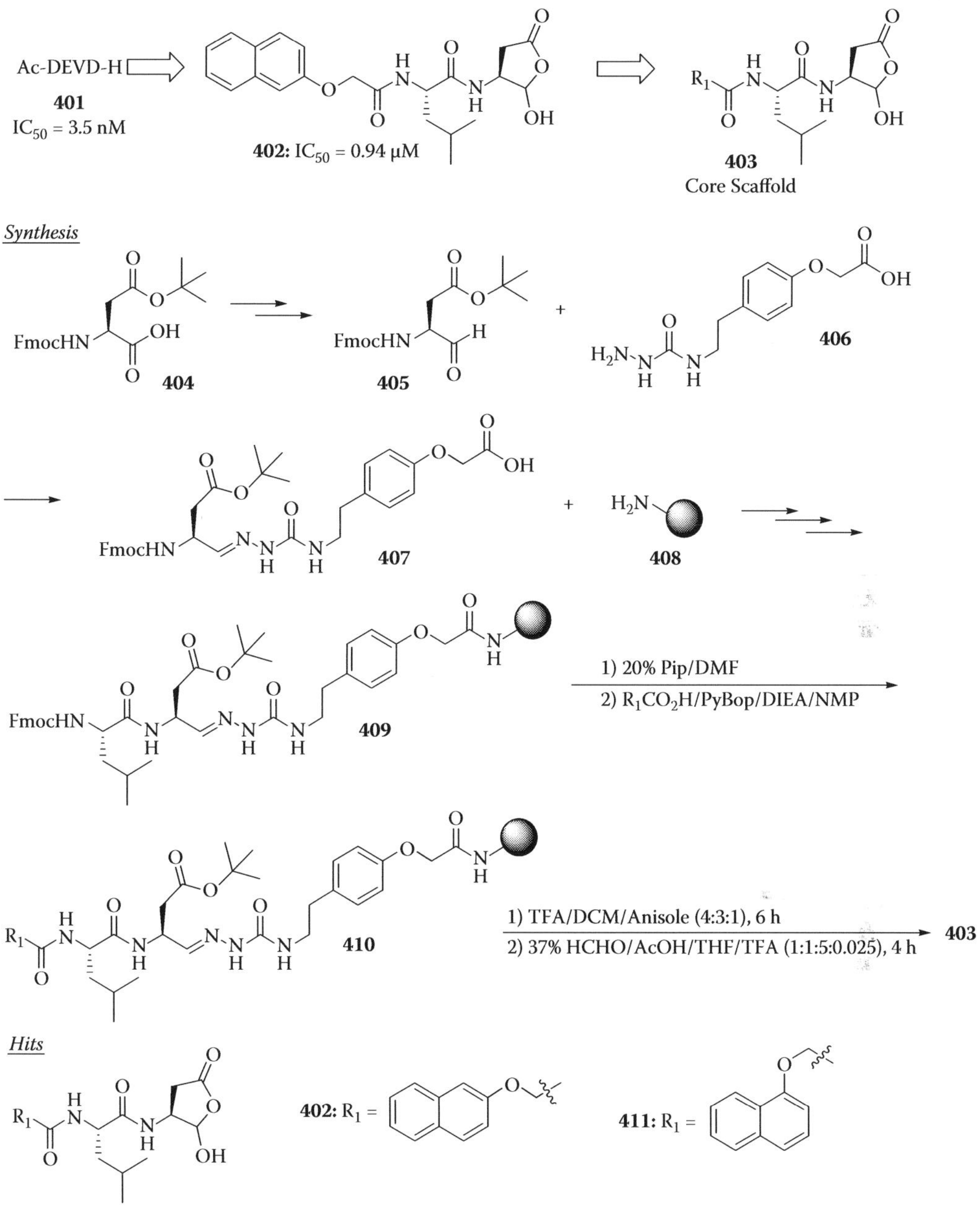

SCHEME 1.46 Caspase-3 targeted library.

how solid-phase chemistry can be used to quickly optimize a lead compound with mild potency and selectivity to identify more potent inhibitors hitting a broad range of caspase targets.

1.13.1.2 Case 2: HCV Inhibitors

(A) Beaulieu et al.[38,39] at Boehringer Ingelheim reported in two consecutive articles how parallel solid-phase synthesis was used for the discovery and optimization of benzimidazole 5-carboxamide analogs as inhibitors of the NS5B polymerase of the hepatitis C virus (HCV).

This work began with two hits found in a high-throughput screening exercise against HCV polymerase. The two active compounds shared a benzimidazole core with identical 5-carboxamide and N-benzimidazole substituents (**412, 413**: $IC_{50} = 12$–$14\ \mu M$). Since close examination of these two compounds indicated that the size of the C2 group contributed little activity, it was decided to use compound **412** as the starting point for optimization. The initial work focused on modification of the 5-carboxamide unit of compound **412**, where it was found that amides did not improve activity, but the carboxylic acid precursor provided more potency (**414**: $IC_{50} = 4.3 \pm 0.6\ \mu M$).

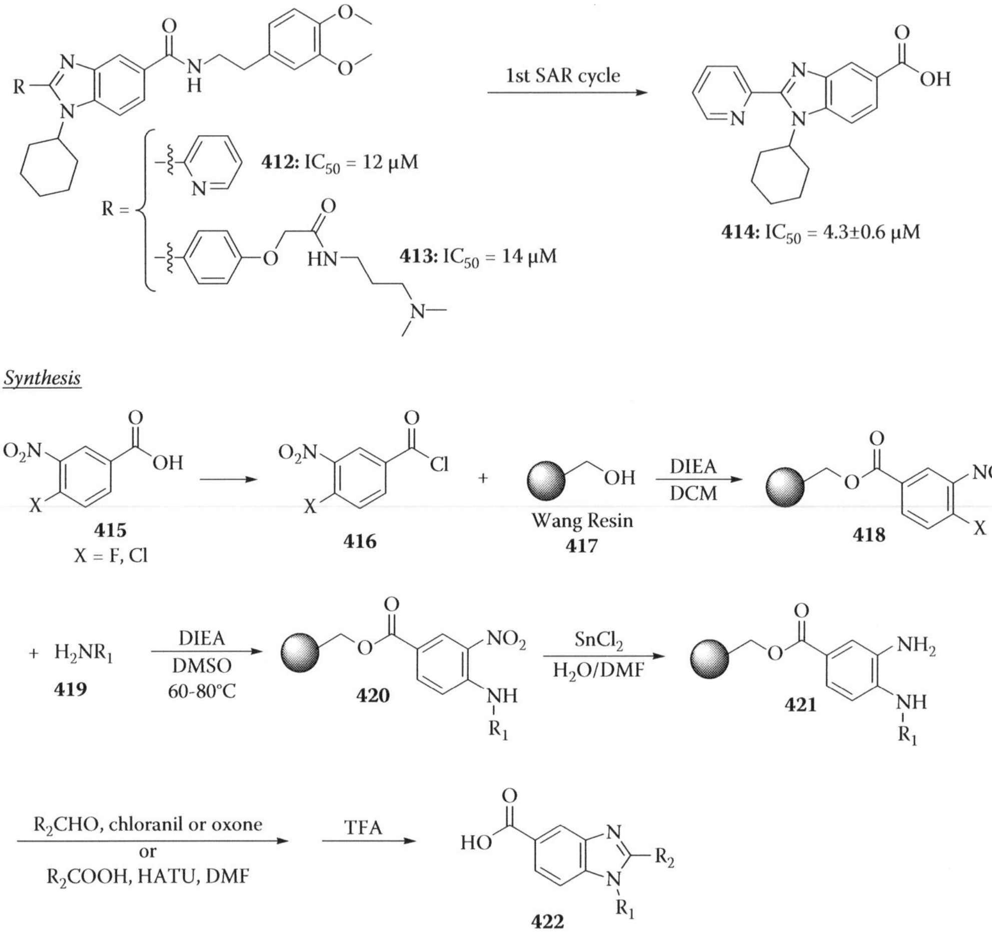

Synthesis

SCHEME 1.47   HCV targeted library: synthesis.

The authors took advantage of reported solid-phase parallel syntheses of benzimidazoles to rapidly optimize the substituents at the N-1 and C-2 positions of the heterocyclic unit of compound **414**. For this the authors immobilized 4-fluoro/chloro-3-nitrobenzoic acid chlorides (**416**) on Wang resin (**417**), displaced the halide group with different amines (**419**) varying in length, aromaticity, and sterics. The nitro group was then reduced and the resulting dianilines (**421**) were either cyclized with picolinaldehyde in the presence of chloranil or oxone, or acylated with picolinic acid followed by TFA cleavage to afford the desired benzimidazoles (**422**). This systematic substitution replacement on N-1 indicated that none of the introduced amines improved activity for which the cyclohexyl group was kept constant for the second round of SAR exploration on C2.

For the next step, a systematic replacement of the 2-pyridyl group on C-2 was pursued. More than 100 analogs were made and, although a clear SAR was not devised, it was found that steric factors had a great impact on improving potency while electronic factors did not. Although a handful of compounds displayed slightly better activity and good selectivity against the polio virus RdRp and the mammalian DdRp II than compound **414**, neither one exhibited inhibition at sub-micromolar levels and they all lacked efficacy in cell-based assays.

For further lead optimization, the authors used traditional solution-phase synthesis, choosing compounds **423** (IC$_{50}$ = 1.6 ± 0.6 µM) and **414** and focusing their efforts initially on the amide group in compound **412**. Amide analogs were made by TBTU coupling of benzimidazole 5-carboxylic precursors with benzyl amines. Placing a carboxylic acid functional group on the benzylic carbon improved both *in vitro* enzyme activity (sub-micromolar levels) as well as water solubility. Coupling **423** and **414** with (*S*)/(*R*)-5-hydroxy-tryptophan afforded more potent inhibitors (**424**: IC$_{50}$ = 140 nM, **425**: IC$_{50}$ = 50 nM) where the (*S*)-isomers were the most potent ones. Furthermore, SAR around the indole unit afforded compound **426**, an oxalic amide analog with single-digit nanomolar activity (IC$_{50}$ = 8 nM) and high selectivity against polio virus RdRp and mammalian DdRp II.

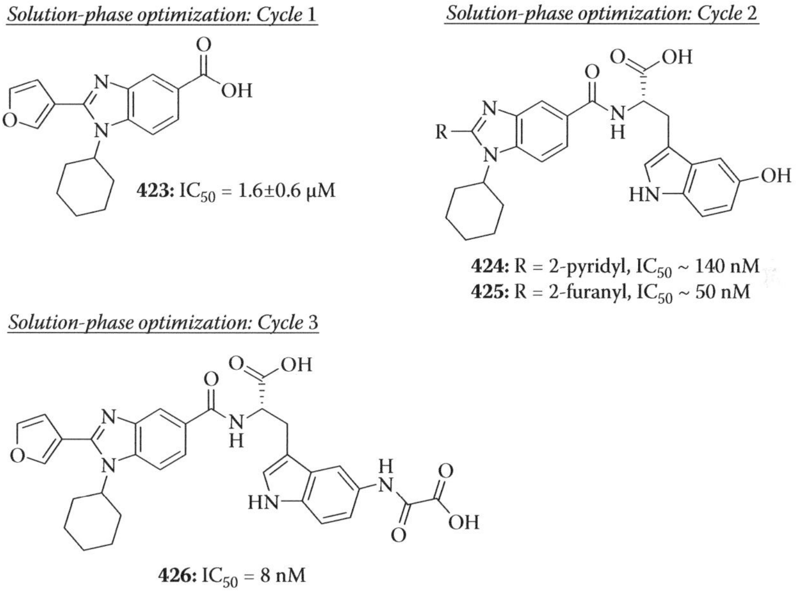

Solution-phase optimization: Cycle 1

423: IC$_{50}$ = 1.6±0.6 µM

Solution-phase optimization: Cycle 2

424: R = 2-pyridyl, IC$_{50}$ ~ 140 nM
425: R = 2-furanyl, IC$_{50}$ ~ 50 nM

Solution-phase optimization: Cycle 3

426: IC$_{50}$ = 8 nM

SCHEME 1.48 HCV targeted library: optimized hits.

(B) Harper et al.[40] at Merck reported the optimization of indole-N-acetamides as allosteric inhibitors of HCV NS5V polymerase. In this program, polymer-supported EDC was used for parallel synthesis and a Waters Fraction-Lynx LC/MS system for the purification of the final indole-N-acetamides. The initial lead compound **427** had a good *in vitro* profile (IC$_{50}$ = 26 ± 11 nM, EC$_{50}$ = 0.8 ± 0.3 µM), offering a good starting point for lead optimization.

Crystallographic data of compound **427** complexed with the enzyme indicated that introduction of liphophilic groups on the C-2 position of the indole core could interact with an empty pocket within the enzyme that could improve affinity. Exploratory SAR confirmed this observation. Also, although substitution on the acetamide group linked to the indole N did not have a significant influence on potency, it did have a profound effect on cell membrane permeability. Following these observations, optimization efforts were focused on the C-2 position of the indole unit in conjunction with the acetamide unit of the N indole to improve simultaneously both potency and cell-based activity. A 10 × 50 exploratory library was constructed using indole-N-acetic acids (**429**) and 50 diverse amines (**430**) chosen to vary compound size and polarity.

Although this library provided **431** with improved potency in both enzyme-based and cell-based *in vitro* assays (IC$_{50}$ = 18 nM, EC$_{50}$ = 0.3 µM), its *in vivo* pharmacokinetic profile in the rat and dog models indicated that this compound has low bioavailability, a short plasma half-life (post IV administration), and high plasma clearance (Rat: %F = 1.9, $t_{1/2}$ = 1.5 ± 0.2 h, Cl$_{(plasma)}$ = 51 ± 9 mL/min/kg;

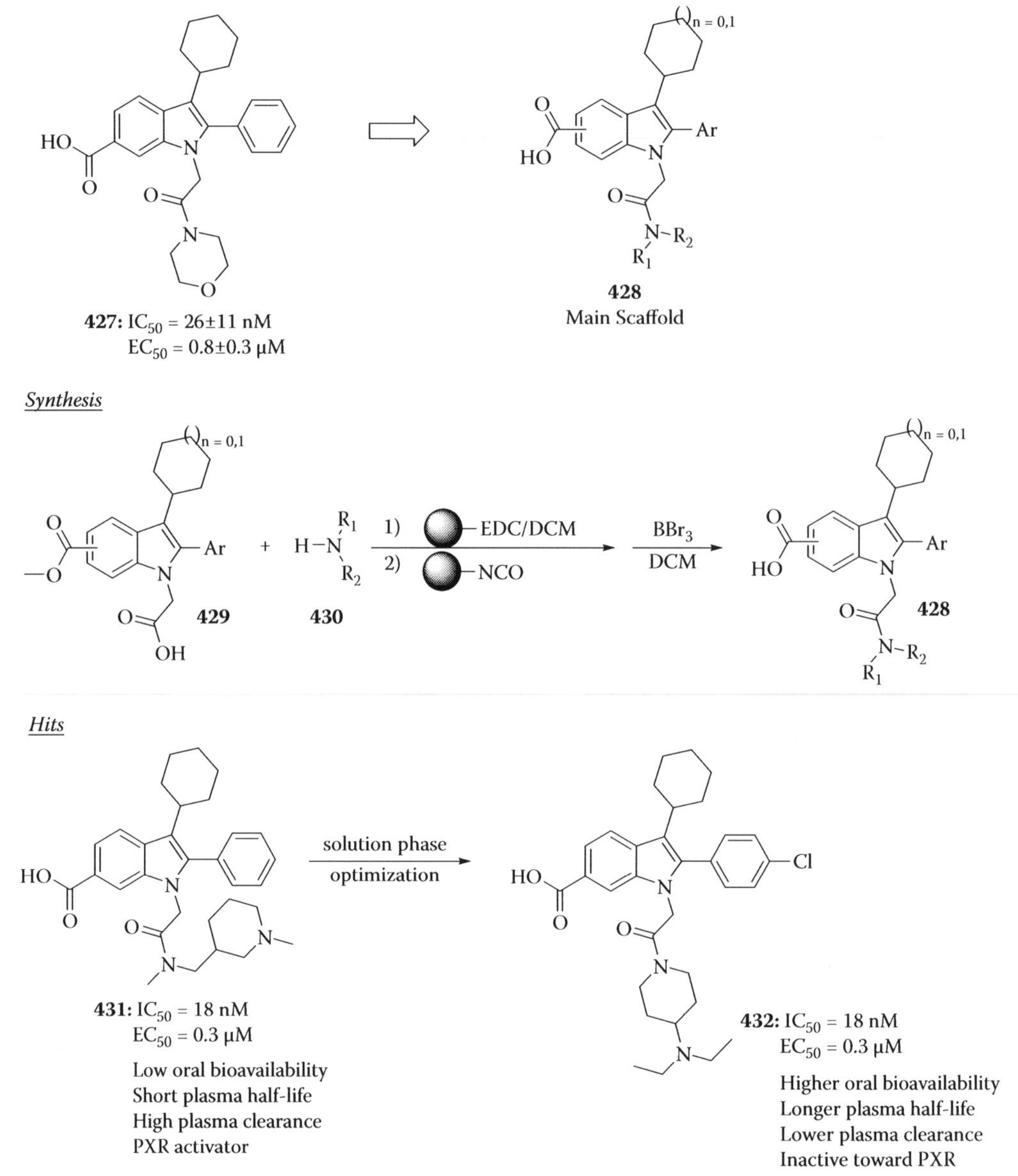

427: IC$_{50}$ = 26±11 nM
EC$_{50}$ = 0.8±0.3 µM

428
Main Scaffold

Synthesis

Hits

solution phase
optimization

431: IC$_{50}$ = 18 nM
EC$_{50}$ = 0.3 µM

Low oral bioavailability
Short plasma half-life
High plasma clearance
PXR activator

432: IC$_{50}$ = 18 nM
EC$_{50}$ = 0.3 µM

Higher oral bioavailability
Longer plasma half-life
Lower plasma clearance
Inactive toward PXR

SCHEME 1.49   HCV targeted library: polymer-assisted synthesis and hits.

Dog: %F = 28 ± 6, $t_{1/2}$ = 1.2 hr, Cl$_{(plasma)}$ = 22 mL/min/kg). Moreover, further testing showed that this compound agonized the pregnane X receptor (PXR), inducing cytochrome-P450 3A4 activity in human liver microsomes, clearly a liability for the use of this compound in humans.

To improve the pharmacokinetics of the compounds, a follow-up synthesis via solution phase afforded compound **432** with a much better *in vivo* profile and with better oral bioavailability and plasma half-life (Rat: %F = 10 ± 3, $t_{1/2}$ = 3.1 ± 1.0 hr, Cl$_{(plasma)}$ = 44 ± 11 mL/min/kg; Dog: %F = 51 ± 19, $t_{1/2}$ = 5.0 ± 0.8 hr, Cl$_{(plasma)}$ = 9 ± 2 mL/min/kg), showing no activity in human XPR assays at 10 µM while maintaining cell-based activity at sub-micromolar levels (EC$_{50}$ = 0.127 ± 23 µM).

This work provides a very good example of the impact that parallel synthesis can have in a lead optimization program.

1.13.1.3 Case 3: HIV Inhibitors

Rano et al.[41] at Merck worked on improving the pharmacokinetics, potency, and metabolic profile of indinavir (**433**). Indinavir is a potent and selective, orally bioavailable HIV protease inhibitor that is metabolized by CYP3A4. To find a lead for a second-generation HIV inhibitor, a solid-phase synthetic route was developed to provide an indinavir-based combinatorial library. It had been established that the 3-pyridyl-methylene group on the piperazine unit increased water solubility and activity against the target. The authors' strategy included keeping the central hydroxyethylene group, varying the potency-driving 3-pyridyl-methylene group to explore the P_3 site of the protease, then varying the benzyl group to probe the P_1' site of the protease and improve the metabolic profile, and finally the modification of the amino indanol group to probe the P_2' site of the protease.

For the 60-member library (5 × 4 × 3), the authors used a split-and-mix spatially addressed approach, beginning with the construction of 5-hydroxyethylene isostere fragments via solution phase (**435**). These fragments, made to probe the P_1' site, were anchored to RAPP TentaGel S-COOH resin (**436**) using EDC coupling conditions. The resins (**437**) were mixed, the allyl group was removed using Pd(PPh)₄, the resins were split into four groups, and each group was treated with different amino indanols (**438**) using EDC/HOBt peptide coupling conditions. The resulting amides (**439**), made to probe the P_2' site, were treated with 30% TFA/DCM to provide unprotected piperazine intermediates that were split into three groups. Two groups were treated with aldehydes (**440**) to alkylate the piperazine unit under reductive alkylation conditions (**441**). The remaining group was acylated with mesyl chloride to afford the corresponding methyl sulfonamide. These last three groups were made to probe the P_3 site of the protease.

The authors used equimolar portions of each pool for screening and found that the 3-pyridyl-methylene group at the P_3 site exhibited the highest potency in both the HIV binding assay and the cell-based spread assay. With this data, the authors used the 20 analogs containing the 3-pyridyl-methylene group on the piperazine unit and tested them as a mixture *in vivo* in dogs (n = 2), administering them orally as a 0.05 M citric acid solution. The analysis of the *in vivo* results indicates that the isobutyl group in the P_1' site provided a higher C_{max} (900–950 ng/mL) and AUC compared to the benzyl group in indinavir (C_{max} ~ 660 ng/mL), although the differences in half-life were only moderate.

This work presents a great example of how combinatorial mixtures can be used to obtain *in vivo* data for lead optimization, SAR elucidation, and derive pharmacokinetic and potency compound profiles with a minimum number of *in vivo* experiments.

1.13.1.4 Case 4: HRV 3CP Inhibitors

(A) Reich et al.[42] at Pfizer Global R&D-La Jolla/Agouron Pharmaceuticals reported a study on the structure-based design of benzamide-based inhibitors of the human rhinovirus 3C protease (HRV 3CP), particularly the HRV-14 3CP serotype. To design non-peptide inhibitors, the authors used the crystal structure of HRV-14 3CP co-crystallized with a peptide aldehyde. Analysis of the crystal structure led the authors to propose that 3-formylbenzamide would provide a scaffold where the aldehyde would be positioned close to the nucleophilic cysteine of the enzyme while a carboxamide group would be oriented toward the S1 pocket of the ezyme, just as observed in the co-crystallized peptide aldehyde. It was found that 3-formylbenzamide only inhibited HRV 3CP weakly (K_i = 104 μM), with moderate antiviral cell activity in H1-HeLa cells (EC_{50} = 15.9 μM) and no cytotoxicity (CC_{50} > 320 μM). Because it was known that HRV 3CP was also inhibited by Michael acceptors, the authors postulated replacing the aldehyde group by an α,β-unsaturated ester moiety. Also, based on the crystal structure of the enzyme-peptide aldehyde, the authors postulated that placing different substituents at position-6 of the phenyl ring would allow access to the S2 subsite of the enzyme with the potential of increasing enzyme-ligand affinity. Surprisingly, initial SAR based on placing different substituents on the α,β-unsaturated ester moiety as well as on position-6 proved

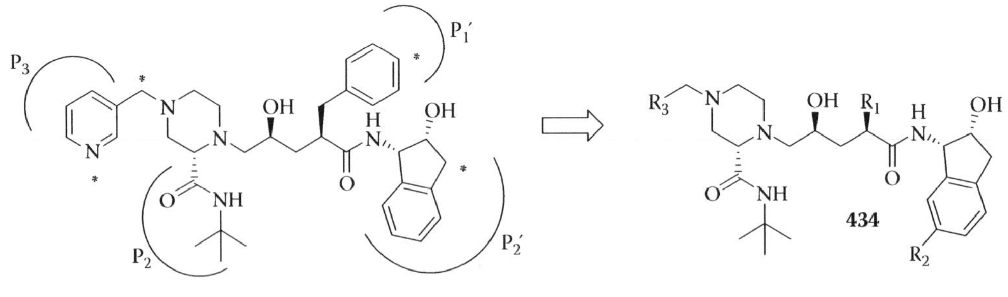

433 (Indinavir)
* = CYP3A4 metabolic sites

Synthesis (mixtures)

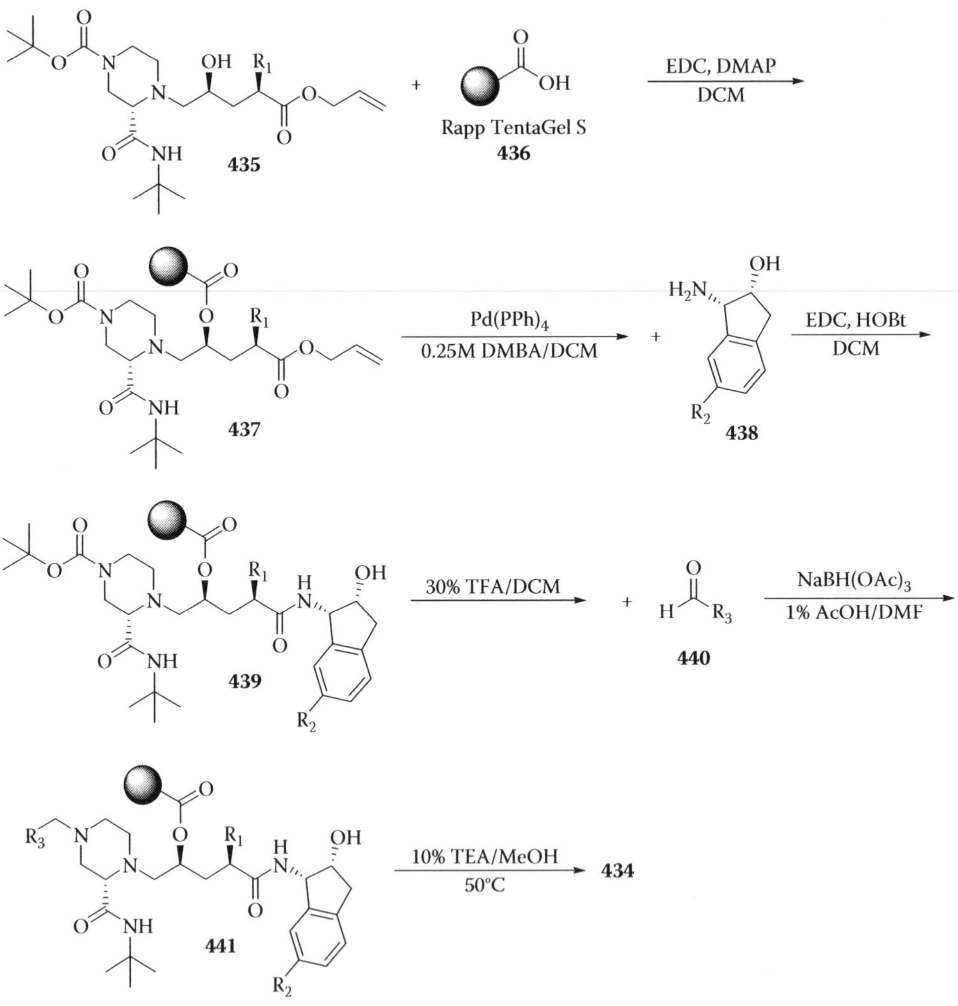

SCHEME 1.50   HIV targeted library.

that the postulated changes did not elicit the anticipated improvement in binding affinity. To find other potential modification sites to improve the activity against HRV 3CP, the authors zoomed into position-5 that, according to the available x-ray structure, would provide access to the S3 and S4 pockets of HRV 3CP.

An analog containing a hydroxymethyl group on position-5 exhibited a $K_i = 54$ µM confirming the theory that substitution at position-5 is tolerable, leading to improved enzyme affinity. Visualizing that a solid-phase synthetic approach could be possible using the benzamide unit as the point of attachment, and that a bromomethyl group at position-5 could allow the introduction of a diverse set of nucleophiles, the authors created a virtual library of approximately 3000 compounds using (E)-ethyl 3-(3-(bromomethyl)-5-carbamoylphenyl)acrylate as the main scaffold and different commercially available primary amines and thiols from the ACD database. The three-dimensional structure of each virtual product was generated and docked against HRV 3CP using a partial-fixed docking approach where the benzamide group was kept fixed as observed in the original x-ray structure. The compounds were ranked first based on low-energy interaction, then by the degree of S3 and S4 space-pocket filling, and finally by ease of synthesis.

A set of 784 compounds was synthesized via solid-phase synthesis using Rink amide loose resin or Chiron Crowns (**442**) where the (E)-3-(bromomethyl)-5-(3-ethoxy-3-oxoprop-1-enyl) benzoic acid (**443**) was anchored using DIC/HOBt coupling conditions. The resulting immobilized bromomethyl benzamide (**444**) was then treated with either primary amines or thiols (**445**) in DMF in the presence of DIEA at 70°C, followed by TFA cleavage and affording the expected aminomethyl- and thioether-benzamides (**446**). This library afforded about a half-dozen compounds with EC_{50} in the single-digit micromolar range and no cell cytotoxicity ($CC_{50} > 100$ µM), with one particular compound (**449**) exhibiting antiviral activity with an EC_{50} of 0.6 µM and a CC_{50} of 79 µM.

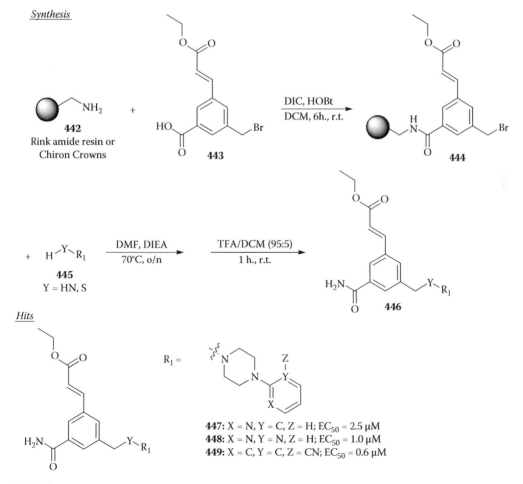

Synthesis

Hits

447: X = N, Y = C, Z = H; $EC_{50} = 2.5$ µM
448: X = N, Y = N, Z = H; $EC_{50} = 1.0$ µM
449: X = C, Y = C, Z = CN; $EC_{50} = 0.6$ µM

SCHEME 1.51 HRV 3CP targeted library.

Although the authors found that the most potent compounds in this campaign had a weaker inactivation constant than expected, this work provides a good example of how x-ray data can be used to propose a scaffold with appropriate functional groups and their position from one another, while providing the desired orientation to carefully probe selected enzyme regions; using computational tools to design, enumerate, and rank/filter potential inhibitors using key structural enzyme features while keeping in consideration the practical aspect of both building blocks availability and ease of synthesis.

(B) Shortly thereafter, Johnson et al.[43] from the same research group reported another structure-based design of small-molecule non-peptidic HRV-14 3CP inhibitors using a polymer-supported parallel synthesis approach. Using as the starting point the Michael acceptor ethyl ester tripeptide **450**, a lead compound developed in-house that had been co-crystallized with HRV-2 3CP, the authors simplified the core structure to an α,β-unsaturated ester moiety where the original carboxamide group was replaced by a lactam (**451**). This core structure offered an amino group as the diversity point to construct in a high-throughput synthesis approach, derivatives to probe the S2, S3, and S4 pockets of the enzyme.

For the structure-based design, the authors used a similar approach just described: A set of commercially available carboxylic acids was selected from the ACD, a virtual library was enumerated, the products minimized, and their three-dimensional coordinates docked against HRV-14 3CP fixing the Michael acceptor group in the same orientation as observed in the HRV-14 3CP-X crystal structure. The virtual compounds were ranked based on the number of hydrogen-bonding interactions, hydrophobic interactions, steric interactions, and by low-energy interaction.

The selected compounds were synthesized via a polymer-supported carbodiimide coupling approach in a 96-well array format using a Charybdis Calypso NXT-75 reaction block system and a Charybdis Iliad PS2 liquid handler. The resulting amides were made with purities greater than 80% and exhibited potent cell-based inhibition with EC_{50} values between 2 and 0.15 µM and no cell cytotoxicity ($CC_{50} > 50$–100 µM). Interestingly, as observed in their previous work, the more potent inhibitors exhibited an observed enzyme inactivation rate much higher than the values expected. The authors explain this as being the result of the compounds possessing a better physicochemical profile leading to a higher cell-wall permeability compared to the lead tripeptide **450**. Further screening of these compounds against a panel of HRV serotypes showed that, despite the observed high inactivation rates, these inhibitors are selective toward HRV-14.

1.13.1.5   Case 5: Neuramidase (NA) Inhibitors

Wang et al.[44] at Abbott Laboratories reported the structure-based design and synthesis of tri- and tetra-substituted pyrrolidines as neuramidase (NA) inhibitors. NA is a glycoprotein expressed on the surface of the influenza virus that catalyzes the cleavage of sialic acid residues, a process that is essential for the replication and activity of the virus. Their work began with the discovery of **456**, a compound that exhibited low inhibition against NA A/Tokyo ($IC_{50} = 50$ µM). The authors compared this compound with **457**, a known potent NA A inhibitor ($IC_{50} \sim 0.5$ nM). With the aid of computer modeling and available crystallographic data of the enzyme, the authors postulated a pharmacophore model for the design of more potent and drug-like inhibitors. It was postulated that NA possesses four key well-conserved regions at the active site responsible for substrate recognition: (1) site 1, which is positively charged and responsible for the recognition of **457**'s carboxylic acid; (2) site 2, which is negatively charged and responsible for the recognition of **457**'s guanidine; (3) site 3, a small hydrophobic pocket that can accommodate **457**'s acetyl group; and (4) site 4, which recognizes and binds to the triglycerol unit of **457**. The authors proposed that

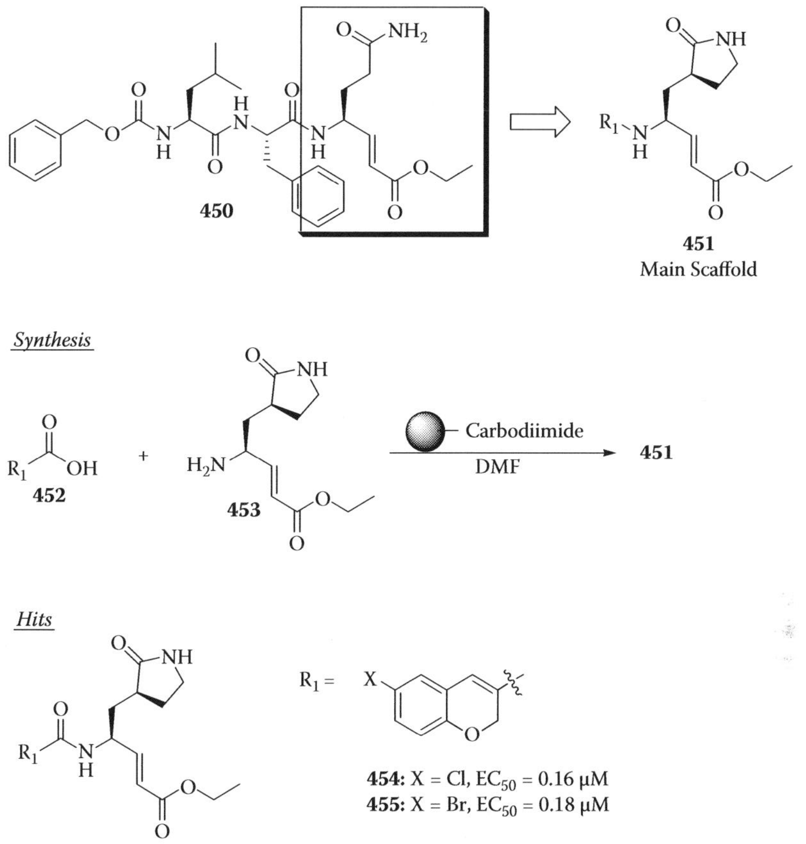

SCHEME 1.52 HRV-14 3CP targeted library.

456 was a close match to the devised four-site pharmacophore, namely a rigid/cyclic frame, a carboxylic acid (site 1), and a free amine (site 2), although they did not comment on how exactly the rest of the scaffold was expected to fit or match the two remaining sites.

The authors synthesized the orthogonally protected pyrrolidine scaffold **459**, a compound whose carboxylic acid could serve as an anchoring point on a resin that, upon release from a solid support, would provide both the amino group (positive charge) and carboxylic acid unit (negative charge) needed for inhibition according to the proposed pharmacophore.

With **459** in hand, a solid-phase methodology was developed using Wang resin (**460**), and a library of trisubstituted pyrrolidines (**458**) was created using an Advanced ChemTech ACT 396 multiple peptide synthesizer. Scaffold **459** was anchored to Wang resin (**460**) via ester bond formation with DIC and DMAP in DMF. The resulting immobilized ester (**461**) was arrayed in batches of 32 and treated with 20% piperidine/DMF, and the free pyrrolidine nitrogen was (1) acylated with various electrophiles to afford resin-bound amides, sulfonamides, or carbamates; or (2) treated with phosgene to treat the resulting carbamyl chloride with amines to afford resin-bound ureas. The intermediates were treated with Pd(PPh$_3$)$_4$ and Bu$_3$SnH in DCM with a catalytic amount of water to remove the allyl protecting group, and the final products (**458**) were released from Wang resin using TFA cleavage conditions.

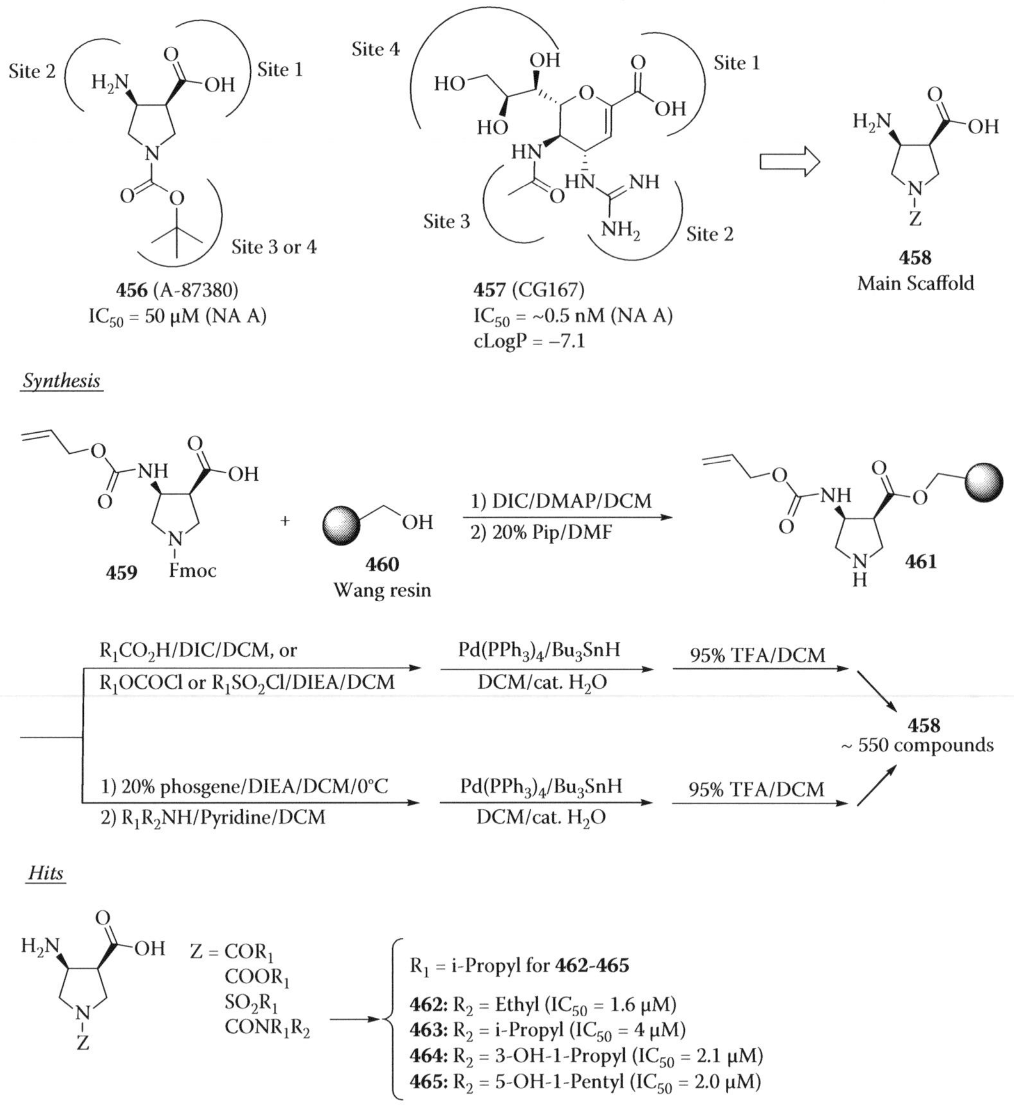

SCHEME 1.53 NA targeted library 1 and hits.

Half a dozen compounds exhibited single-digit micromolar NA A inhibition where, interestingly, all the compounds were disubstituted urea derivatives sharing an isopropyl group (**462** to **465**). The second urea substituent showed it could accommodate diverse functionalities such as pyridyl, carboxylate, and alcohol groups. Although co-crystallization of **464** with N9 A/Tern NA proved that the carboxylate group interacted with the enzyme at site 1, it was observed that the isopropyl substituent of the urea was not placed toward hydrophobic site 3 as expected, but instead toward site 4. This induced an unexpected conformational change in the enzyme, resulting in the creation of a new hydrophobic pocket. This conformational change the orientation of **464** such that the hydroxypropyl group of the urea was exposed toward the solvent.

With the knowledge that the isopropyl urea unit is placed in site 4 of the enzyme, the authors proceeded to explore site 3. To achieve this, they synthesized **467** where, according to x-ray data, a 2C-aminomethyl group *trans* to the 3-amino and 4-carboxylate groups could be derivatized, positioning the introduced substituents in the right region in space to probe site 3. For these tetrasubstituted pyrrolidines, a solution-phase strategy similar to their solid-phase approach was used. About 70 tetrasubstituted pyrrolidines (**468**) were synthesized in parallel, of which those containing an acetamido or trifluoroacetamido group were active, the latter (**469**) with an $IC_{50} = 280$ nM. This observation is in line with what was known about site 3 — that it is a small hydrophobic site.

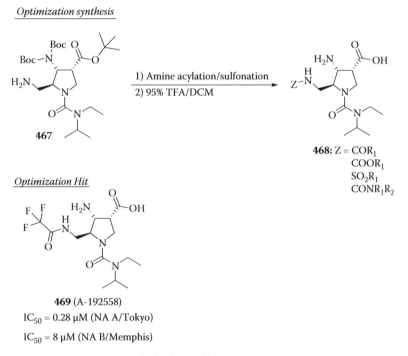

SCHEME 1.54   NA targeted library: optimization and hits.

Although compound **469** was 180 times more potent than the initial lead compound **456**, this compound lacked selectivity inhibiting NA B/Memphis with an IC_{50} of 8 μM.

1.13.1.6   Case 6: Anti-Poxvirus Agents

Pirrung et al.[45] at the Levine Science Research Center used combinatorial pools for the screening and optimization of isatin-β-thiosemicarbazones as anti-poxvirus agents (**472**). Based on the anti-poxvirus compound isatin-β-thiosemicarbazone hits (**470**, **471**), the strategy followed involved the solid-phase synthesis of N-alkylated isatin-β-thiosemicarbazones (**472**). Such analogs not only needed to be potent, but also to possess a favorable physicochemical profile for cell-based activity. The indexed combinatorial library approach included the use of two pool sets: (1) one set keeping the isatin constant while being combined with selected secondary amines, and (2) one set keeping the secondary amine constant while being combined with the available isatins.

With ten isatins in hand and ten selected secondary amines, this indexed combinatorial approach involved only the synthesis of twenty isatin mixtures. Compounds (**476**) were anchored onto a trityl thiosemicarbazone resin (**475**) via imine-bond formation, and the immobilized isatins (**477**) were then split into two sets. Set 1 was used for the N-alkylation of each immobilized isatin with a mixture of ten formaminals (**478**) derived from the selected secondary amines. For set 2, the immobilized isatins were mixed, divided in ten subsets, and each subset N-alkylated with a single formaminal. Final products (**472**) were released from the polymer support using 10% triisopropyl-silane in DCM followed by the addition of 30% TFA in DCM.

Each of the resulting 20 pools was screened for the inhibition of plaque formation in vaccinia-infected and cowpox-infected human foreskin fibroblast (HFF) cells. From set 1 it was observed that the pool containing 5-bromo-6-methyl-isatin exhibited the most potent EC_{50} inhibition, and from set 2 four amines were identified with the most potent EC_{50} inhibition. From these results, four analogs were synthesized as singles based on the most potent isatin from set 1 and the four more potent amines from set 2. It was found that these four analogs were up to 45 times more potent (**480**–**483**: $EC_{50} = 6.8$ to 0.6 µM) than reported anti-poxvirus agents **470** ($EC_{50} = 14$ µM) and **471** ($EC_{50} = 3.3$ µM), and they also had a safe nontoxic cell profile ($CC_{50} > 180$ µM). It is worth pointing out that compound **481** ($EC_{50} = 6.8$ µM), the single compound whose building blocks were expected to produce the most potent compound from set 1 and set 2, was not the most potent agent as expected; instead, it was compound **483** with an $EC_{50} = 0.6$ µM.

The indexed combinatorial library approach allows for the identification of the building blocks that contribute the most to a particular effect in a compound mixture, in this case potency, while minimizing the number of analogs that need to be made. However, this approach is not without risks, as this method assumes that the overall observed effect derives from only one compound in a mixture and not from the synergistic effect of two or more potent compounds or the combination of several weaker inhibitors. Interestingly, the authors did not report the use of computational chemistry to predict and select compounds with desired cell-permeability molecular properties such as cLogP, yet this exercise found potent, selective, and cell-permeable anti-poxvirus candidates. A quick cLogP calculation (using ChemDraw 9.0) of these potent compounds clearly indicates that they are right at the border of the acceptable Lipinski's rule-of-5 limit (cLogP = 4.9–5.5) with the exception of compound **481**, whose cLogP (~2.4) positions this compound as a good follow-up candidate for further lead optimization.

1.13.1.7   Case 7: Tissue Factor/Factor VIIa (TF/VIIa) Complex Inhibitors

Parlow et al.[46] at Pharmacia Corporation reported the structure-base design and polymer-assisted solution-phase synthesis of pyrazinones as selective small-molecule inhibitors of the tissue factor/factor VIIa (TF/VIIa) complex. The coagulation cascade involves several proteins that are activated at certain points in the coagulation pathway and form active complexes that, in turn, proteolytically activate other targets, ultimately generating the formation of thrombin (IIa) and a fibrin clot. Deregulation of this process can lead to over-production of clots, resulting in vascular thrombosis and ultimately causing conditions such as heart attacks, stroke, pulmonary embolism, etc. Tissue factor (TF) is a membrane-bound protein responsible for the coagulation initiation process in the event of blood vessel damage. Factor VIIa resides in the plasma and when

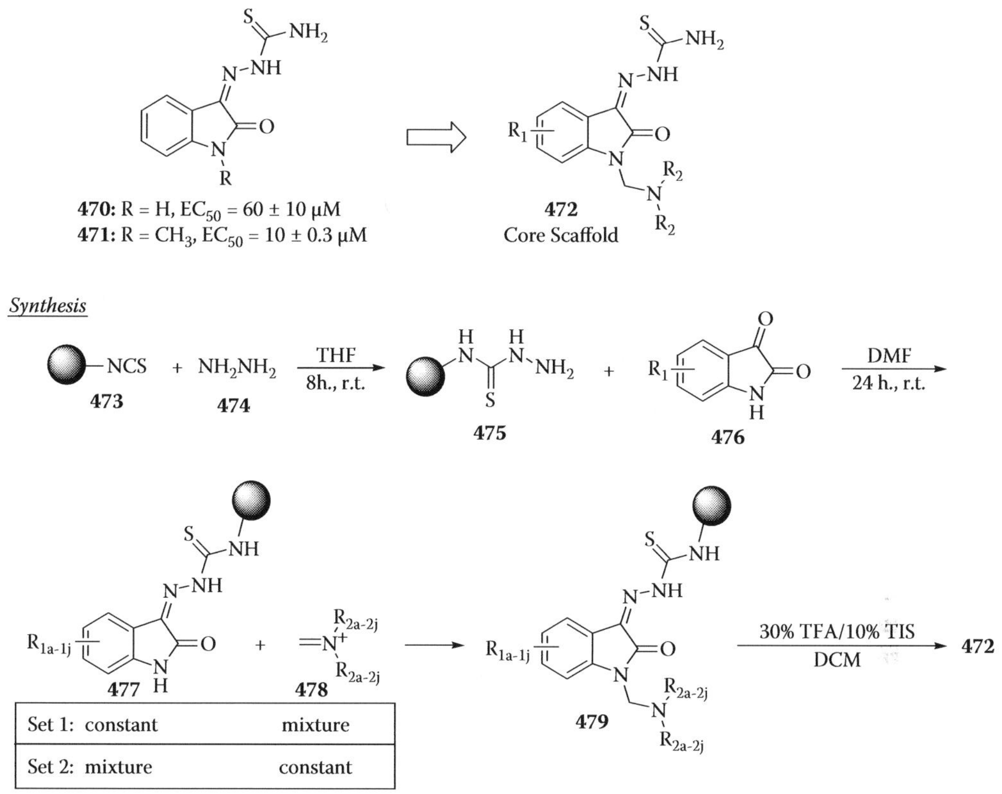
470: R = H, EC50 = 60 ± 10 μM
471: R = CH3, EC50 = 10 ± 0.3 μM
472
Core Scaffold
Synthesis
473
474
THF
8h., r.t.
475
476
DMF
24 h., r.t.
477
478
479
30% TFA/10% TIS
DCM
472
Set 1: constant mixture
Set 2: mixture constant

Hits

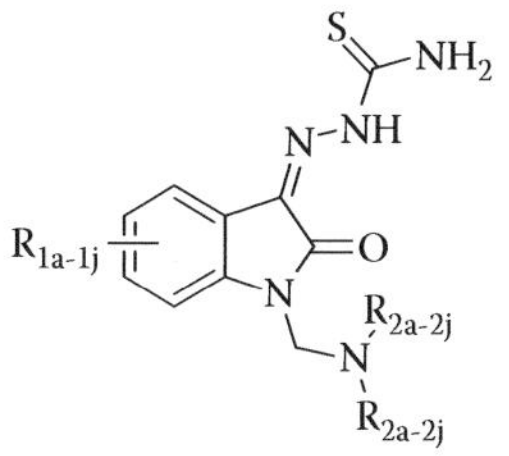

compound	Isatin (Set 1)	Amine (Set 2)	vaccinia Copenhagen EC_{50} (μM), CC_{50} (μM)	
480	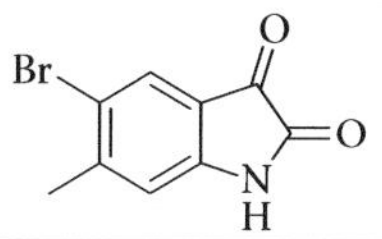	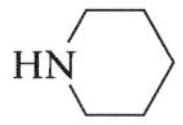	1.0 ± 0.8	> 244 ± 0
481		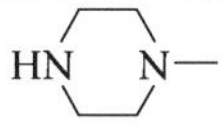	6.8 ± 8.7	> 235 ± 0
482		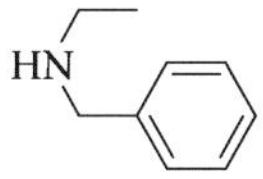	0.8 ± 0.1	> 201 ± 23
483		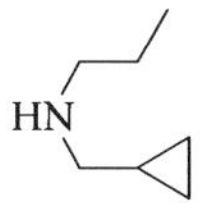	0.6 ± 0.5	> 161 ± 67

SCHEME 1.55 Anti-poxvirus targeted library.

it encounters TF, it forms a complex that activates factors IX and X, triggering the coagulation cascade. The selective inhibition of the TF/VIIa complex offers a means to regulate the coagulation process as long as undesirable side effects such as hemophilia are avoided.

An internal program focusing on the synthesis of tripeptide-α-ketothiazoles rendered hit **484** as a potent nanomolar TF/VIIa inhibitor exhibiting 500-fold selectivity against thrombin ($IC_{50} = 0.2$ μM for VIIa, $IC_{50} = 100$ 2 μM for IIa). This compound was co-crystallized with the TF/VIIa complex, giving insight into how **484** interacts with the TF/VIIa complex. It was observed that the arginine sidechain reaches deep into the S_1 pocket, forming four strong hydrogen bond interactions. The pyridyl group reaches into the relatively open, negatively charged S_2 pocket. Factor VIIa is the only coagulation protease that has a negatively charged S_2 pocket, a key detail that the authors will exploit for the design of selective TF/VIIa inhibitors. With these features at hand, several six-member heterocyclic systems were designed and docked against TF/VIIa, identifying pyrazinone scaffold **485** as a promising core for the synthesis of TF/VIIa complex inhibitors.

The synthetic approach involved first the solution-phase synthesis of the pyrazinone core bearing desired substituents for the interactions with the S_2 pocket. For the library synthesis, discrete compounds were made in multiple reaction block vessels spatially arranged using polymer-supported reagents for the rapid purification of intermediates and excess reagents. The process involves a modified Strecker reaction of glycine benzyl ester **486** with TMSCN and an aldehyde following the cyclization of the intermediate with oxalyl chloride to afford dichloropyrazinones **487** (S_2 pocket). These compounds were then reacted with 3 equiv of amines (**488**) to afford the aminopyrazinones **489** (S_3 pocket), which were then converted into their respective carboxylic acid analogs by either saponification or hydrogenolysis (**490**). Acids **490** obtained via saponification were purified by first immobilizing the acid on polymer-supported trisamine resin (**491**), the immobilized ammonium carboxylate intermediates (**492**) were washed to remove excess reagents, and then the acids **490** were released upon treatment with 4 N HCl/dioxane.

For the amide-coupling reaction (S_1 pocket) purified acids **490** were treated with HOBT (**494**) in the presence of polymer-supported carbodiimide (**493**). Amines (**496**) were then added to the 1-benzotriazole ester intermediates (**495**) to form the desired crude amides (**497**). To finally remove excess **490**, **494**, and **495**, the authors used both polymer-supported trisamine resin (**491**) and polymer-supported aldehyde (**498**). Filtration of all resins followed by solvent evaporation afforded pure products **497**.

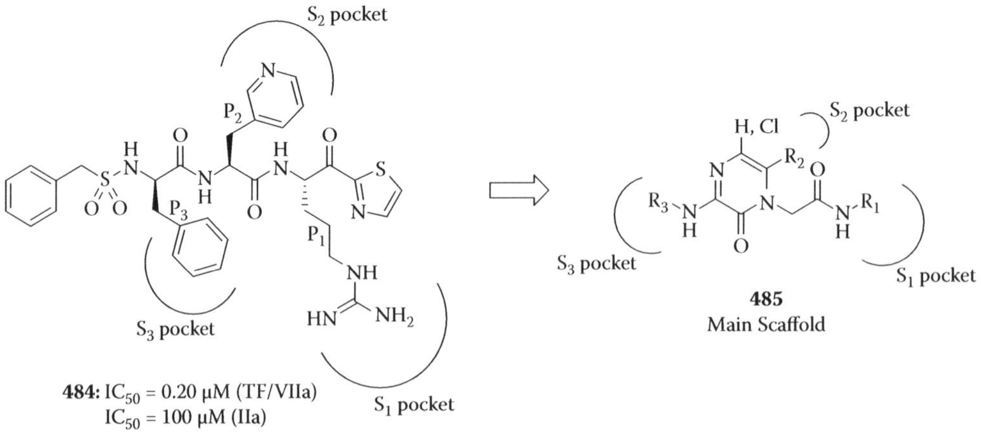

484: $IC_{50} = 0.20$ μM (TF/VIIa)
$IC_{50} = 100$ μM (IIa)

485
Main Scaffold

SCHEME 1.56 TF/VIIa targeted library synthesis.

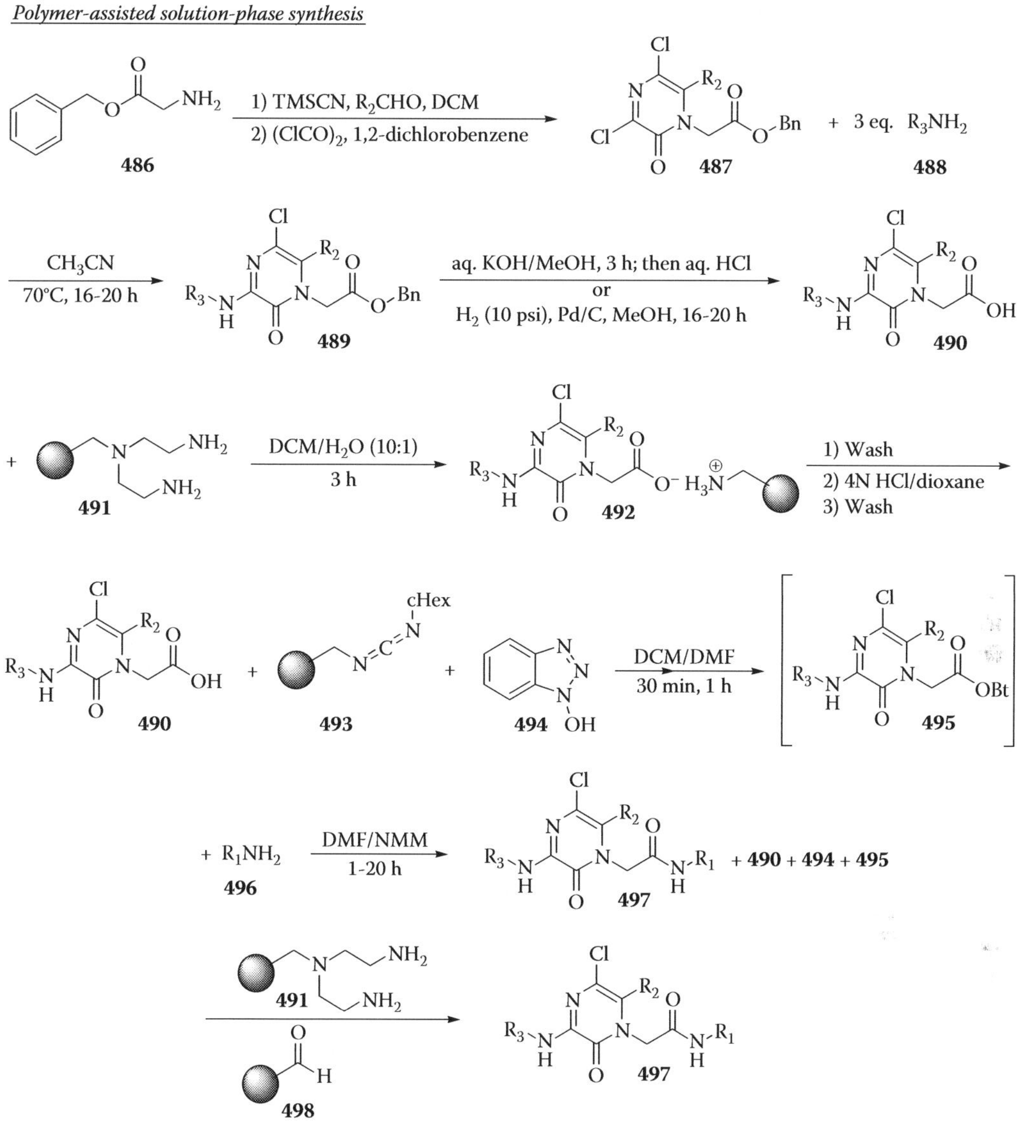

Polymer-assisted solution-phase synthesis

SCHEME 1.56 TF/VIIa targeted library synthesis (continued).

For this study, the authors decided to elucidate the SAR, first focusing on the S_1 pocket, then on the S_3 pocket, and finally the S_2 pocket. The SAR on the S_1 pocket indicated that a *para*-benzamidine group was crucial for activity (**499, 500**: $IC_{50} = 0.3–0.6$ µM for VIIa). The 3-pyridylamidine group also was well tolerated with a slight loss in activity (**501**: $IC_{50} = 1.4$ µM for VIIa), but the 2-pyridylamine resulted in a fivefold loss in activity. Changing a phenethylamidine group for the phenylamidine group was detrimental for activity as it was for nitrogen-containing heterocyclic systems. Positioning the amidine group at the 3-position of the phenyl ring was also deleterious for activity. Despite the single-digit to sub-micromolar potency of **499, 500**, and **501** against VIIa, it was observed that no significant selectivity against IIa was exhibited.

The next optimization phase focused on the S_3 pocket. Here, the chain length between the phenyl unit of the phenethylamine group and the pyrazinone was systematically varied, finding that a benzyl group not only retained potency against VIIa, but also increased tenfold selectivity against IIa (**503**: $IC_{50} = 0.77$ μM for VIIa). Removal of all methylene linkers or additional ones resulted in loss of potency (**502, 505, 506**). Similarly, replacement of the benzyl or phenethylamine with small tertiary amines was conterproductive for activity. Substituting a sulfur atom for the nitrogen of the benzylamine also resulted in a decrease in activity. Interestingly, replacing the phenylamine group with small alkyl amines not only was well tolerated, but it also resulted in higher selectivity against IIa (10- to 20-fold) (**507, 508, 509**). Replacement of the phenylamine group with small polar groups (alkyl alcohols, alkyl ethers, unsubstituted amine-containing alkyls) was well tolerated, but tertiary amines were not.

S_1 pocket: SAR and Hits

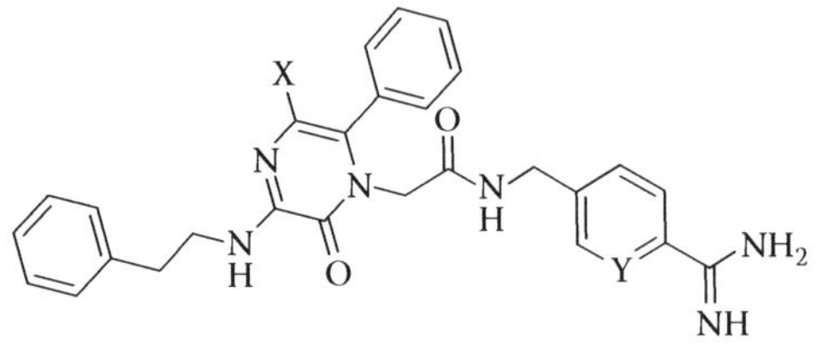

499: X = H, Y = C, $IC_{50} = 0.34$ μM (VIIa), $IC_{50} = 0.30$ μM (IIa)
500: X = Cl, Y = C, $IC_{50} = 0.63$ μM (VIIa), $IC_{50} = 0.16$ μM (IIa)
501: X = Cl, Y = N, $IC_{50} = 1.40$ μM (VIIa), $IC_{50} = 0.15$ μM (IIa)

S_3 pocket: SAR and Hits

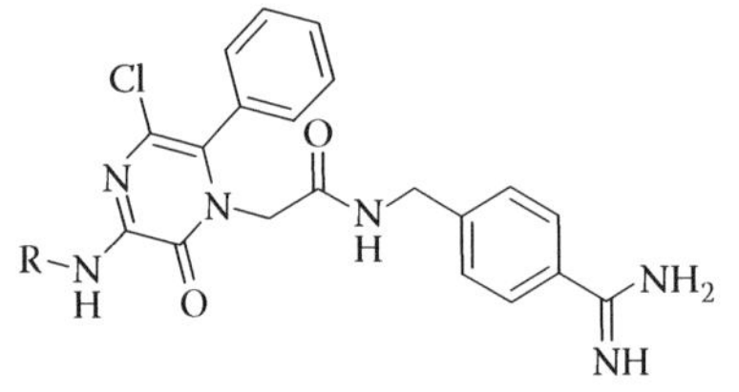

502: R = Ph, $IC_{50} = 11.2$ μM (VIIa), $IC_{50} > 30$ μM (IIa)
503: R = CH_2Ph, $IC_{50} = 0.77$ μM (VIIa), $IC_{50} = 7.63$ μM (IIa)
504: R = CH_2CH_2Ph, $IC_{50} = 0.63$ μM (VIIa), $IC_{50} = 0.16$ μM (IIa)
505: R = $(CH_2)_3Ph$, $IC_{50} = 12.6$ μM (VIIa), $IC_{50} = 23.6$ μM (IIa)
506: R = $(CH_2)_4Ph$, $IC_{50} = 4.1$ μM (VIIa), $IC_{50} = 7.1$ μM (IIa)
507: R = Me, $IC_{50} = 0.7$ μM (VIIa), $IC_{50} = 20.4$ μM (IIa)
508: R = Et, $IC_{50} = 0.4$ μM (VIIa), $IC_{50} = 11.3$ μM (IIa)
509: R = cycBu, $IC_{50} = 0.40$ μM (VIIa), $IC_{50} = 4.6$ μM (IIa)

S_2 pocket: SAR and Hits   *Most potent and selective hit*

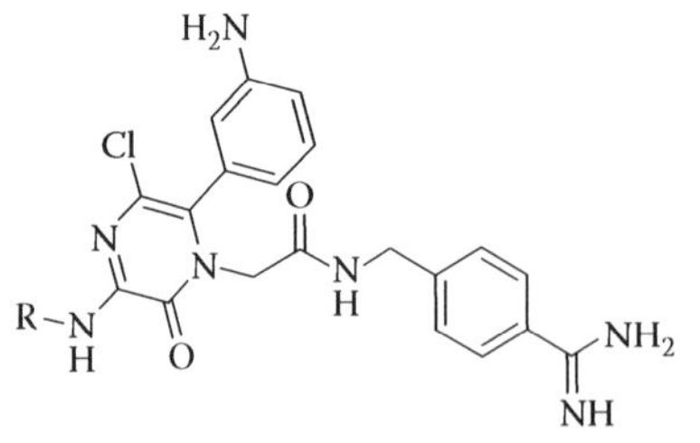

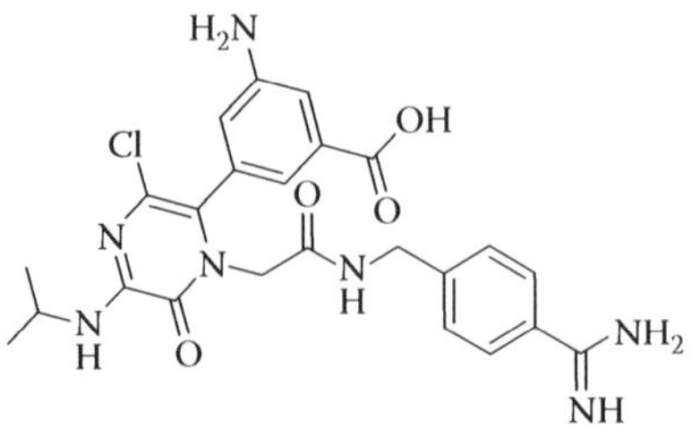

510: R = Me, $IC_{50} = 0.06$ μM (VIIa), $IC_{50} > 30$ μM (IIa)
511: R = Et, $IC_{50} = 0.04$ μM (VIIa), $IC_{50} > 28.4$ μM (IIa)
512: R = cycBu, $IC_{50} = 0.02$ μM (VIIa), $IC_{50} = 8.0$ μM (IIa)

513: $IC_{50} = 16$ nM (VIIa)
$IC_{50} > 100$ μM (IIa)
$IC_{50} > 100$ μM (Xa)

SCHEME 1.57   TF/VIIa targeted library: SAR and hits.

The last optimization phase focused on the phenyl ring that interacts at the S_2 pocket. Here, the authors found that anilines favored dramatically activity reaching nanomolar potency with high selectivity against IIa (>500-fold) (**510, 511, 512**: $IC_{50} = 0.06–0.02$ µM for VIIa). Several substituents were used instead of the *meta*-aminophenyl moiety varying size, length, and electronic properties but all analogs could only reach single-digit micromolar activity at best. Having established that the *meta*-aminophenyl moiety was crucial for activity and selectivity, further studies focused on the effects of substituents on the aniline functionality. The results indicated that mono- and dimethylation, acetylation, and sulfonation, although well tolerated, always resulted in a slight loss of activity, with more severe consequences for selectivity. So the authors proceeded to explore the synthesis of other *meta*-aminophenyl derivatives to find that compound **513** not only retained nanomolar potency against the TF/VIIa complex, but also exhibited a dramatic increase in selectivity (6250-fold) against both thrombin (IIa) and Factor Xa (**513**: $IC_{50} = 16$ nM for VIIa, $IC_{50} > 100$ mM for both IIa and Xa).

1.13.2 Metalloproteases

1.13.2.1 Case 1: Endothelin Converting Enzyme (ECE-1) Inhibitors

(A) Kitas et al.[47] at F. Hoffmann-La Roche Ltd. reported a solid-phase synthesis of 1,2,4-triazoles linked to a 4-mercaptopyrrolidine core as endothelin converting enzyme (ECE) inhibitors. ECE-1 is one of the two known homologs of the Zn-endopeptidase endothelin converting enzyme (ECE). ECE interacts with big-ET, cleaving and releasing endothelin-1 (ET-1), a 21-amino-acid peptide found in high plasma levels in people with hypertension, congestive heart failure, and pulmonary hypertension conditions. Consequently, preventing ECE from interacting with big-ET offers a potential route to treat the above-mentioned conditions.

Starting with compound **514** ($IC_{50} = 0.5$ µM), the authors focused on the triazole-linked [2S,4R]-4-mercapto-proline as the main scaffold (**515**) and developed a solid-phase synthesis using a 4-(α,α-diphenylhydroxymethyl)-benzoic acid as a linker bound to BHA-NH$_2$ resin (**516**). The proline portion was introduced by treating resin **516** with [2S,4R]-4-sulfanyl-1-(Fmoc)-pyrrolidine-2-carboxylic acid (**517**) in 10% TFA/DCM. The carboxylic acid of the resulting resin **518** was first treated with trimethylsilyl chloride, the Fmoc removed, and the amine acylated (**519**). The carboxylic acid was then treated with TPTU, the ester intermediates treated with hydrazine, the resulting hydrazides reacted with different isothiocyanates to afford hydrazinecarbothioamides (**520**). These intermediates were S-alkylated with alkyl halides and the resulting carbamohydrazonothioates (**521**) cyclized in the presence of 0.06 M TFA to afford the desired triazole core (**522**). The desired products **515** were released from the solid support with 60% TFA and 10% TIPS in DCE.

The SAR found indicated that, from the acylating agents used on the nitrogen of the proline core, aromatic sulfonamides were in general the most favorable group, although the 2-naphthyl group in lead **514** was still found to provide the most potency. Replacing the methyl group on the triazole-linked thioether core with aliphatic alkyl groups was fairly well tolerated without a significant increase in activity. Activity was significantly improved when different substitution patterns were explored on the phenyl group of the N-triazole, finding that a fluoro atom at the *para* position increased activity threefold (**523**, $IC_{50} = 150$ nM).

(B) Brands et al.[48] at BAYER HealthCare AG reported the solid-phase systematic optimization of a non-obvious indole-based Zn-chelating compound lead **524** ($IC_{50} = 0.22$ µM) discovered from an HTS campaign. Using ethyl-5-amino-1H-indole-2-carboxylate (**527**) as the initial building block, the authors built a solid-phase combinatorial library via a split-and-pool approach using IRORI's radiofrequency-tagged Kans. The approach began by immobilizing indole **527** on an acid-labile aldehyde linker (**519**) via reductive amination to afford immobilized indole **528**. The amino unit was acylated with acyl chlorides, then the indole nitrogen was derivatized by alkylation with LiHMDS and primary bromides. The next step consisted of the saponification of the ethyl ester, where

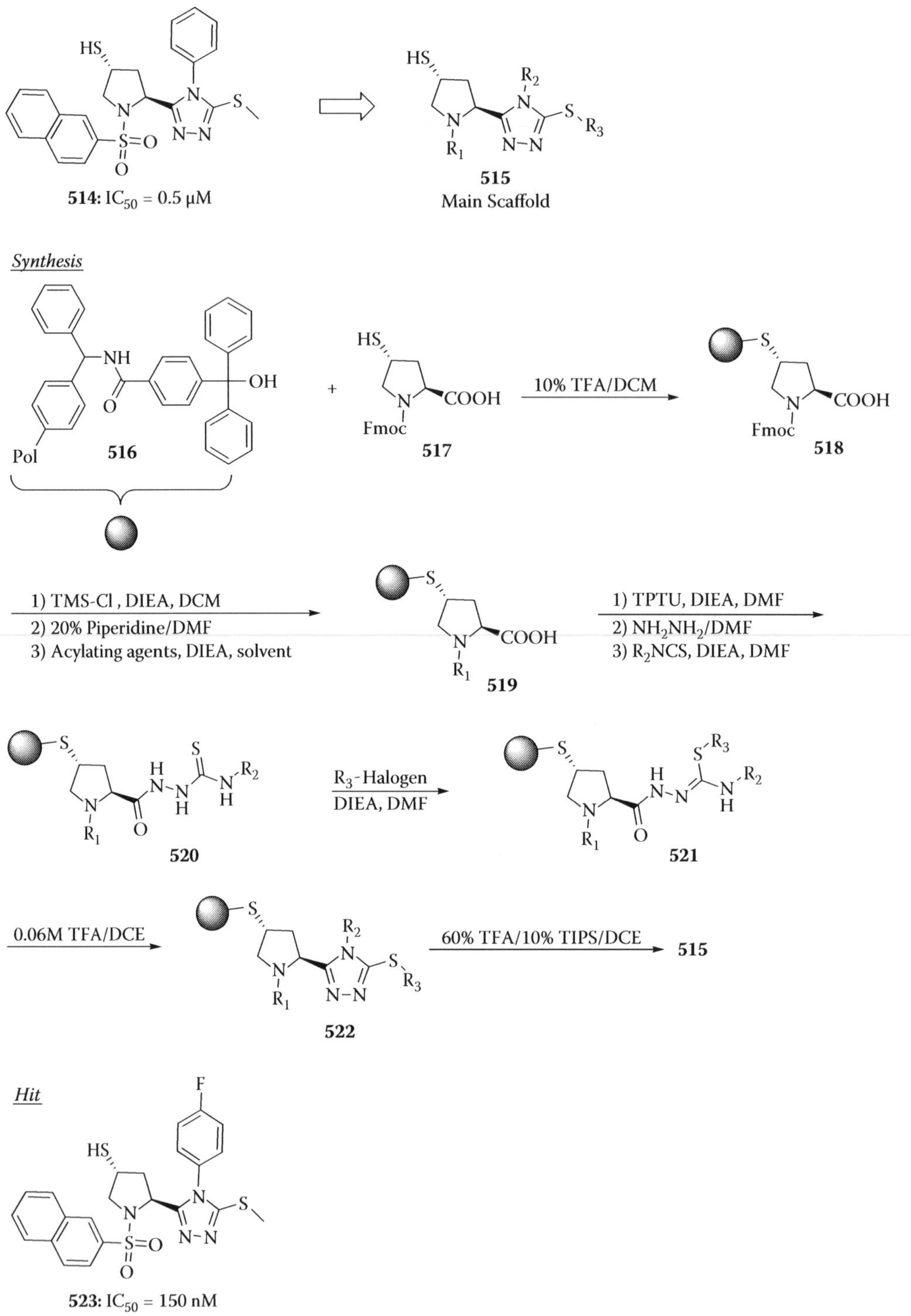

514: IC$_{50}$ = 0.5 μM

515
Main Scaffold

Synthesis

516

517

518

519

520

521

522

Hit

523: IC$_{50}$ = 150 nM

SCHEME 1.58 ECE-1 targeted library 1.

the resulting carboxylic acid intermediates (**531**) were either coupled with anilines using HATU as the coupling reagent or with amino phenols using PPh$_3$/NBS. The functionalized indole intermediates (**532**) were then cleaved from the resin with TFA/DCM to afford desired products (**525**).

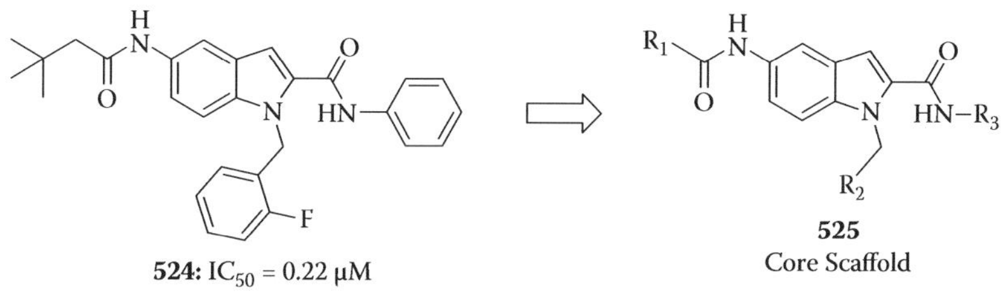

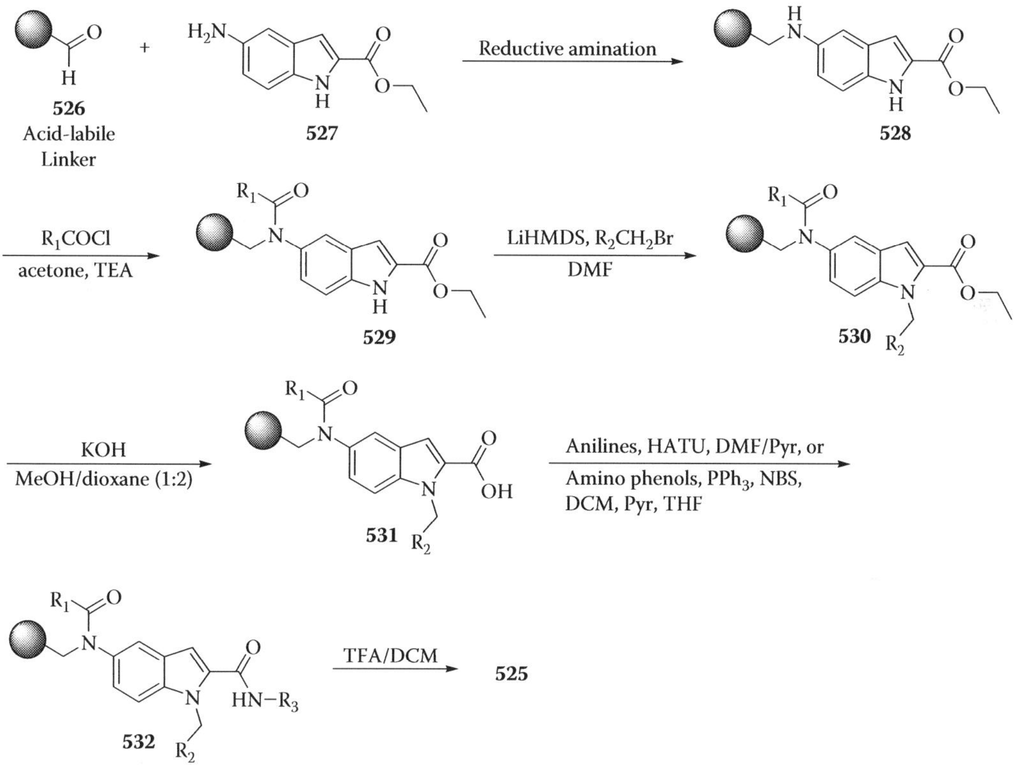

SCHEME 1.59 ECE-1 targeted library 2.

The initial SAR centered on the benzyl group on the indole nitrogen. It was observed that either elimination of the benzylic methylene carbon (phenyl group, **533:** IC_{50} = 5.9 μM) or addition of another methylene group (phenethyl group, **534:** IC_{50} = 5.1 μM) resulted in loss of activity against ECE-1. Substitution exploration on the aromatic portion of the benzyl group found that replacing the *ortho*-fluoro group with a trifluoromethyl group increased activity slightly (**535:** IC_{50} = 0.17 μM).

With this SAR in hand, the authors proceeded toward the optimization of the neopentyl amide group by elongating the aliphatic chain, replacing the neopentyl group with bulkier cyclic aliphatic groups, changing the amide for a thioamide and sulfonamide, and reverse amide group. All these changes were detrimental for activity, except for the incorporation of bulkier cyclic aliphatic groups, albeit the activity did not increase dramatically (**536:** IC_{50} = 0.12 μM). Positioning the neopentyl amide group on the 4-position of the indole also resulted in a dramatic loss of activity.

Because none of the above changes provided a significant improvement in activity, the authors decided to keep the original substituents of the original lead **524** and proceeded toward the optimization

Optimization of Benzyl Group (R_2)   Optimization of Neopentyl Amide Group (R_1)

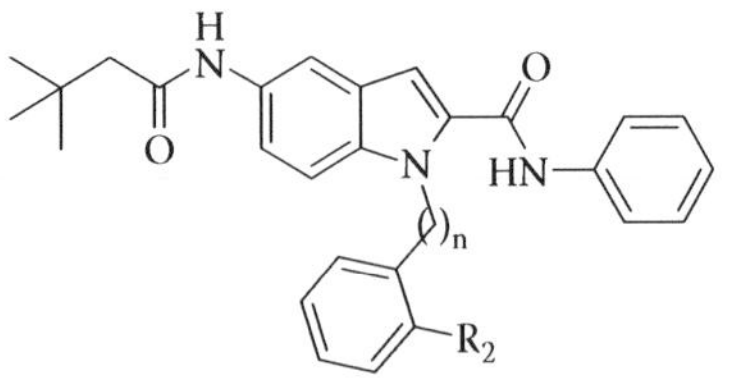

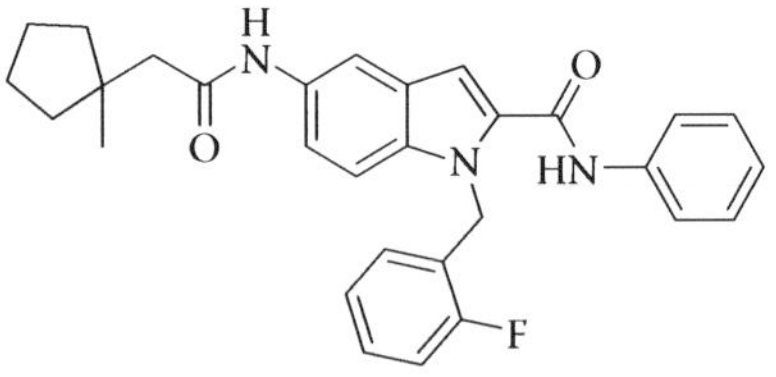

524: $n = 1$, $R_2 = F$, $IC_{50} = 0.22$ μM **536:** $IC_{50} = 0.12$ μM
533: $n = 0$, $R_2 = H$, $IC_{50} = 5.9$ μM
534: $n = 2$, $R_2 = H$, $IC_{50} = 5.1$ μM
535: $n = 1$, $R_2 = CF_3$, $IC_{50} = 0.17$ μM

Optimization of Phenyl Amide Group (R_3)

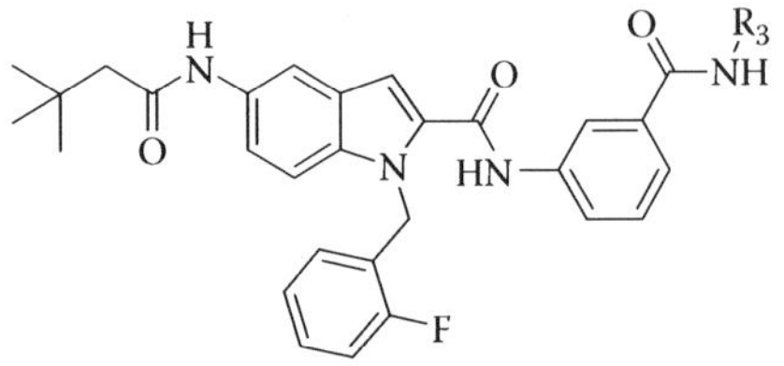

537: $R_3 = H$, $IC_{50} = 0.22$ μM
538: $R_3 = Ph(4\text{-}SOCH_3)$, $IC_{50} = 0.066$ μM
539: $R_3 = Ph(4\text{-}SONH_2)$, $IC_{50} = 0.039$ μM

SCHEME 1.60 ECE-1 targeted library 2: optimization and hits.

of the phenyl amide group. Deletion of the carbonyl group (amine analog) was deleterious for activity. Replacement of the phenyl group with different unsubstituted heterocyclic groups did not result in any major improvement in activity. However, when additional functionalities were present, activity improved. Eventually, it was found that the addition of aryl amides at the 3-position of the original phenyl amide group was tolerated and afforded double-digit nanomolar ECE-1 inhibitors (**538:** $IC_{50} = 0.066$ μM; **539:** $IC_{50} = 0.039$ μM).

1.13.2.2   Case 2: MMP-2 Inhibitors

Salvino et al.[49] at Rhone Poulenc Rorer reported the development and use of a five-step solid-phase approach for the synthesis and optimization of arylsulfone hydroxamates (**553**) as selective MMP or PDE4 inhibitors. Using Wang resin (**540**) as the solid support, Salvino's team first anchored 2-(diethoxyphosphoryl)acetic acid (**541**) by converting it into its acid chloride analog in pyridine. Then the resulting resin (**543**) was treated with aldehydes (**544**) in the presence of lithium bis(trimethylsilyl)amide (**545**) to afford acrylates (**546**). Thiols (**547**) were then incorporated via Michael addition, the thioethers (**548**) oxidized with mCPBA to their corresponding sulfones (**549**), and the desired carboxylic acids (**550**) released with 30% TFA/DCM. Coupling of **550** to hydroxylamine Wang resin followed by TFA cleavage afforded desired hydroxamic acids **553**.

The SAR indicated that 3,4-dialkoxy-phenyl sulfones afforded PDE4 inhibitors with MMP inhibition activity. It was also observed that a variety of substituents were tolerated at R_1, where the length of the alkyl sidechain found to be important for activity affording very potent PDE4 inhibitors (**544**, PDE4 $IC_{50} = 1.3$ nM). Further screening against an MMP isozyme panel revealed that several compounds exhibited selectivity against MMP-2. The SAR around the phenyl sulfone group indicated that the 4-methoxy-phenyl group exhanced selectivity towards MMP-2. Keeping this group constant, further optimization studies at R_1 showed that replacing the alkyl phenyl group with alkyl groups was tolerated, particularly cyclohexyl-based analogs improved activity up to 25-fold, affording sub-nanomolar MMP-2 inhibition (**557**, MMP-2 $IC_{50} = 0.8$ nM).

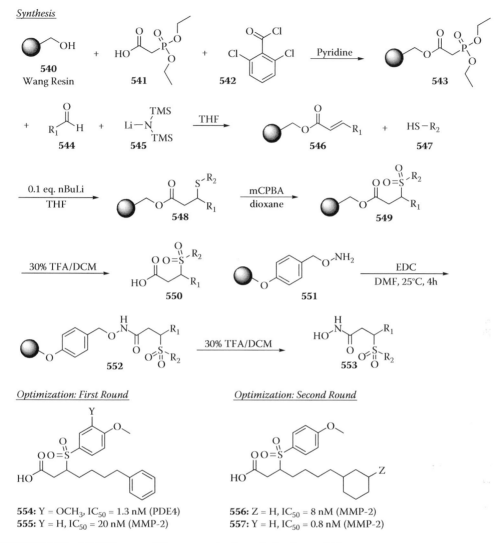

Synthesis

SCHEME 1.61 MMP-2 targeted library: synthesis, optimization, and hits.

1.13.2.3 Case 3: Procollagen C-Proteinase (PCP) Inhibitors

Dankwardt et al.[50] at Roche Bioscience presented the solid-phase synthesis of sulfonamide hydroxamates as Procollagen C-Proteinase (PCP) inhibitors. PCP interacts with procollagen, cleaving C-propeptide, the latter responsible for the production of collagen fribrils. Because PCP over-expression has been associated with fibrotic diseases such as arthritis, surgical adhesions, and adult respiratory distress syndrome, inhibition of PCP would prevent or regulate collagen production as a viable route for disease treatment.

The lead structure **548** had a sulfonamide hydroxamic acid scaffold **549**. The authors then devised a solid-phase synthetic methodology where amino-protected amino acids (**551**) were immobilized via ester-bond formation on a hydroxyl-containing solid support (Argonaut Technologies' ArgoGel-OH resin, **550**). The amino group of the immobilized amino acids (**552**) was then unprotected, and the resulting amines (**553**) were treated with 4-methoxy-benzene sulfonyl chloride (**554**) to afford the desired polymer-supported sulfonamides (**555**). The sulfonamides were then N-alkylated either with alcohols via Mitsunobu conditions or with alkyl halides. The resulting N-aklyl-sulfonamide ester intermediates (**556**) were then treated with hydroxylamine to cleave the compounds, forming the desired sulfonamide hydroxamic acids (**549**).

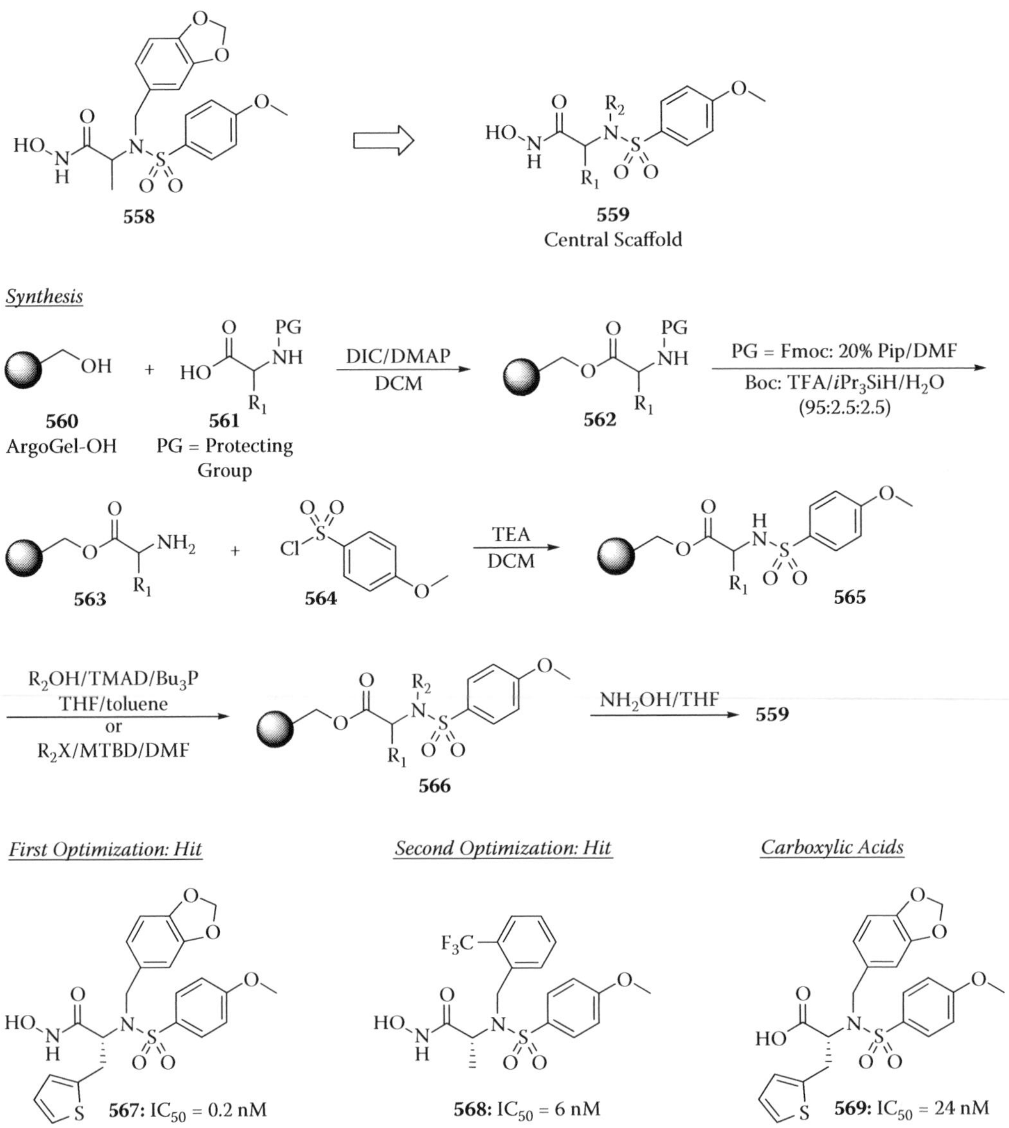

SCHEME 1.62   PCP targeted library: synthesis, optimization, and hits.

The authors noted that previously made analogs containing an N-piperonyl-sulfonamide substituent exhibited high potency; therefore the first optimization cycle focused on the amino acid region, keeping the N-piperonyl-sulfonamide portion constant. The initial round indicated that only D-analogs were active, so full optimization efforts centered on the use of D-amino acids. A 69-member library was made and it was found that Orn- and Dpr-carbamate protected analogs (e.g., CBz, Fmoc) and heteroaromatic acids were well tolerated, particularly 2-thienylalanine (2-Thi) affording PCP picomolar inhibition (**567**, $IC_{50} = 0.2$ nM).

For the second optimization cycle, the authors made a 161-member compound library where D-alanine was kept constant (R_1) mainly for convenience and ease of synthesis. Here it was found that alkylation with substituted benzylic (R_2-Ph-CH$_2$) or methylene-heteroaryl (het-CH$_2$) was very well tolerated. Interestingly, it was observed that in the case of benzyl analogs, no particular substitution pattern (*o-*, *m-*, or *p-*) was preferred. Despite achieving high potency (**568**, $IC_{50} = 6$ nM), none of the compounds provided sub-nanomolar potency.

For the third optimization cycle, the authors built a 29-member library, keeping the D-2-thie-nylalanine group constant and varying the benzyl alkylating agents on the sulfonamide group. Unfortunately, none of the compounds exhibited close or better potency than **567**. Interestingly, the authors also reported that during the course of this work, many carboxylic acids were isolated from the final hydroxamic products. These carboxylic acids were also assayed against PCP and several active compounds were found with double-digit nanomolar potency (**569**, $IC_{50} = 24$ nM).

1.13.2.4 Case 4: Ras Farnesyl Transferase Inhibitors

Saha et al.[51] at Johnson & Johnson reported the solid-phase synthesis of triazole-based Ras farnesyl transferase inhibitors. Farnesyl transferase (FTase) is a zinc metalloenzyme that catalyzes the bind-ing of a farnesyl group to the thiol of a cysteine residue of several proteins involved in cell signaling. The Ras oncogene protein contains a tetrapeptide sequence known as the CAAX motif that, upon farnesylation, promotes Ras translocation to the internal cell membrane. It has been determined that of the mutations of the three known Ras isozymes (that is, H-Ras, N-Ras, and K-Ras) it is the latter that is found highly expressed in human cancer cells (e.g., pancreas, lung, colon). Molecules capable of disrupting the Ras farnesylation process could provide an anticancer effect.

Using reported FTase inhibitors (**570** to **572**) as a starting point, the authors identified that both the imidazole and 4-CN-phenyl groups were key pharmacophore sites that were needed for activ-ity. With this information the authors focused their efforts on the synthesis of 3,4-disubstituted triazoles (**573**). The solid-phase approach involved the immobilization of triazole aldehyde **574** on 2-chlorotrityl resin. The immobilized aldehyde (**575**) was then alkylated with amines under reduc-tive amination conditions. The resulting amines (**576**) were then reacted with either carboxylic acids using standard peptide coupling conditions, acylated with sulfonyl chlorides, or further N-alkylated under reductive alkylation conditions to afford their corresponding amides, sulfonamides, and ter-tiary amines (**577**, **578,** and **579**, respectively). The 4-nitrogen of the triazole unit of the resulting intermediates was then N-alkylated with 4-CN-benzyl alcohol (**580**) in the presence of Tf_2O and DIEA at -78 °C. The resulting 1,2,4-triazol-4-ium intermediates (**581** to **583**) where then cleaved with 10% TFA/DCM to produce the desired triazoles (**584** to **586**).

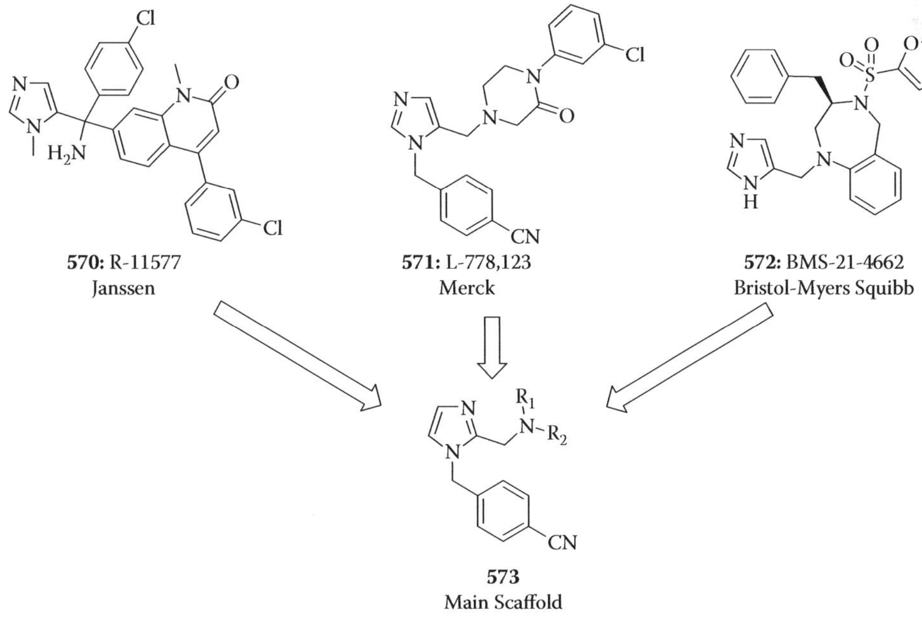

570: R-11577
Janssen

571: L-778,123
Merck

572: BMS-21-4662
Bristol-Myers Squibb

573
Main Scaffold

SCHEME 1.63 Ras FTase targeted library.

Synthesis

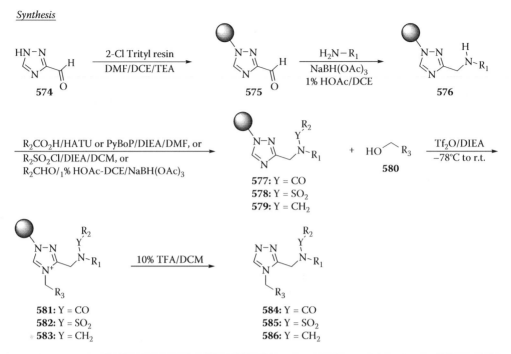

SCHEME 1.63 Ras FTase targeted library (continued).

The initial library afforded several potent amide and sulfonamide derivatives, except for the tertiary amine series where none of the analogs exhibited activity against FTase. The initial SAR indicated that amides were more potent than sulfonamides. One of the hits (**587:** $IC_{50} = 26$ nM) contained a 4-bromo-phenetyl amine sidechain and a quinoline-2-carboxylic amide unit. The first optimization round consisted of keeping constant the 4-bromo-phenetyl amine and changing the nature of the amides. A range of aromatic and heteroaromatic acids was used but none exhibited higher activity than **587**. The next variation was for the amine sidechain, keeping the quinoline-2-carboxylic amide unit constant. Here it was found that replacing the bromide with a fluoride was well tolerated, improving the activity slightly (**588:** $IC_{50} = 19$ nM). When the sidechain between the amino group and the phenyl group was shortened (phenethyl to benzyl), potency increased fivefold compared to **587** (**589:** $IC_{50} = 5$ nM). Addition of another fluorine atom at the *ortho*-position of the benzyl group increased activity almost another fold (**590:** $IC_{50} = 3.5$ nM).

For the second optimization round, the authors explored small changes in **590**. It was observed that replacing imidazole for the triazole unit resulted in a dramatic 53-fold decrease in activity. Replacing the triazole unit with a tetrazole group was particularly detrimental for activity ($IC_{50} > 1$ µM). Replacing the *ortho*-carbon atom of the 4-cyano-benzyl group with a nitrogen atom did not improve potency against FTase (**591:** $IC_{50} = 5.5$ nM). However, when the hetero-nitrogen atom of the quinoline-2-carboxylic amide unit was replaced with a carbon (naphthalene), a slight improvement in activity was observed (**592:** $IC_{50} = 2.2$ nM). It was when a methyl substituent was added to the triazole group that sub-nanomolar activity was achieved (**593:** $IC_{50} = 0.9$ nM). At the end, the overall optimization process improved the activity of the initial discovery hit **587** almost 30-fold, producing **593** with picomolar inhibition potency against FTase.

First Optimization Cycle: Hits	*Second Optimization Cycle: Hits*

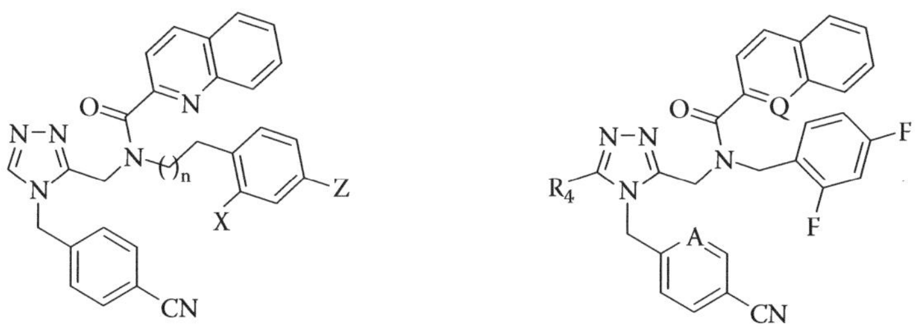

587: n = 1, X = H, Z = Br, IC_{50} = 26 nM
588: n = 1, X = H, Z = F, IC_{50} = 19 nM
589: n = 0, X = H, Z = F = F, IC_{50} = 5 nM
590: n = 0, X = Z = F, IC_{50} = 3.5 nM

591: R_4 = H, Q = A = N, IC_{50} = 5.5 nM
592: R_4 = H, Q = A = C, IC_{50} = 2.2 nM
593: R_4 = CH$_3$, Q = N, A = C, IC_{50} = 0.9 nM

SCHEME 1.64 Ras FTase targeted library: optimization and hits.

1.14 TRANSPORTERS

1.14.1 Case 1: Human Glycine type-2 Transporter (hGlyT$_2$)

Ho et al.[52] at Pharmacopeia Inc. published work done on the optimization of 2-aminomethylbenz-amides as selective and potent inhibitors of the human glycine type-2 transporter (hGlyT$_2$). Glycine levels in the CNS (central nervous system) are mediated by the glycine type-1 and type-2 transporters (GlyT$_1$ and GlyT$_2$, respectively) in conjunction with the strychnine-sensitive glycine receptor (ssGlyR).

The authors began this work based on compound **594**, a hit with an IC$_{50}$ of 1.8 μM against GlyT$_2$ that was discovered from an internal HTS campaign. For the lead optimization phase, the authors developed a solid-phase route using two different resins (ArgoGel™-NH$_2$ and Tentagel-S, **596**) with two different linkers, one acid labile and the other photolabile. Their strategy involved first the immobilization of amines (**597**) to explore the 4-phenylbutan-1-amine unit of **594,** followed by acylation with chloromethylbenzoyl chloride or a 3- or 4-formylbenzoyl chloride (**599**), whereupon treatment with amines (**601**) could be incorporated either by displacing the chloride of the chlorom-ethylbenzamides or by reductive amination on the 3- or 4-formylbenzamides. The release of the desired products (**603**) then occurred via photolysis or acidolysis.

With this methodology in hand, the authors first proceeded to corroborate the optimal positional substitution in **594** by making the *meta-* and *para*-analogs finding that the *ortho*-analog provides the highest potency of the three regioisomers. With this piece of information, the next step was to investigate the 4-methoxyphenylmethylamine component of **594**, finding that the 4-hydroxyphen-ethylamine analog **604** improved potency by around tenfold with an IC$_{50}$ of 170 nM. Positioning the hydroxyl group at the 2- or 3-position resulted in a loss of potency, with the same observations when the 4-hydroxyl group in **604** was replaced by an amino, a fluoro, or a trifluoromethyl group. Selective methylation of the amine and amide groups in **604** also led to loss of activity in both cases. With this SAR, the authors proceeded to explore the phenylbutyl moiety of **604**. The chain length was varied from one to three carbon units and in all instances resulted in much less potency. To cor-roborate that a 4-carbon chain length offers the best activity, different substituents were placed and probed at different positions in the phenyl unit to find that a chloro group on the 3- and 4-positions improved activity twofold (IC$_{50}$ = 99 and 77 nM, respectively) compared to **604**. It was then found

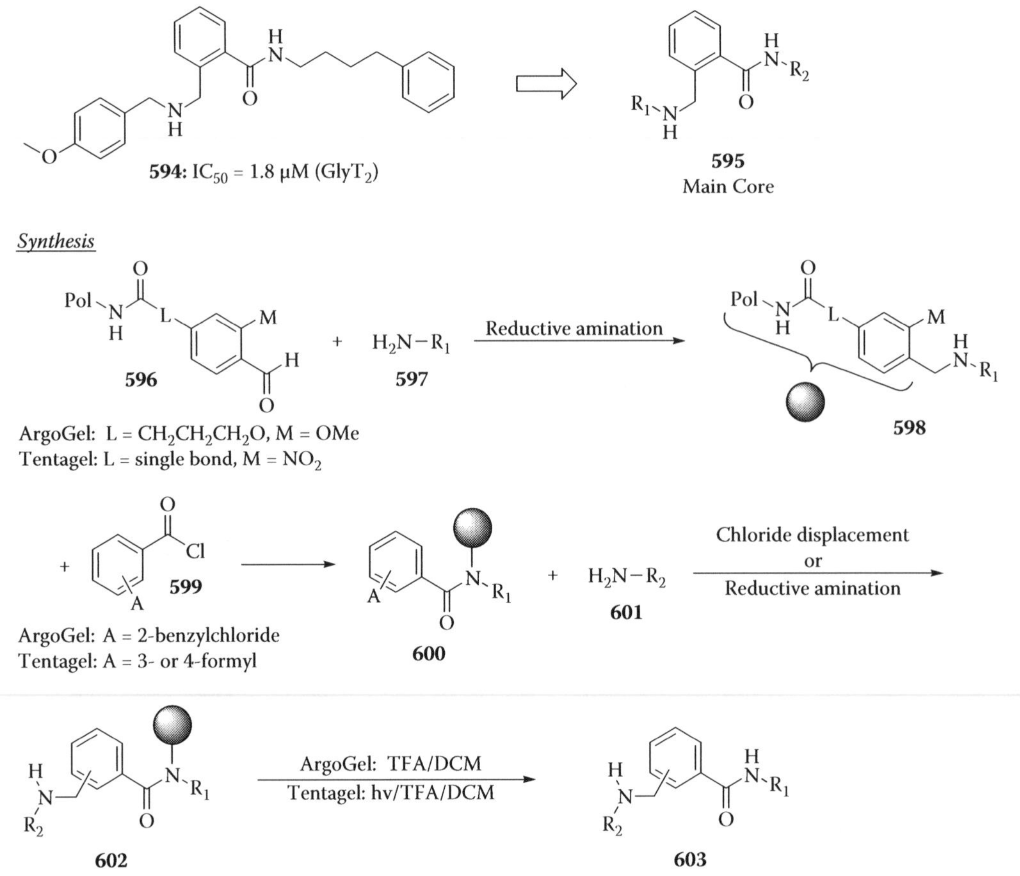

SCHEME 1.65 GlyT$_2$ targeted library synthesis.

that replacing the benzyl-phenethyl single bond with an alkene group improved potency further where the *trans*-isomer (**605**) was the most potent analog (IC$_{50}$ = 56 nM). The selectivity of **605** was tested against GlyT$_1$, ssGlyR, dopamine transporter (DAT), norepinephrine (noradrenalin) transporter (NET), and 5-hydroxytryptamine transporter (5HTT), and it was determined that **605** exhibited good selectivity against GlyT$_1$, ssGlyR, and 5HTT but low selectivity against DAT and NET.

Based on this SAR, the authors proceeded to optimize **604** and **605** by probing the effect of imposing structural constraints on these two compounds. A second library was constructed where the benzyl amine group was linked endocyclic or exocyclic to 5-, 6-, 7-, and 8-member rings fused to the phenyl group. When possible, the phenolic group was also placed on different positions of the phenyl ring. This second library afforded several constrained analogs with an IC$_{50}$ of less than 50 nM. The authors selected three compounds and tested for selectivity against the transporter panel mentioned above and identified that compounds **606** and **607** (IC$_{50}$ ~ 30 nM, 60-fold more potent than **594**) were highly selective toward hGlyT$_2$ (>84-fold selective). Of these two compounds, **607** would be the first choice to explore further (i.e., efficacy, toxicity profile, etc.) as it contains a more rigid benzannulated five-member ring and has no chiral centers, which would simplify its synthesis on a larger scale.

1.14.2 CASE 2: P-GLYCOPROTEIN MODULATORS

Sarshar et al.[53] at Ontogen Corporation reported the solid-phase synthesis of 2,4,5-trisubstituted imidazoles as P-glycoprotein modulators. P-glycoprotein (Pgp) is a transporter protein found in many

Library 1: Hits

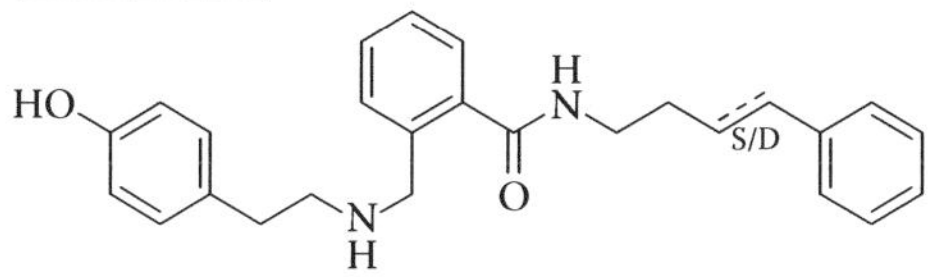

604 (single bond): IC$_{50}$ = 170 nM (GlyT$_2$)
605 (trans-double bond): IC$_{50}$ = 56 nM (GlyT$_2$)

Library 2: Hit

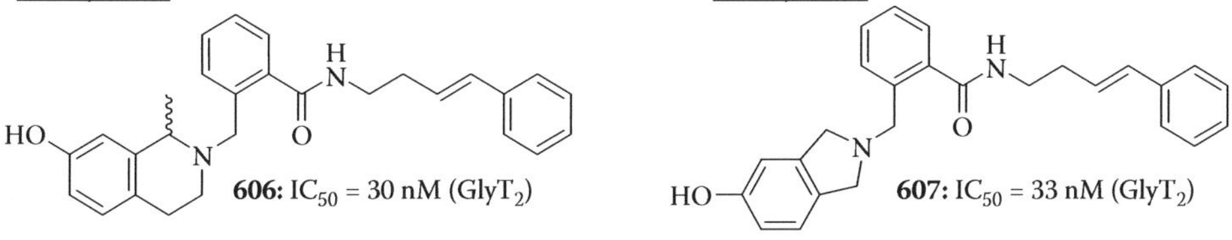

606: IC$_{50}$ = 30 nM (GlyT$_2$)

Library 3: Hit

607: IC$_{50}$ = 33 nM (GlyT$_2$)

SCHEME 1.66 GlyT$_2$ targeted library: optimization and hits.

tissues acting as a drug efflux pump. Drug efflux has a direct impact on drug absorbance levels of compounds crossing the cell membrane. High levels or over-expression of Pgp has been associated with drug resistance (e.g., cancer), so down-regulation of Pgp could enhance drug absorbance.

The authors proceeded with a well-known 4-multicomponent imidazole synthesis where aldehydes and amines are cyclized when mixed with vicinal diones and an NH$_3$ source. In this case, the authors developed two approaches. In the first approach, aldehyde-containing compounds were immobilized on Wang resin (**608**) to provide N-alkyl imidazoles (**612**). In the second approach, amine-containing derivatives were immobilized (**613**) to afford non N-alkylated imidazoles (**615**). In each case, the immobilized reagent reacted with vicinal diones (**700**), NH$_4$OAc (**701**), and either an amine (**609**, route 1) or an aldehyde (**614**, route 2) in acetic acid at 100°C for 4 hr. The resulting polymer-bound imidazoles (**612**, **615**) were then cleaved with 20% TFA in DCM to produce the desired products (**616**).

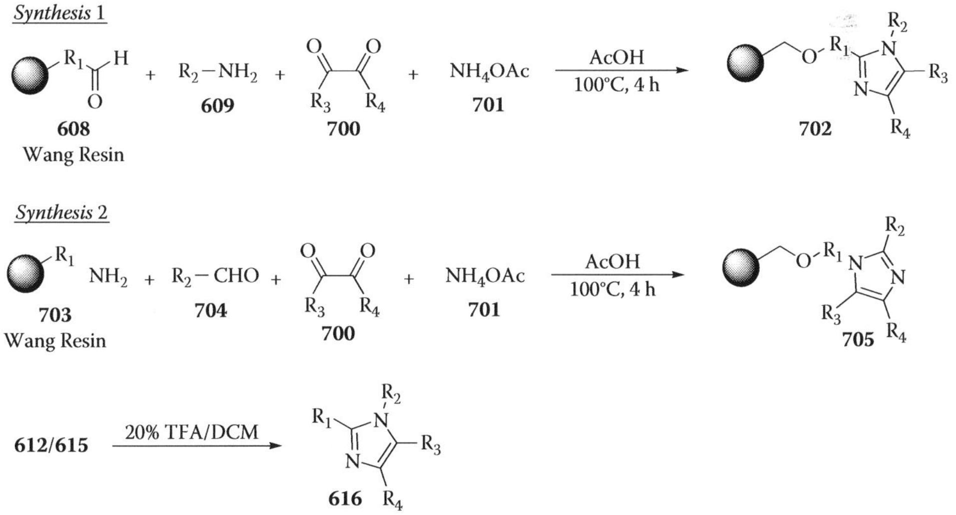

Synthesis 1

Synthesis 2

SCHEME 1.67 Pgp targeted library synthesis: routes 1 and 2.

With this methodology in hand, the authors produced a 500-member library where the products were screened in various MDR potentiation assays using the CEM/VLB1000 cell line. This approach quickly produced an early SAR on the series. It was determined that a carboxylic acid unit or an alcohol group at position-2, combined with a methoxy or dimethylamino group at the 4- and 5-positions were essential in eliciting activity. Also, with the aid of a preliminary solution-phase made library, the authors further devised that N-unsubstituted imidazoles and a cinnamic methyl ester at the 2-position were favored for activity (**617**: ED_{50} = 0.3 μM).

Because the presence of an ester group could lead to undesirable compound metabolism by esterases, the authors used the current SAR at hand and explored different groups and esters as a substitute for the cinnamic methyl ester group. It was observed that the alkene part was needed for activity, none of the prepared esters exhibited better activity, and that replacing the methyl ester with an ethoxy group afforded an allylic ethyl ether with a fourfold improved activity (**618**: ED_{50} = 0.08 μM).

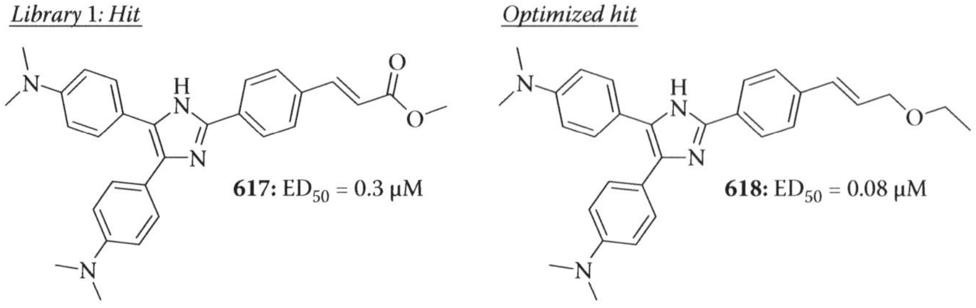

Library 1: Hit *Optimized hit*

617: ED_{50} = 0.3 μM   **618**: ED_{50} = 0.08 μM

SCHEME 1.68 Pgp targeted library: hits.

1.15 SUMMARY AND CONCLUSIONS

Combinatorial chemistry has matured to a point where its impact in research has evolved from a mere approach to synthesize large compound collections to a more established discipline to be integrated in a more meaningful way in medicinal chemistry for the optimization of biologically active compounds. The initial approach of using combinatorial chemistry to synthesize as many compounds as possible just based on chemical feasibility led the industry toward the synthesis of large compound libraries that failed to deliver hits suitable for lead optimization, mainly due to their high molecular weight and lipophilicity. The pharmaceutical industry has learned the hard way that to find an effective solution to a problem not only is it necessary to use the right tool at the right step/time, but also synergystically with other tools. In the case of pharmaceutical R&D, this translates into how best parallel synthesis can be used to have a positive and significant impact on the process of taking a hit compound all the way into development.

The integration of medicinal chemistry guidelines changed the focus of combinatorial chemistry from "what can be made" to "what should be made." The introduction of physicochemical parameters associated with drug-like compounds has translated into the synthesis of smaller-sized, better-designed compound libraries, which not only shortens synthesis cycle campaigns, but also increases the quality of the final compounds. When additional target information derived from crystallographic enzyme-ligand studies was taken into consideration, it proved useful both in optimizing lead compounds for a better target-ligand interaction and for proving/disproving theories around the biological target.

In medicinal chemistry, it is well known that potency and selectivity against a target quite often do not meet the criteria required for a compound to graduate into a drug candidate. The physicochemical properties of a compound play a key role in reaching a balance between water solubility, lipophilicity, and the acidic/basic/neutral nature of the compound. These factors, in turn, influence the pharmacokinetic profile of a compound when administered into a living dynamic system (e.g., compound stability, plasma concentration, etc.). Eventually, intrinsic pharmacokinetic factors will dictate the overall pharmacokinetic compound profile leading to the determination of the efficacy of the compound, dosage, and safety profile.

The examples discussed in this review clearly show how the successful integration of combinatorial chemistry with traditional medicinal chemistry and computational chemistry provides a far superior outcome as opposed to treating combinatorial chemistry as the magic bullet to solve all the challenges in a given research project.

REFERENCES

1. Merrifield, R.B. Solid phase peptide synthesis. I. The synthesis of a tetrapeptide. J. Am. Chem. Soc., 1963, 85, 2149–2154.
2. Bunin, B.A.; and Ellman, J.A. A general and expedient method for the solid-phase synthesis of 1,4-benzodiazepine derivatives. J. Am. Chem. Soc., 1992, 114, 10997–10998.
3. Lipinski, C.A.; Lombardo, F.; Dominy, B.W.; and Feeney, P.J. Experimental and computational approaches to estimate solubility and permeability in drug discovery and development settings. Adv. Drug Deliv. Rev., 1997, 23, 3–25.
4. Lipinski, C.A. Drug-like properties and the causes of poor solubility and poor permeability. J. Pharmacol. Toxicol. Methods, 2000, 44, 235–249.
5. Lipinski, C.A.; Lombardo, F.; Dominy, B.W.; and Feeney, P.J. Experimental and computational approaches to estimate solubility and permeability in drug discovery and development settings. Adv. Drug Deliv. Rev., 2001, 46, 3–26.
6. Veber, D.F.; Johnson, S.R.; Cheng, H.Y.; Smith, B.R.; Ward, K.W.; and Kopple, K.D. Molecular properties that influence the oral bioavailability of drug candidates. J. Med. Chem., 2002, 45, 2615–2623.
7. Teague, S.J.; Davis, A.M.; Leeson, P.D.; and Oprea, T. The design of leadlike combinatorial libraries. Angew. Chem. Int. Ed. Engl., 1999, 38, 3743–3748.
8. Sanders, W.J.; Nienaber, V.L.; Lerner, C.G.; McCall, J.O.; Merrick, S.M.; Swanson, S.J.; Harlan, J.E.; Stoll, V.S.; Stamper, G.F.; Betz, S.F.; Condroski, K.R.; Meadows, R.P.; Severin, J.M.; Walter, K.A.; Magdalinos, P.; Jakob, C.G.; Wagner, R.; and Beutel, B.A. Discovery of potent inhibitors of dihydroneopterin aldolase using CrystaLEAD high-throughput x-ray crystallographic screening and structure-directed lead optimization. J. Med. Chem., 2004, 47, 1709–1718.
9. Nienaber, V.L.; Richardson, P.L.; Klighofer, V.; Bouska, J.J.; Giranda, V.L.; and Greer, J. Discovering novel ligands for macromolecules using x-ray crystallographic screening. Nat. Biotechnol., 2000, 18, 1105–1108.
10. Kundu, B.; Rastogi, S.K.; Batra, S.; Raghuwanshi, S.K.; and Shukla, P.K. Combinatorial approach to lead optimization of a novel hexapeptide with antifungal activity. Bioorg. Med. Chem. Lett., 2000, 10, 1779–1781.
11. Saha, A.K.; Liu, L.; Marichal, P.; and Odds, F. Novel antifungals based on 4-substituted imidazole: solid-phase synthesis of substituted aryl sulfonamides towards optimization of in vitro activity. Bioorg. Med. Chem. Lett., 2000, 10, 2735–2739.
12. Saha, A.K.; Liu, L.; Simoneaux, R.L.; Kukla, M.J.; Marichal, P.; and Odds, F. Novel antifungals based on 4-substituted imidazole: a combinatorial chemistry approach to lead discovery and optimization. Bioorg. Med. Chem. Lett., 2000, 10, 2175–2178.
13. Altorfer, M.; Ermert, P.; Fassler, J.; Farooq, S.; Hillesheim, E.; Jeanguenat, A.; Klumpp, K.; Maienfisch, P.; Martin, J.A.; Merrett, J.H.; Parkes, K.E.B.; Obrecht, J.-P.; Pitterna, T.; and Obrecht, D. Applications of parallel synthesis to lead optimization. Chimia, 2003, 57, 262–269.
14. Chitkul, B.; and Bradley, M. Optimising inhibitors of trypanothione reductase using solid-phase chemistry. Bioorg. Med. Chem. Lett., 2000, 10, 2367–2369.
15. Dhanak, D.; Christmann, L.T.; Darcy, M.G.; Keenan, R.M.; Knight, S.D.; Lee, J.; Ridgers, L.H.; Sarau, H.M.; Shah, D.H.; White, J.R.; and Zhang, L. Discovery of potent and selective phenylalanine derived CCR3 receptor antagonists. Part 2. Bioorg. Med. Chem. Lett., 2001, 11, 1445–1450.
16. Guo, T.; Adang, A.E. P.; Dong, G.; Fitzpatrick, D.; Geng, P.; Ho, K.K.; Jibilian, C.H.; Kultgen, S.G.; Liu, R.; McDonald, E.; Saionz, K.W.; Valenzano, K.J.; van Straten, N.C.R.; Xie, D.; and Webb, M.L. Small molecule biaryl FSH receptor agonists. 2. Lead optimization via parallel synthesis. Bioorg. Med. Chem. Lett., 2004, 14, 1717–1720.
17. Guo, T.; Adang, A.E.; Dolle, R.E.; Dong, G.; Fitzpatrick, D.; Geng, P.; Ho, K.K.; Kultgen, S.G.; Liu, R.; McDonald, E.; McGuinness, B.F.; Saionz, K.W.; Valenzano, K.J.; van Straten, N.C.; Xie, D.; and Webb, M.L. Small molecule biaryl FSH receptor agonists. 1. Lead discovery via encoded combinatorial synthesis. Bioorg. Med. Chem. Lett., 2004, 14, 1713–1716.

18. Chen, Z.; Miller, W.S.; Shan, S.; and Valenzano, K.J. Design and parallel synthesis of piperidine libraries targeting the nociceptin (N/OFQ) receptor. Bioorg. Med. Chem. Lett., 2003, 13, 3247–3252.

19. Koberstein, R.; Aissaoui, H.; Bur, D.; Clozel, M.; Fischli, W.; Jenek, F.; Mueller, C.; Nayler, O.; Sifferlen, T.; Treiber, A.; and Weller, T. Tetrahydroisoquinolines as orexin receptor antagonists: strategies for lead optimization by solution-phase chemistry. Chimia, 2003, 57, 270–275.

20. Berger, M.; Albrecht, B.; Berces, A.; Ettmayer, P.; Neruda, W.; and Woisetschlager, M. S(+)-4-(1-phenylethylamino)quinazolines as inhibitors of human immunoglobulin E synthesis: potency is dictated by stereochemistry and atomic point charges at N-1. J. Med. Chem., 2001, 44, 3031–3038.

21. Choi, C.; Li, J.H.; Vaal, M.; Thomas, C.; Limburg, D.; Wu, Y.Q.; Chen, Y.; Soni, R.; Scott, C.; Ross, D.T.; Guo, H.; Howorth, P.; Valentine, H.; Liang, S.; Spicer, D.; Fuller, M.; Steiner, J.; and Hamilton, G.S. Use of parallel-synthesis combinatorial libraries for rapid identification of potent FKBP12 inhibitors. Bioorg. Med. Chem. Lett., 2002, 12, 1421–1428.

22. Gopalsamy, A.; Yang, H.; Ellingboe, J.W.; Kees, K.L.; Yoon, J.; and Murrills, R. Parallel solid-phase synthesis of vitronectin receptor (avb3) inhibitors. Bioorg. Med. Chem. Lett., 2000, 10, 1715–1718.

23. Castanedo, G. M.; Sailes, F.C.; Dubree, N.J.; Nicholas, J.B.; Caris, L.; Clark, K.; Keating, S.M.; Beresini, M.H.; Chiu, H.; Fong, S.; Marsters, J.C., Jr.; Jackson, D.Y.; and Sutherlin, D.P. Solid-phase synthesis of dual alpha4beta1/alpha4beta7 integrin antagonists: two scaffolds with overlapping pharmacophores. Bioorg. Med. Chem. Lett., 2002, 12, 2913–2917.

24. Bauser, M.; Delapierre, G.; Hauswald, M.; Flessner, T.; D'Urso, D.; Hermann, A.; Beyreuther, B.; De, V.J.; Spreyer, P.; Reissmuller, E.; and Meier, H. Discovery and optimization of 2-aryl oxazolo-pyrimidines as adenosine kinase inhibitors using liquid phase parallel synthesis. Bioorg. Med. Chem. Lett., 2004, 14, 1997–2000.

25. Zhao, Z.; Leister, W.H.; Robinson, R.G.; Barnett, S. F.; Feo-Jones, D.; Jones, R. E.; Hartman, G.D.; Huff, J.R.; Huber, H.E.; Duggan, M.E.; and Lindsley, C.W. Discovery of 2,3,5-trisubstituted pyridine derivatives as potent Akt1 and Akt2 dual inhibitors. Bioorg. Med. Chem. Lett., 2005, 15, 905–909.

26. Fancelli, D.; Berta, D.; Bindi, S.; Cameron, A.; Cappella, P.; Carpinelli, P.; Catana, C.; Forte, B.; Giordano, P.; Giorgini, M.L.; Mantegani, S.; Marsiglio, A.; Meroni, M.; Moll, J.; Pittala, V.; Roletto, F.; Severino, D.; Soncini, C.; Storici, P.; Tonani, R.; Varasi, M.; Vulpetti, A.; and Vianello, P. Potent and selective Aurora inhibitors identified by the expansion of a novel scaffold for protein kinase inhibition. J. Med. Chem., 2005, 48, 3080–3084.

27. Kim, K.S.; Kimball, S.D.; Misra, R.N.; Rawlins, D.B.; Hunt, J.T.; Xiao, H.Y.; Lu, S.; Qian, L.; Han, W.C.; Shan, W.; Mitt, T.; Cai, Z.W.; Poss, M. A.; Zhu, H.; Sack, J.S.; Tokarski, J.S.; Chang, C.Y.; Pavletich, N.; Kamath, A.; Humphreys, W.G.; Marathe, P.; Bursuker, I.; Kellar, K.A.; Roongta, U.; Batorsky, R.; Mulheron, J.G.; Bol, D.; Fairchild, C.R.; Lee, F.Y.; Webster, K.R. Discovery of aminothiazole inhibitors of cyclin-dependent kinase 2: synthesis, x-ray crystallographic analysis, and biological activities. J. Med. Chem., 2002, 45, 3905–3927.

28. Hardcastle, I.R.; Cockcroft, X.; Curtin, N.J.; El-Murr, M.D.; Leahy, J.J.; Stockley, M.; Golding, B.T.; Rigoreau, L.; Richardson, C.; Smith, G.C.; and Griffin, R.J. Discovery of potent chromen-4-one inhibitors of the DNA-dependent protein kinase (DNA-PK) using a small-molecule library approach. J. Med. Chem., 2005, 48, 7829–7846.

29. Naerum, L.; Norskov-Lauritsen, L.; and Olesen, P.H. Scaffold hopping and optimization towards libraries of glycogen synthase kinase-3 inhibitors. Bioorg. Med. Chem. Lett., 2002, 12, 1525–1528.

30. Chen, P.; Norris, D.; Iwanowicz, E.J.; Spergel, S.H.; Lin, J.; Gu, H.H.; Shen, Z.; Wityak, J.; Lin, T.A.; Pang, S.; De Fex, H.F.; Pitt, S.; Shen, D.R.; Doweyko, A. M.; Bassolino, D.A.; Roberge, J.Y.; Poss, M.A.; Chen, B.C.; Schieven, G.L.; and Barrish, J.C. Discovery and initial SAR of imidazoquinoxalines as inhibitors of the Src-family kinase p56(Lck). Bioorg. Med. Chem. Lett., 2002, 12, 1361–1364.

31. Dumas, J.; Sibley, R.; Riedl, B.; Monahan, M.K.; Lee, W.; Lowinger, T.B.; Redman, A.M.; Johnson, J.S.; Kingery-Wood, J.; Scott, W.J.; Smith, R.A.; Bobko, M.; Schoenleber, R.; Ranges, G.E.; Housley, T.J.; Bhargava, A.; Wilhelm, S.M.; and Shrikhande, A. Discovery of a new class of p38 kinase inhibitors. Bioorg. Med. Chem. Lett., 2000, 10, 2047–2050.

32. Smith, R.A.; Barbosa, J.; Blum, C.L.; Bobko, M.A.; Caringal, Y.V.; Dally, R.; Johnson, J.S.; Katz, M.E.; Kennure, N.; Kingery-Wood, J.; Lee, W.; Lowinger, T. B.; Lyons, J.; Marsh, V.; Rogers, D.H.; Swartz, S.; Walling, T.; and Wild, H. Discovery of heterocyclic ureas as a new class of raf kinase inhibitors: identification of a second generation lead by a combinatorial chemistry approach. Bioorg. Med. Chem. Lett., 2001, 11, 2775–2778.

33. Coppola, G.M.; Kukkola, P.J.; Stanton, J.L.; Neubert, A.D.; Marcopulos, N.; Bilci, N.A.; Wang, H.; Tomaselli, H.C.; Tan, J.; Aicher, T.D.; Knorr, D.C.; Jeng, A.Y.; Dardik, B.; and Chatelain, R.E. Perhydro-quinolylbenzamides as novel inhibitors of 11beta-hydroxysteroid dehydrogenase type 1. J. Med. Chem., 2005, 48, 6696–6712.

34. Collins, J.L.; Fivush, A.M.; Watson, M.A.; Galardi, C.M.; Lewis, M.C.; Moore, L.B.; Parks, D.J.; Wilson, J.G.; Tippin, T.K.; Binz, J.G.; Plunket, K.D.; Morgan, D.G.; Beaudet, E.J.; Whitney, K.D.; Kliewer, S.A.; and Willson, T.M. Identification of a nonsteroidal liver X receptor agonist through parallel array synthesis of tertiary amines. J. Med. Chem., 2002, 45, 1963–1966.

35. Haque, T.S.; Tadesse, S.; Marcinkeviciene, J.; Rogers, M.J.; Sizemore, C.; Kopcho, L.M.; Amsler, K.; Ecret, L.D.; Zhan, D.L.; Hobbs, F.; Slee, A.; Trainor, G.L.; Stern, A.M.; Copeland, R.A.; and Combs, A.P. Parallel synthesis of potent, pyrazole-based inhibitors of Helicobacter pylori dihydroorotate dehydrogenase. J. Med. Chem., 2002, 45, 4669–4678.

36. Fritzen, E.L.; Brightwell, A.S.; Erickson, L.A.; and Romero, D.L. The solid phase synthesis of tetrahydroisoquinolines having cdc25B inhibitory activity. Bioorg. Med. Chem. Lett., 2000, 10, 649–652.

37. Linton, S.D.; Karanewsky, D.S.; Ternansky, R.J.; Wu, J.C.; Pham, B.; Kodandapani, L.; Smidt, R.; Diaz, J.L.; Fritz, L.C.; and Tomaselli, K.J. Acyl dipeptides as reversible caspase inhibitors. 1. Initial lead optimization. Bioorg. Med. Chem. Lett., 2002, 12, 2969–2971.

38. Beaulieu, P.L.; Bos, M.; Bousquet, Y.; Fazal, G.; Gauthier, J.; Gillard, J.; Goulet, S.; LaPlante, S.; Poupart, M.A.; Lefebvre, S.; McKercher, G.; Pellerin, C.; Austel, V.; and Kukolj, G. Non-nucleoside inhibitors of the hepatitis C virus NS5B polymerase: discovery and preliminary SAR of benzimidazole derivatives. Bioorg. Med. Chem. Lett., 2004, 14, 119–124.

39 Beaulieu, P.L.; Bos, M.; Bousquet, Y.; DeRoy, P.; Fazal, G.; Gauthier, J.; Gillard, J.; Goulet, S.; McKercher, G.; Poupart, M.A.; Valois, S.; and Kukolj, G. Non-nucleoside inhibitors of the hepatitis C virus NS5B polymerase: discovery of benzimidazole 5-carboxylic amide derivatives with low-nanomolar potency. Bioorg. Med. Chem. Lett., 2004, 14, 967–971.

40. Harper, S.; Avolio, S.; Pacini, B.; Di, F.M.; Altamura, S.; Tomei, L.; Paonessa, G.; Di, M.S.; Carfi, A.; Giuliano, C.; Padron, J.; Bonelli, F.; Migliaccio, G.; De, F.R.; Laufer, R.; Rowley, M.; and Narjes, F. Potent inhibitors of subgenomic hepatitis C virus RNA replication through optimization of indole-N-acetamide allosteric inhibitors of the viral NS5B polymerase. J. Med. Chem., 2005, 48, 4547–4557.

41. Rano, T.A.; Cheng, Y.; Huening, T.T.; Zhang, F.; Schleif, W.A.; Gabryelski, L.; Olsen, D.B.; Kuo, L.C.; Lin, J.H.; Xu, X.; Olah, T.V.; McLoughlin, D.A.; King, R.; Chapman, K.T.; and Tata, J.R. Combinatorial diversification of indinavir: in vivo mixture dosing of an HIV protease inhibitor library. Bioorg. Med. Chem. Lett., 2000, 10, 1527–1530.

42. Reich, S.H.; Johnson, T.; Wallace, M.B.; Kephart, S.E.; Fuhrman, S.A.; Worland, S.T.; Matthews, D.A.; Hendrickson, T.F.; Chan, F.; Meador, J., III; Ferre, R.A.; Brown, E.L.; DeLisle, D.M.; Patick, A.K.; Binford, S.L.; and Ford, C.E. Substituted benzamide inhibitors of human rhinovirus 3C protease: structure-based design, synthesis, and biological evaluation. J. Med. Chem., 2000, 43, 1670–1683.

43. Johnson, T.O.; Hua, Y.; Luu, H.T.; Brown, E.L.; Chan, F.; Chu, S.S.; Dragovich, P.S.; Eastman, B.W.; Ferre, R.A.; Fuhrman, S.A.; Hendrickson, T.F.; Maldonado, F.C.; Matthews, D.A.; Meador, J.W., III; Patick, A.K.; Reich, S.H.; Skalitzky, D.J.; Worland, S.T.; Yang, M.; and Zalman, L.S. Structure-based design of a parallel synthetic array directed toward the discovery of irreversible inhibitors of human rhinovirus 3C protease. J. Med. Chem., 2002, 45, 2016–2023.

44. Wang, G.T.; Chen, Y.; Wang, S.; Gentles, R.; Sowin, T.; Kati, W.; Muchmore, S.; Giranda, V.; Stewart, K.; Sham, H.; Kempf, D.; and Laver, W.G. Design, synthesis, and structural analysis of influenza neuraminidase inhibitors containing pyrrolidine cores. J. Med. Chem., 2001, 44, 1192–1201.

45. Pirrung, M.C.; Pansare, S.V.; Sarma, K.D.; Keith, K.A.; and Kern, E.R. Combinatorial optimization of isatin-beta-thiosemicarbazones as anti-poxvirus agents. J. Med. Chem., 2005, 48, 3045–3050.

46. Parlow, J.J.; Case, B.L.; Dice, T.A.; Fenton, R.L.; Hayes, M.J.; Jones, D.E.; Neumann, W.L.; Wood, R.S.; Lachance, R.M.; Girard, T.J.; Nicholson, N.S.; Clare, M.; Stegeman, R.A.; Stevens, A.M.; Stallings, W.C.; Kurumbail, R.G.; and South, M.S. Design, parallel synthesis, and crystal structures of pyrazinone antithrombotics as selective inhibitors of the tissue factor VIIa complex. J. Med. Chem., 2003, 46, 4050–4062.

47. Kitas, E.A.; Loffler, B.M.; Daetwyler, S.; Dehmlow, H.; and Aebi, J.D. Synthesis of triazole-tethered pyrrolidine libraries: novel ECE inhibitors. Bioorg. Med. Chem. Lett., 2002, 12, 1727–1730.

48. Brands, M.; Erguden, J.K.; Hashimoto, K.; Heimbach, D.; Schroder, C.; Siegel, S.; Stasch, J.P.; and Weigand, S. Novel, selective indole-based ECE inhibitors: lead optimization via solid-phase and classical synthesis. Bioorg. Med. Chem. Lett., 2005, 15, 4201–4205.

49. Salvino, J.M.; Mathew, R.; Kiesow, T.; Narensingh, R.; Mason, H.J.; Dodd, A.; Groneberg, R.; Burns, C.J.; McGeehan, G.; Kline, J.; Orton, E.; Tang, S.Y.; Morrisette, M.; and Labaudininiere, R. Solid-phase synthesis of an arylsulfone hydroxamate library. Bioorg. Med. Chem. Lett., 2000, 10, 1637–1640.

50. Dankwardt, S.M.; Abbot, S.C.; Broka, C.A.; Martin, R.L.; Chan, C.S.; Springman, E.B.; Van Wart, H.E.; and Walker, K.A. Amino acid derived sulfonamide hydroxamates as inhibitors of procollagen C-proteinase. 2. Solid-phase optimization of side chains. Bioorg. Med. Chem. Lett., 2002, 12, 1233–1235.

51. Saha, A.K.; Liu, L.; Simoneaux, R.; DeCorte, B.; Meyer, C.; Skrzat, S.; Breslin, H.J.; Kukla, M.J.; and End, D.W. Novel triazole based inhibitors of Ras farnesyl transferase. Bioorg. Med. Chem. Lett., 2005, 15, 5407–5411.

52. Ho, K.K.; Appell, K.C.; Baldwin, J.J.; Bohnstedt, A.C.; Dong, G.; Guo, T.; Horlick, R.; Islam, K.R.; Kultgen, S.G.; Masterson, C.M.; McDonald, E.; McMillan, K.; Morphy, J.R.; Rankovic, Z.; Sundaram, H.; and Webb, M. 2-(Aminomethyl)-benzamide-based glycine transporter type-2 inhibitors. Bioorg. Med. Chem. Lett., 2004, 14, 545–548.

53. Sarshar, S.; Zhang, C.; Moran, E.J.; Krane, S.; Rodarte, J.C.; Benbatoul, K.D.; Dixon, R.; and Mjalli, A.M. 2,4,5-Trisubstituted imidazoles: novel nontoxic modulators of P-glycoprotein mediated multidrug resistance. Part 1. Bioorg. Med. Chem. Lett., 2000, 10, 2599–2601.

2 Application of Parallel Synthesis to the Optimization of Inhibitors of the ZipA-FtsZ Protein–Protein Interaction

Lee D. Jennings

CONTENTS

2.1   INTRODUCTION

2.1.1   FEATURES OF PROTEIN-PROTEIN INTERFACES

The design of small molecules able to disrupt protein-protein interactions is of great current interest to the bioorganic and pharmaceutical research community. Protein-protein interactions are ubiquitous in physiological processes and have a role in cell division, activation of signaling pathways, and immune response. A survey of the literature on protein-protein binding sites that are of interest as targets for therapeutic intervention reveals that the protein surfaces are varied in size, shape, and depth. The integrin superfamily of adhesion receptors binds a short linear amino acid sequence (RGD or KQAGVD) in an interfacial region between the α and β subunits [1,2]. Interleukin-2 (IL-2) binds a small molecule inhibitor of the IL-2/IL-2Rα receptor interaction through a conformational change of the protein binding surface that creates a small-molecule binding site [3]. The human growth hormone receptor, typical of many proteins with protein ligands, binds human growth hormone at a "hot-spot" where a small set of primary hydrophobic contacts in a relatively small area dominate affinity [4,5]. Similarly, the constant fragment (Fc) of immunoglobulin G binds different proteins using a consensus binding site, which makes conformational changes to adapt to its various binding partners [6]. The consensus binding region is characterized by a high degree of solvent accessibility, predominantly nonpolar character, and low hydrogen bonding ability.

Due to the common features of protein binding sites, the design of small-molecule inhibitors of protein-protein interactions is often difficult. Opportunities for making hydrogen bonds, which are very efficient for providing favorable binding energetics, are limited. Moreover, the functional group presentation on the surface of a protein is divergent, rather than convergent [7]. As a consequence, small ligand binding is often relatively inefficient as only one side of a ligand is able to make interactions with the functionality presented on the protein surface. Contemporary approaches for the design of non-peptide inhibitors of protein–protein interactions include α-helix mimics [8], mimics of extended β-sheets [9], bridged cyclic ligands [10], the *de novo* design of ligands [11], the construction of ligands through a fragment assembly approach [12], and the optimization of leads discovered through the screening of compound libraries [13]. Researchers in all the above-mentioned projects used information about the structure of the protein-ligand interaction obtained by x-ray crystallographic or NMR studies.

2.1.2   ZIPA IS A TARGET FOR INHIBITION OF BACTERIAL CELL-WALL BIOSYNTHESIS

Our particular protein-protein interaction of interest involves the bacterial protein ZipA and the prokaryotic tubulin analog FtsZ. Given good growing conditions, a bacterium grows slightly in size or length and a new cell wall grows through the center forming two daughter cells. In *Escherichia coli*, the membrane-bound protein ZipA appears to play an essential role in the formation and dynamics of the septal ring [14]. ZipA is localized at the site of cell division at a very early stage in the division cycle and tethers the FtsZ protofilaments to the membrane during the infolding of the septum. Consequently, the inhibition of the ZipA-FtsZ interaction should arrest cell division and lead to cell death. The ability to selectively arrest bacterial cell division would establish a new mechanism of antibacterial activity.

The structure of the C-terminal FtsZ binding domain of ZipA (ZipA$_{185-328}$) [15], as well as that of the complex between ZipA$_{185-328}$ and the C-terminal fragment of FtsZ (FtsZ$_{367-383}$), has been elucidated both by x-ray crystallography and by NMR [16]. In this complex, FtsZ$_{367-383}$ forms an α-helix and fills a shallow solvent-exposed hydrophobic cavity on the surface of ZipA. The α-helix conformation of the FtsZ$_{367-383}$ peptide directs only six amino acid sidechains toward significant interaction with ZipA. Sequential substitution of each of these six amino acid residues with alanine indicates that just residues Ile374, Phe377, and Leu378 account for virtually all of the binding affinity to ZipA$_{185-328}$. These results indicate that binding energetics are dominated by burial of hydrophobic residues and also suggest that, because the number of buried residues is only three, a small molecule could reasonably be expected to interfere with this binding.

During the exploratory phase of this project, a number of approaches were undertaken to identify leads. Testing of a collection of structurally diverse, lead-like compounds for inhibition of ZipA-FtsZ association provided us with indoloquinolizinone **1**, dihydrocarbazole **2**, and acridine **3** [17], and an HTS screening campaign performed subsequently gave us aminopyrimidine lead **4** (Structure 2.1). We describe here our approaches to optimizing these leads using parallel synthesis and structure based design.

STRUCTURE 2.1

2.2 CASE STUDY I: INDOLOQUINOLIZINONES

2.2.1 ORIGIN OF THE INDOLOQUINOLIZINONE LEAD

In the early stages of this project, leads were identified by focused screening of a limited set of compounds that were either identified by molecular modeling or were members of a collection of compounds designed to exemplify a diverse set of molecular scaffolds. Initially, compounds were assayed by detection of the displacement of $ZipA_{185-328}$ from the immobilized C-terminal 16 amino acid fragment of the FtsZ protein by surface plasmon resonance (SPR) [18]. Later on, as a new assay became available to us, IC_{50} values were routinely determined by a fluorescence polarization assay using a fluoresceine-labeled analog of the C-terminal domain of FtsZ [19].

Indoloquinolizinone (**1**) was found to inhibit ZipA-FtsZ binding by SPR (39% inhibition at 1 mM conc.). Not surprisingly, however, **1** did not show antibacterial activity at concentrations up to 200 μM. To find out how **1** interfered with binding of FtsZ, co-crystals of the inhibitor bound to ZipA were obtained and the crystal structure was solved at 2.0 Å resolution (Figure 2.1a). It was found that **1** does, in fact, occupy the hydrophobic cavity on the surface of ZipA necessary for the binding of the FtsZ peptide and, as expected, the strongest contribution to binding appeared to be the hydrophobic effect, as the molecule displaces about 600 Å² worth of low-entropy water from the site. A smaller enthalpic contribution to binding stems from van der Waals interactions and a π-stacking interaction between the hexahydroquinolizinone ring system and the phenylalanine sidechain that forms part of the bottom of the concavity. Additionally, the carbonyl oxygen makes a hydrogen bond with a bound water molecule (primarily associated with the backbone carbonyl of Asn247) that is also present in the FtsZ peptide-ZipA structure.

A series of searches of our corporate compound collection for compounds similar to **1** yielded a number of compounds of interest. Among these were the dihydrocarbazole **2** and the acridine **3**, which, while less potent inhibitors than **1**, also yielded high-quality x-ray structures of their complex with ZipA. The indole nucleus of **2** bound in an essentially identical position to that of **1** but had the added feature of the amine sidechain accessing a neighboring pocket of ZipA that is not involved in the binding of FtsZ. The acridine ring system of **3** also bound to a similar region of ZipA but the longer sidechain not only entered the neighboring pocket, but also made a hydrogen-bonding interaction with His46.

Building on the structural information, and taking into account the opportunities available for elaboration of the indoloquinolizine core, we focused on targets that could access the additional

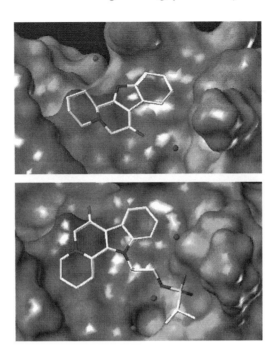

FIGURE 2.1 Crystallographic orientation of (a) compound **1** and (b) compound **7b**. (*Source:* Reproduced with permission from *Bioorg. Med. Chem. Lett.,* 2004, 14, 1427–1431.)

hydrogen bonding opportunity seen with **3**. We devised straightforward ways to link hydrogen-bond-accepting groups to the indoloquinolizine ring through tethers of various lengths. It was hoped that small hydrophobic groups attached to the different hydrogen-bond-accepting groups would shield the developing hydrogen bonds from water. Our initial plan was to do a multivariate array using parallel synthesis. A multivariate, or combinatorial, array holds the possibility of discovering synergistic effects between different substituent groups (*R*-groups) of the molecule that might not be found if only one part is varied while the other parts are fixed.

2.2.2 SYNTHESIS OF ANALOGS OF 1 AND BIOLOGICAL RESULTS

The indoloquinolizine scaffold was resynthesized and was alkylated with acrylonitrile, methyl bromoacetate, methyl acrylate, and methyl 4-bromobutyrate (Schemes 2.1 and 2.2) [18]. The alkylation of the indoloquinolizine nitrogen was unexpectedly difficult due to unfavorable sterics forcing out-of-plane bending of the newly formed C-N bond. Nitrile intermediate **6** was reduced by hydrogenation over Raney nickel to give the resulting amine **7**. Amine **7** was acylated in parallel with acid chlorides and sulfonyl chlorides to give amides and sulfonamides. Ester intermediates formed by the alkylation of the indoloquinolizine ring system by methyl bromoacetate, methyl acrylate, and methyl 4-bromobutyrate were subjected to ester hydrolysis and coupled in parallel with amines. Products were individually purified by reverse-phase HPLC (high-pressure liquid chromatography) and checked for purity and integrity by LC-MS (liquid chromatography coupled mass spectroscopy).

The product molecules were assayed for inhibition of ZipA-FtsZ interaction using SPR [18]. Table 2.1 shows that despite significant variation in the shape and length of the sidechains attached to the indoloquinolizinone scaffold, the inhibition values of the different analogs made were essentially the same and no significant improvement in the inhibition of the interaction of ZipA with FtsZ

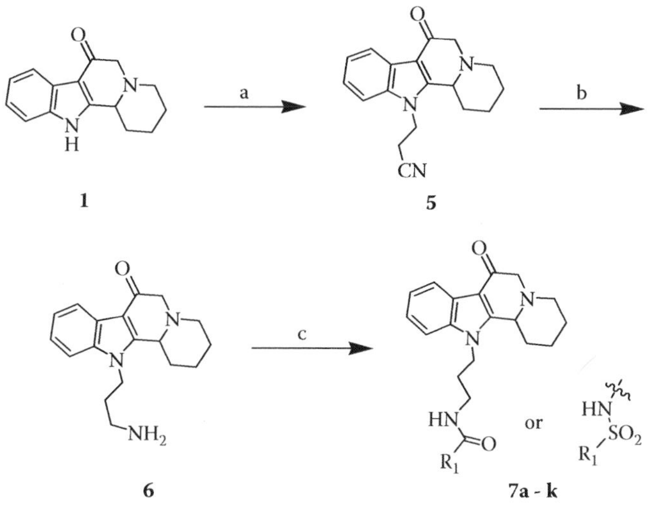

SCHEME 2.1 Synthesis of Compounds **7a–k**. (Conditions: (a) acrylonitrile, LiHMDS, toluene, 150°C, 17 min, microwave irradiation; (b) H$_2$, 50 psi, Raney nickel, NH$_3$, MeOH (65%); (c) R^{1}COCl or R^{1}SO$_2$Cl, Na$_2$CO$_3$, CH$_2$Cl$_2$-H$_2$O.)

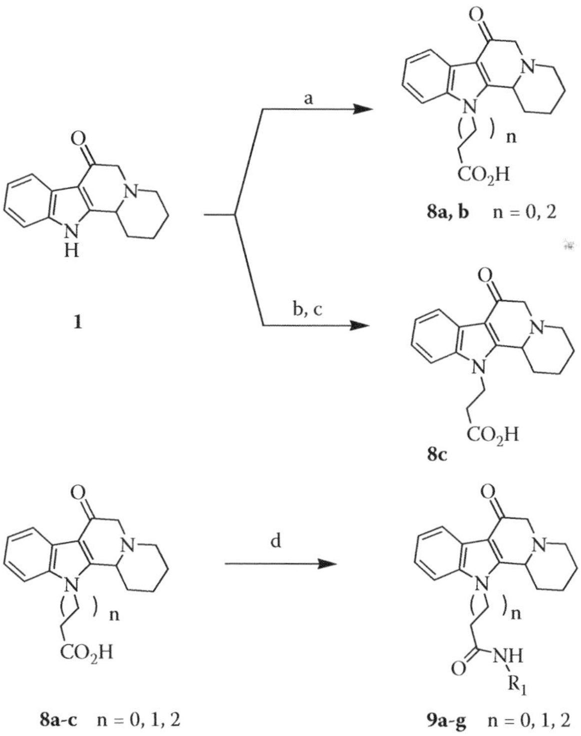

SCHEME 2.2 Synthesis of compounds **9a–g**. (Conditions: (a) methyl bromoacetate or methyl 4-bromobutyrate, KOH, DMSO, 180°C, 2 to 17 min, microwave irradiation; (b) methyl acrylate, KOt-Bu, toluene, 150°C, 17 min, microwave irradiation; (c) 6N HCl; (d) R^{1}NH$_2$, EDCI, DIEA, CH$_2$Cl$_2$.)

TABLE 2.1

Inhibition of ZipA-FtsZ$_{367-383}$ Association by Indoloquinolizinones 1–9g.

Ex.	R$_1$	% Inhibition at 1 mM	Ex.	R$_1$	% Inhibition at 1 mM
1	H	39	7h		25
6		47	7I		30
7a		38	7j		8
7b		46	7k		38
7c		33	9a		24
7d		25	9b		31
7e		11	9c		23
7f		12	9d		39
7g		41	9e		21
7h		25	9f		32
7I		30	9g		22

was observed. Consequently, we looked to a co-crystallization study with one of these analogs to understand the binding and to help determine how best to make progress on the optimization of these ligands.

The co-crystal of **7b** with ZipA was obtained and its diffraction pattern solved. The crystal structure showed that, while **7b** occupies the same hydrophobic area on ZipA as **1**, the scaffold is inverted in the binding domain and the sulfonamide chain extends in the exact opposite direction it was expected to go [Figure 2.1(b)]. The key van der Waals interaction between the hexahydroquinolizine and the active site Phe269 is maintained. Additionally, the sulfonamide nitrogen atom makes a hydrogen bond with the same bound water molecule noticed in the ZipA-**1** structure, although this time from a different direction. The *iso*-propylsulfonyl group acts to shield the water molecule, isolating it from the bulk water that surrounds the protein. In retrospect, this binding mode might have been anticipated but because the hydrophobic effect is nonspecific and dominant for the interaction of **7b** with ZipA, **7b** was able to find an alternate, more potent orientation within the binding site. While disappointed that we were not able to correctly predict the binding mode of our new molecules, we were encouraged by the consistency by which our molecules made use of the bound water molecule associated with Asn247 for making hydrogen bonds. Given these considerations, locking inhibitors into a specific orientation appeared possible, thus opening the door to structure-based potency optimization.

2.3 CASE STUDY II: 3-(2-INDOLYL)PIPERIDINES AND 2-PHENYL INDOLES

2.3.1 A New Approach to the Optimization of Lead 1 for Potency

In a practical analysis, our ability to make derivatives of our lead **1** was limited. We had exploited derivatization of the indole nitrogen in the first phase of our investigation of this lead. We could have functionalized the core scaffold using the carbonyl group as a handle but we rejected this approach, anticipating that the sorts of linkages that we could form would either be rigid (e.g., alkenes) and negatively affect water solubility or else would create new chiral centers (e.g., ethers). Instead, we decided on a strategy of reconfiguring the lead by deleting the C ring and repositioning or deleting the nitrogen atom in the piperidine ring. This approach provided more synthetically accessible sites for functionalization as well as greater synthetic flexibility.

The desired synthetic targets were obtained by way of the synthetic routes described in Schemes 2.3 through 2.6. In general, our strategy was to make a limited set of two, four, or seven intermediates and to functionalize the intermediates in a combinatorial manner using parallel synthesis in the last step. Scaffolds were designed with advice from the computational chemistry group. Building blocks (e.g., amines or acid chlorides) used in making the final products were selected using an experimental design process described previously [20]. At this stage, a new fluorescence assay for measuring the inhibition of FtsZ binding to ZipA was ready [19], and from this point on compounds were tested for their ability to inhibit ZipA-FtsZ interaction using that assay. Additionally, all compounds were tested for antibacterial activity in an *in vitro* microbial growth inhibition assay. Selected compounds were subjected to biophysical binding experiments using NMR and tested for inhibition of bacterial cell division using microscopy [20].

2.3.2 Synthesis of 3-(2-Indolyl)piperidines and 2-Phenyl Indoles and Biological Results

Substituted 2-phenyl indoles were prepared as shown in Scheme 2.3. 2-(2-Phenyl-1*H*-indol-1-yl)propanoic acids **12a–e** and acetic acid analogs **14a–b**, which were judged by docking experiments to be well accommodated in the FtsZ binding site of ZipA, were synthesized and coupled with amines to make a library of 116 compounds. None of these compounds offered any advantage over the lead **1**. (% Inhibition of ZipA-FtsZ interaction at 0.5 mM concentration as high as 44%; IC_{50} values were not determined.) However, we noticed that 2-phenyl indoles with amide substituents with basic amine functional groups were generally more active than other analogs. We would use this observation in subsequent library designs.

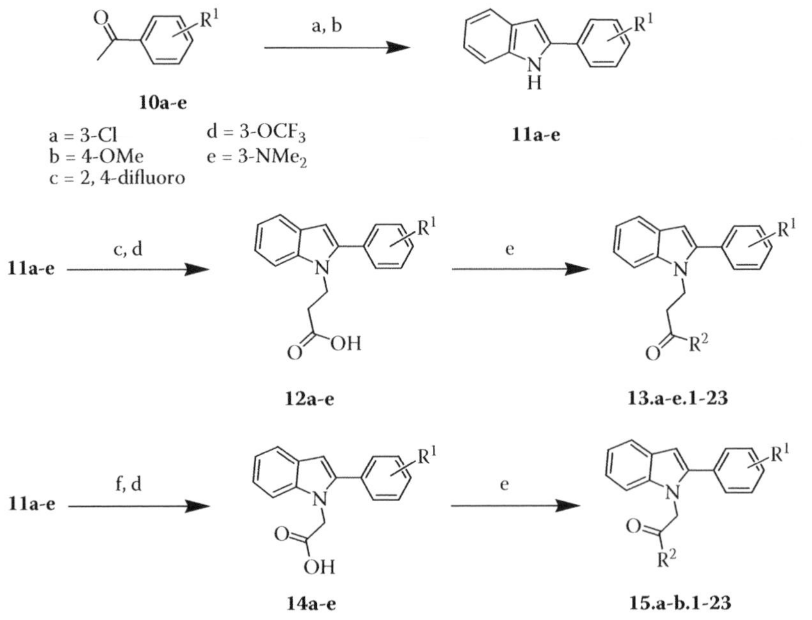

a = 3-Cl   d = 3-OCF$_3$
b = 4-OMe   e = 3-NMe$_2$
c = 2, 4-difluoro

SCHEME 2.3 Synthesis of 2-phenyl-1*H*-indoles **13.a-3.1**–**23** and **15.a-b-1**–**23**. (Conditions: (a) phenyl hydrazine, EtOH; (b) polyphosphoric acid, 130°C; (c) methyl acrylate, potassium *tert*-butoxide, toluene; (d) NaOH, MeOH; (e) amine R^2, EDCI, HOBt, *i*-Pr$_2$EtN; (f) methyl bromoacetate, NaH, DMF.)

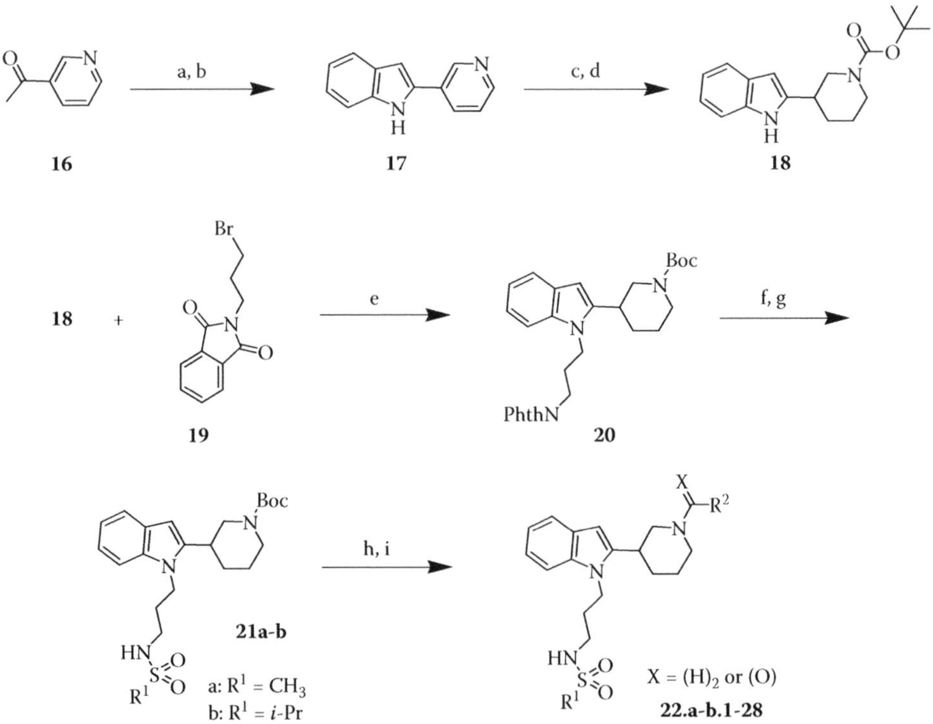

SCHEME 2.4 Synthesis of substituted 2-(piperidin-3-yl)-1*H*-indoles **22.a-b.1**–**28**. (Conditions: (a) phenyl hydrazine, EtOH; (b) polyphosphoric acid, 130°C; (c) H$_2$, Pt/C, AcOH; (d) di(*tert*-butyl)dicarbonate; (e) NaH, DMF, cat. TBAI; (f) H$_2$NNH$_2$.H$_2$O, EtOH; (g) R^{1}SO$_2$Cl, NEt$_3$, CH$_2$Cl$_2$; (h) trifluoroacetic acid, CH$_2$Cl$_2$; (i) either R^{2}CH$_2$Br, Na$_2$CO$_3$, acetone, cat. (*n*-butyl)$_4$N$^+$I$^-$, or R^{2}COCl, aq. Na$_2$CO$_3$, CH$_2$Cl$_2$.)

N-(3-indol-1-yl-propyl)-alkylsulfonamides **22.a.1–28** and **22.b.1–20** (Scheme 2.4) were designed to (1) cover the protein binding site, (2) display the sulfonamide moiety that we knew from experience was well-placed to make a hydrogen bond with the previously identified bound water molecule, and (3) display new hydrophobic sidechains that we hoped would increase binding potency through favorable van der Waals interactions. There were no active compounds in this array. (IC_{50} values for the three most potent library members at 0.5 μM concentration were determined to all be >2000 μM.) However, using the information gained from screening the 2-phenyl-1H-indoles, we modified the synthetic route to afford a library of indoles having the basic 3-aminopropyl substituent (Scheme 2.5). This series contained one of the most active indole ZipA inhibitors in this study, **23.a.4**. (IC_{50} = 296 μM.)

Alkylcarboxamide analogs **26.a.1–b.22** were prepared as shown in Scheme 2.6. Amines used in the preparation of the four alkylcarboxamide intermediates **25a–d** were a subset of the 23 diverse

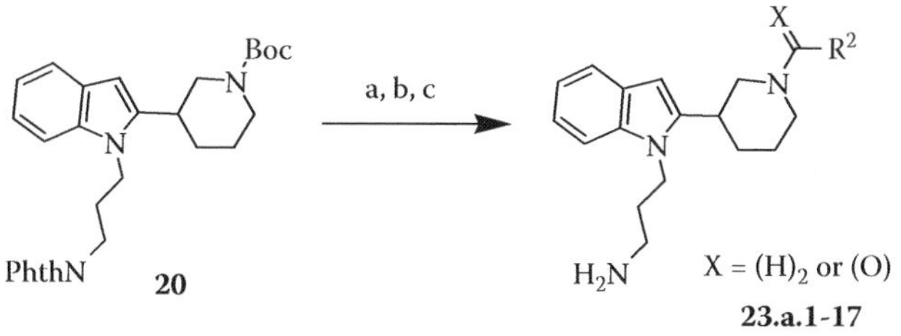

SCHEME 2.5 Synthesis of substituted 1-(3-aminopropyl)-1H-indoles 23.a.1–17. (Conditions: (a) trifluoroacetic acid, CH_2Cl_2; (b) either R^2CH_2Br, Na_2CO_3, acetone, cat. TBAI, or R^2COCl, aq. Na_2CO_3, CH_2Cl_2; (c) $H_2NNH_2.H_2O$, EtOH.)

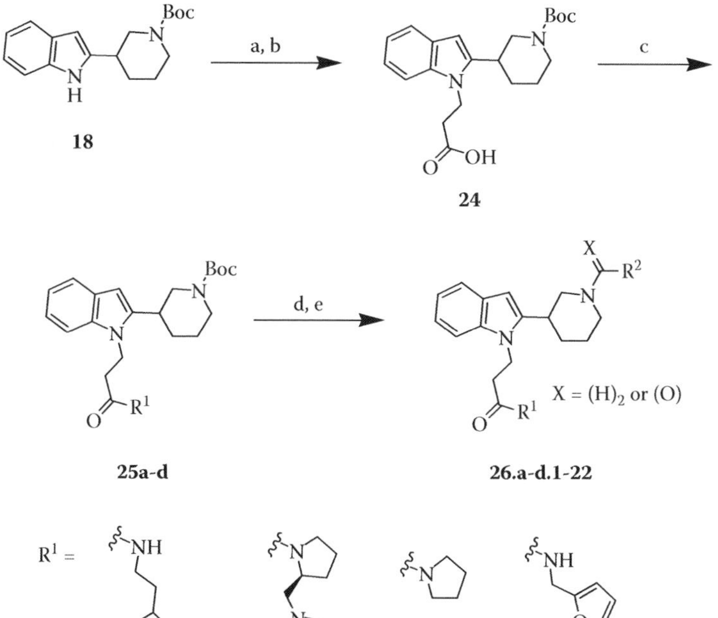

SCHEME 2.6 Synthesis of substituted 3-(indol-1-yl)-propionamides **26.a-d.1–22**. (Conditions: (a) methyl acrylate, potassium *tert*-butoxide, toluene; (b) aq. NaOH, MeOH; (c) amine **a, b, c,** or **d**, EDCI, HOBt, *i*-Pr$_2$EtN, CH_2Cl_2; (d) trifluoroacetic acid, CH_2Cl_2; (e) either R^2CH_2Br, Na_2CO_3, acetone, cat. TBAI or R^2COCl, aq. Na_2CO_3, CH_2Cl_2.)

amines used in the synthesis of aryl indoles **13.a.1–23** and **13.b.1–23,** and were chosen based on the results from the screening of the members of the aryl indoles library (see Scheme 2.3) for inhibition of ZipA-FtsZ binding. The combination of the amide substituent formed using amine **b** (Scheme 2.6) and 4-biphenylcarbonyl group (**26.b.15**) gave us our most potent inhibitor in this lead series (IC_{50} = 271 μM).

Parallel synthesis enabled a team of three chemists to prepare 265 compounds, exemplifying two core scaffolds [2-phenyl-1*H*-indoles and 2-(piperidin-3-yl)-1*H*-indoles], in a short period of time. By employing four different synthetic routes and using an experimental design approach to building-block selection, we were able to make compounds that varied in molecular weight, lipophilicity, and shape. Although the final targets are relatively large (MW = 437 to 589), all were experimentally determined to be at least moderately soluble in water at pH 7.4 (19 to 50 μg/mL) and highly permeable in an artificial membrane permeability assay (1.3 to 7.0 Pe × 10^{-6} cm/sec). While our best compounds, **23.a.4** and **26.b.15** (Structure 2.2), did not meet our program goal of an inhibitor of ZipA-FtsZ with a K_D ~ 10 μM and *in vitro* microbial growth inhibition at ≤16 μg/mL, we were able to identify features (a water-solubilizing group and a 4-biphenylcarbonyl substituent group) that provided increased potency over the lead. Moreover, using a novel cell elongation assay, we were able to find evidence that small molecule inhibitors of ZipA-FtsZ could indeed inhibit cell division in both *Bacillus subtilis* and *Escherichia coli*. We were still hopeful that high-throughput screening would provide us with leads that we could elaborate, using the information gathered up to this point, into potent ZipA-FtsZ inhibitors with *in vitro* antimicrobial activity.

23.a.4
IC_{50} = 268 mM
MIC (μg/mL) *E. coli* (imp) 32
MIC (μg/mL) *E. coli* (wt) >200

26.b.15
IC_{50} = 271 mM
MIC (μg/mL) *E. coli* (imp) >200
MIC (μg/mL) *E. coli* (wt) >200

STRUCTURE 2.2

2.4 CASE STUDY III: TRIAZOLOPYRIDAZINES

2.4.1 ORIGIN OF THE TRIAZOLOPYRIDAZINE LEAD

High-throughput screening of our compound collection using the fluorescence polarization assay provided approximately 30 reproducible and interesting hits. Of these, amino pyrimidine **4** was most potent (IC_{50} = 58 μM). NMR binding studies confirmed that **4** interfered with FtsZ binding by binding to ZipA in the solvent-exposed cavity used to bind FtsZ and an x-ray structure of the complex of **4** with ZipA was subsequently obtained and used to guide further development of this series. Unfortunately, follow-up testing of **4** showed nonspecific toxicity toward both bacteria and eukaryotes (represented by *Candida albicans*). Additionally, amine-substituted pyrimidines are heavily covered in the patent literature in the context of kinase inhibition, antitumor, and antiviral activity and designing analogs that fell outside of prior claims would be extremely challenging. It was thought that a search of the structures of compounds in Wyeth's corporate compound collection using a shape-based computational procedure would be an effective approach to finding new leads.

A database of low-energy conformers of a subset of compounds in Wyeth's corporate database was searched using molecular shape comparison with **4** [21]. A detailed computational study of the protein-ligand interactions of 380 best matches led to the selection of 29 molecules for experimental

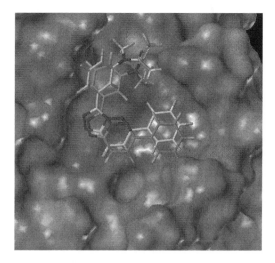

FIGURE 2.2   The x-ray structure of **28** overlaid with its ROCS determined pose. (*Source:* Reproduced with permission from *J. Med. Chem.*, 2005, 48, 1489–1495.)

testing in the ZipA-FtsZ assay. The three most interesting and active molecules discovered in this exercise were **27** through **29** (Structure 2.3). Although these leads were weaker than the original HTS hit **4** (IC$_{50}$ = 554 to 344 μM), they did not exhibit cytotoxicity and were not encumbered by prior art. To get guidance for synthetic elaboration, a crystal structure of **28** in complex with ZipA was obtained. As anticipated by modeling of the protein-ligand interaction, **28** was closely aligned with **4** in the FtsZ binding concavity of ZipA with the triazolopyridazine ring stacked above the Phe269 that forms the floor of the binding pocket and the N-methylacetamide group extending over a ledge formed by the edge of the FtsZ binding concavity (Figure 2.2).

27	**28**	**29**
IC$_{50}$ = 554 μM	IC$_{50}$ = 371 μM	IC$_{50}$ = 344 μM

STRUCTURE 2.3

Our strategy for the optimization of **27** was to simultaneously investigate variation of the substitution pattern on one of the phenyl group substituents on the triazolopyridazine ring, potentially improving the enthalpy gain from van der Waals interaction with the binding pocket, and introduce different water solubilizing groups on the other phenyl group, in an effort to improve solubility.

2.4.2 Synthesis of Analogs of Triazolopyrimidine Lead 27 and Biological Results

The synthetic route used for the preparation of analogs of **27** is shown in Scheme 2.7. Seven intermediate triazolopyridazines **34a–g** were prepared from chloropyridazine **33** and different carboxylic acid hydrazides, shown in Figure 2.3, which were selected based on modeling of the anticipated synthetic products. Addition of water-solubilizing groups to the substituent phenyl group was accomplished via the Heck coupling methodology using alkenes, acrylate esters, and acrylamides [22] in a microwave reactor [23]. Boc-protection of primary and secondary amines

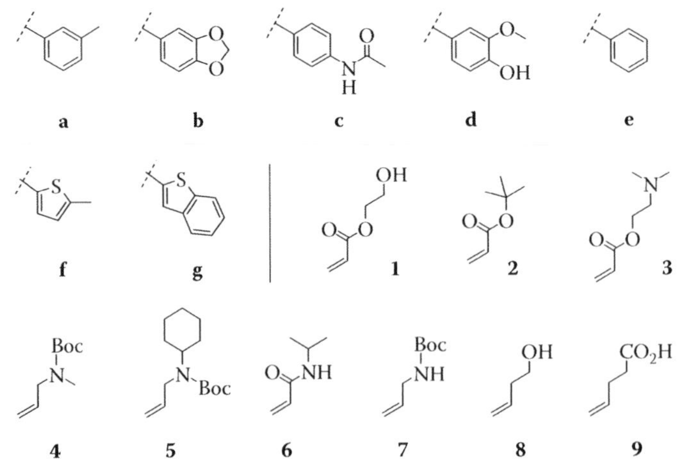

FIGURE 2.3   Structures of substituent groups on triazolopyridazines **35.a–g.1–9.**

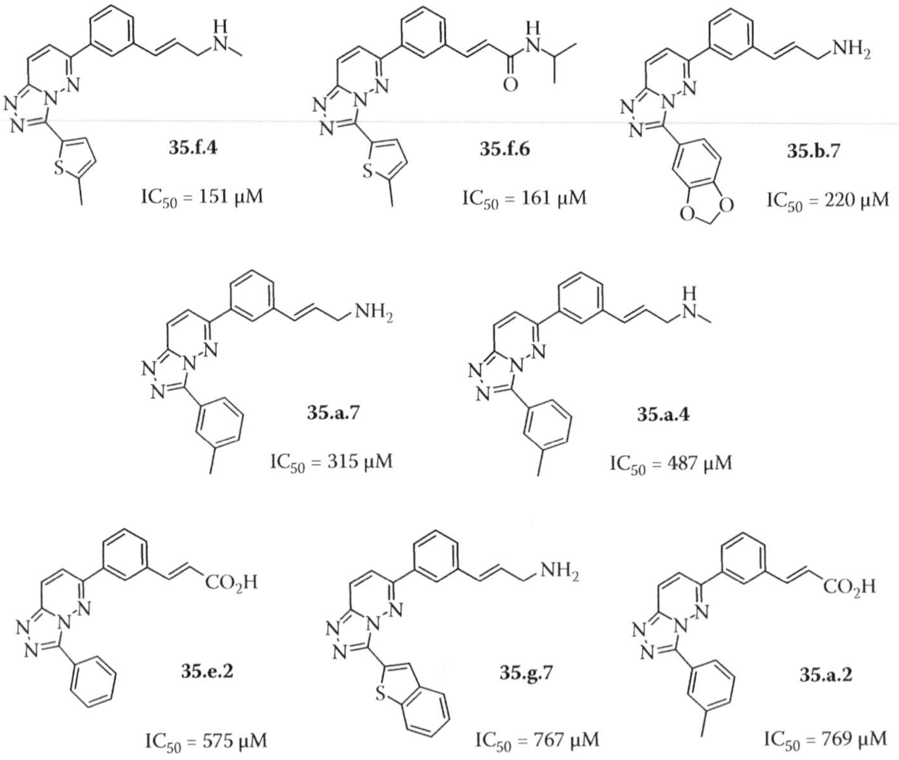

FIGURE 2.4   Most potent triazolopyridazine inhibitors of ZipA-FtsZ$_{367-383}$ association.

was found necessary for successful coupling of substituted allylic amines. Carbamate and *t*-butyl ester protecting groups were removed in a subsequent step using TFA and Et$_3$SiH at room temperature. This process enabled the introduction of a diverse set of substituent groups using uniform reaction conditions and 27 compounds were made. (Not every alkene was reacted with each intermediate.) The most potent inhibitors of ZipA-FtsZ interaction are shown in Figure 2.4. (All other compounds had IC$_{50}$ >2000 μM.) With the exception of **35.f.6, 35.e.2,** and **35.a.2,** the most potent inhibitors all had basic amine substituents. The 3-methyl substitution pattern on the southern triazolopyridazine substituent (5-methyl on the thienyl substituent) was clearly preferred, as was predicted by modeling. MICs were >25 μM for all compounds against all species in the panel; *Escherichia coli* growth was not inhibited by any of the compounds.

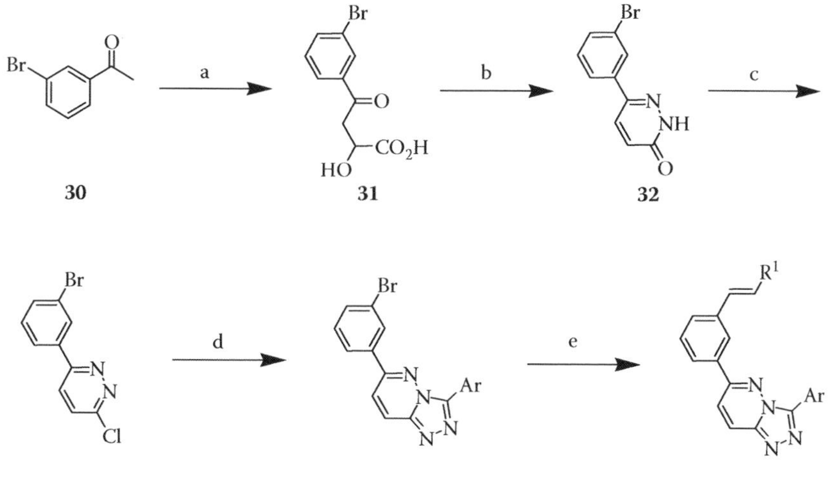

SCHEME 2.7 Synthesis of substituted triazolopyridazines **35.a-g.1-9**. (Conditions: (a) glyoxylic acid, 110°C, 2.5 hr; (b) hydrazine hydrate, NH_4OH, H_2O, 80°C, 2 hr; (c) $POCl_3$, 90°C, 2 hr; (d) aryl carboxylic acid hydrazide, *n*-butanol, reflux, overnight; (e) alkene, $Pd(OAc)_2$, $P(o-tol)_3$, NEt_3, DMF, 180°C, 11 sec.)

2.5 SUMMARY AND CONCLUSION

Typical of many of the protein-protein interactions characterized to date, features of the ZipA-FtsZ interface make the design of small, drug-like inhibitors of the ZipA-FtsZ interaction particularly difficult. The FtsZ binding interface is not significantly larger than typical substrate or cofactor binding sites found in many enzymes or proteins, but is relatively flat and is wide open to water. The binding pocket contrasts strikingly with typical enzyme active sites that are enclosed on three or more sides and feature a convergent presentation of protein functionality. Designing molecules with sufficient binding affinity is therefore a challenge because binding energy must compensate for a relatively high cost of desolvation, and only one side of an inhibitor is able to make interactions with the target. The need to maintain a favorable pharmaceutical profile, solubility, specificity, and cell penetration still applies, further complicating our efforts. Despite the number of approaches taken to finding leads, and the efficient approach taken in making analogs of the different leads, we were unable to find a potent inhibitor possessing a reasonable level of antimicrobial activity without toxicity or interfering patent claims. However, we were able to make improvements to our leads in terms of potency and solubility, and gained a certain degree of understanding about the SAR (Structure Activity Relation) of these series. More significantly, we were able to demonstrate that compounds from these series bind to ZipA at the FtsZ binding site and obtained evidence that small-molecule inhibitors of the ZipA-FtsZ interaction could indeed inhibit cell division in both *Bacillus. subtilis* and *Escherichia coli* [20]. Finally, our research organization gained increased appreciation for the challenges inherent with a biological target that involves a protein-protein interaction.

ACKNOWLEDGMENTS

The author would like to acknowledge that the advances recounted in this chapter were made possible by contributions by the members of the ZipA Exploratory Project Team: team leaders Steven

Haney, Alexey Ruzin, and Alan Sutherland, and contributors Atul Agarwal, Juan Alvarez, Weidong Ding, Elizabeth Dushin, Russell Dushin, Eric Feyfant, Kenneth Foreman, Charles Ingalls, Cynthia Kenny, Scott Kincaid, Pornpen Labthavikul, Yuanhong Li, Soraya Moghazeh, Lidia Mosyak, Peter Petersen, Thomas Rush, Mohani Sukhdeo, Desiree Tsou, Margareta Tuckman, Karen Wheless, and Yan Zhang.

REFERENCES

1. Xiong, J.-P.; Stehle, T.; Zhang, R.; Joachimiak, A.; Frech, M.; Goodman, S.L.; and Arnaout, M.A. *Science,* 2002, 296, 151–155.
2. Feuston, B.P.; Culberson, J.C.; Duggan, M.E.; Hartman, G.D.; Leu, C.-T.; and Rodan, S.B. *J. Med. Chem.,* 2002, 45, 5640–648.
3. Arkin, M.R.; Randal, M.; Delano, W.L.; Hyde, J.; Luong, T.N.; Oslob, J.D.; Raphael, D.R.; Taylor, L.; Wang, J.; McDowell, R.S.; Wells, J.A.; and Braisted, A.C. *Proc. Nat. Acad. Sci.,* 2003, 100, 1603–1608.
4. Cunningham, B.C.; and Wells, J.A. *Science,* 1989, 244, 1081–1084.
5. Atwell, S.; Ultsch, M.; De Vos, A.M.; and Wells, J. A. *Science,* 1997, 278, 1125–1128.
6. DeLano, W.L.; Ulstch, M.H.; De Vos, A.M.; and Wells, J.A. *Science,* 2000, 287, 1279–1283.
7. Peczuh, M.W.; and Hamilton, A.D. *Chem. Rev.,* 2000, 100, 2479–2494.
8. (a) Kutzki, O.; Park, H.S.; Ernst, J.T.; Orner, B.P.; Yin, H.; and Hamilton, A.D. *J. Am. Chem. Soc.,* 2002, 124, 11838–11839. (b) Yin, H.; and Hamilton, A.D. *Bioorg. Med. Chem. Lett.,* 2004, 14, 1375–1379.
9. (a) Bowman, M.J.; and Chmielewski, J. *Bioorg. Med. Chem. Lett.,* 2004, 14, 1395–1398. (b) Hwang, Y.S.; and Chmielewski, J. *J. Med. Chem.,* 2005, 48, 2239–2242.
10. Li, T.; Saro, D.; and Spaller, M.R. *Bioorg. Med. Chem. Lett.,* 2004, 14, 1385–1388.
11. Tan, C.; Wei, L.; Ottensmeyer, F.P.; Goldfine, I.; Maddux, B.A.; Yip, C.C.; Batey, R.A.; and Kotra, L.P. *Bioorg. Med. Chem. Lett.,* 2004, 14, 1407–1410.
12. Braisted, A.C.; Oslob, J.D.; Delano, W.L.; Hyde, J.; McDowell, R.S.; Waal, N.; Yu, C.; Arkin, M.R.; and Raimundo, B.C. *J. Am. Chem. Soc.,* 2003, 125, 3714–3715.
13. Hardcastle, I.R.; Ahmed, S.U.; Atkins, H.; Calvert, A.H.; Curtin, N.J.; Farnie, G.; Golding, B.T.; Griffin, R.J.; Guyenne, S.; Hutton, C.; Källblad, P.; Kemp, S.J.; Kitching, M.S.; Newell, D.R.; Norbedo, S.; Northen, J.S.; Reid, R.J.; Saravanan, K.; Willems, H.M.G.; and Lunec, J. *Bioorg. Med. Chem. Lett.,* 2005, 15, 1515–1520.
14. Hale, C.A.; and de Boer, P.A.J. *Cell,* 1997, 88, 175–185.
15. Moy, F.R.; Glasfeld, E.; Mosyak, L.; and Powers, R. *Biochemistry,* 2000, 39, 9146–9156.
16. (a) Mosyak, L.; Zhang, Y.; Glasfeld, E.; Haney, S.; Stahl, M.; Seehra, J.; and Somers, W.S. *EMBO J.* 2000, 19, 3179–3191. (b) Moy, F.J.; Glasfeld, E.; and Powers, R. *J. Biomolec. NMR,* 2000, 17, 275–276.
17. Tsao, D.H.H.; Sutherland, A.G.; Jennings, L.D.; Li, Y.; Rush, T.S., III; Alvarez, J.C.; Ding, W.; Dushin, E.G.; Dushin, R.G.; Haney, S.A.; Kenny, C.H.; Malakian, A.K.; Nilakantan, R.; and Mosyak, L.; *Bioorg. Med. Chem.,* 2006, 14, 7953–7961.
18. Jennings, L.D.; Foreman, K.W.; Rush, T.S., III; Tsao, D.H.H.; Mosyak, L.; Li, Y.; Sukhdeo, M.N.; Ding, W.; Dushin, E.G.; Kenny, C.H.; Moghazeh, S.L.; Petersen, P.J.; Ruzin, A.V.; Tuckman, M.; and Sutherland, A.G. *Bioorg. Med. Chem. Lett.,* 2004, 14, 1427–1431.
19. Kenny, C.H.; Ding, W.; Kelleher, K.; Benard, S.; Dushin, E.G.; Sutherland, A. G.; Mosyak, L.; Kriz, R.; and Ellestad, G. *Anal. Biochem.,* 2003, 323, 224–233.
20. Jennings, L.D.; Foreman, K.W.; Rush, T.S., III; Tsao, D.H.H.; Mosyak, L.; Kincaid, S.L.; Sukhdeo, M.N.; Sutherland, A.G.; Ding, W; Kenny, C.H.; Sabus, C.L.; Liu, H.; Dushin, E.G.; Moghazeh, S.L.; Labthavikul, P.; Petersen, P. J.; Tuckman, M.; Haney, S.A.; and Ruzin, A.V. *Bioorg. Med. Chem.,* 2004, 12, 5115–5131.
21. Rush, T.S, III; Grant, J.A.; Mosyak, L.; and Nicholls, A. *J. Med. Chem.,* 2005, 48, 1489–1495.
22. A solution of 0.2 mmol of aryl bromide, 0.45 mmol alkene, 5 mg Pd(OAc)$_2$, 13 mg P(o-tol)$_3$, and 50 µL NEt$_3$ in 3.5 mL DMF in a SmithProcessVial™ was heated to 180 °C by microwave for 11 min. The reaction solution was filtered through Celite, concentrated, redissolved in DMSO, and purified by reverse phase HPLC (RP-HPLC).
23. Smith Synthesizer™ (Personal Chemistry AB, Uppsala, Sweden), now marketed by Biotage AB as the Initiator™Sixty.

3 Case Studies of Parallel Synthesis in Hit Identification, Hit Exploration, Hit-to-Lead, and Lead Optimization Programs

Alexander Ernst and Daniel Obrecht

Dedicated to Professor Sir Jack Baldwin, F.R.S.

CONTENTS

3.1 INTRODUCTION

The discovery of safe and efficacious biologically active compounds is a structured multi-step process in the pharmaceutical and agrochemical industries.[1] It typically starts with the screening of chemical libraries or compound collections, an established method to provide entry points ("hits") to new discovery chemistry programs. This hit identification phase leads to the subsequent optimization efforts ("hit exploration," "hit-to-lead," and "lead optimization") and ultimately to the selection of a compound that meets the criteria to enter clinical trials or field trials, respectively (Figure 3.1).

Chemistry plays an important role in this lead-finding and optimization process, and the availability of chemistry resources has become a bottleneck in the discovery process. Hence, increasingly, pharmaceutical and agrochemical companies involve external partners in these processes to obtain access to innovative compounds and to shorten the lead-time in the optimization cycles. Thereby, they are sourcing products and services covering all stages of the drug discovery chemistry process, that is, general libraries for hit identification and the synthesis of focused libraries for hit exploration, hit-to-lead, and lead optimization.

To provide an effective and attractive support, such compound library producers and service providers must deliver a distinct skill set in a reliable manner. Leaders in the field distinguish themselves through outstanding operational skills, a set of privileged assets, and a valuable network of special relationships (Figure 3.2). Their ability to work effectively, under time pressure, to solve problems in a timely fashion and to provide innovative solutions enables them to be valuable partners in the discovery process. Flexibility and responsiveness to customer needs, in combination with the ability to complete discovery projects within the agreed budget and timelines, allow them to provide valuable support that complements the in-house activities of the pharmaceutical and agrochemical companies.

A number of reports from industrial research groups in the area of targeted chemical libraries for lead identification and optimization have been published and are summarized in a recent review

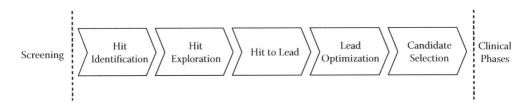

FIGURE 3.1 Value chain in the drug discovery phase.

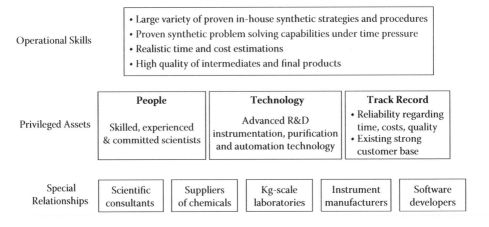

FIGURE 3.2 Capabilities.

article.[2] However, for confidentiality reasons, only a very small fraction of outsourced programs in this area are published. The authors are thus pleased to have the opportunity to present in this chapter three case studies that illustrate different synthetic approaches for the efficient and effective synthesis of focused libraries using parallel synthesis carried out at Polyphor, a Swiss drug discovery chemistry company (www.polyphor.com).

3.2 CASE STUDY I: A NOVEL SEQUENTIAL THREE-COMPONENT REACTION TOWARD INSECTICIDAL THIAZOLES

3.2.1 INTRODUCTION: SEQUENTIAL MULTI-COMPONENT REACTION (SMCR)

Among the synthetic strategies that are especially useful for parallel synthesis, multi-component reactions (MCRs)[3,4] have generated a lot of attention among chemists in academia and industry. The classical MCRs include, among others, *Passerini*,[5] *Ugi*,[6] *Hantzsch*,[7] and *Bucherer-Bergs*,[8] which all have in common that the components react in one-pot to the final products. Mixing three components A, B, and C in a classical MCR generally yields one type of final product A-B-C (Figure 3.3); and the sequence of component addition generally does not change the course of the reaction as the thermodynamically most stable products are irreversibly formed from a reactive intermediate.

Although the classical one-pot MCRs are very valuable for parallel library synthesis and many highly useful applications especially applying the *Passerini* and *Ugi* reactions were described,[4] we were looking for reactive components that would allow *sequential multi-component reactions* (SMCRs) to generate different scaffolds depending on the sequence of component addition (see Reference 3; Chapter 1.9.5). Such an approach would be highly attractive because the same set of building blocks could generate a whole range of scaffolds and thus allow synthesizing libraries with high scaffold diversity. In this case, mixing of three components A, B, and C could give theoretically six different types of products (e.g., A-B-C, A-C-B, B-A-C, B-C-A, C-A-B, and C-B-A) as schematically shown in Figure 3.4. This concept, however, is only feasible if the intermediates (A-B, B-A, A-C, C-A, B-C, and C-B) are formed quasi-irreversibly, yet live long enough to react further with the third component.

3.2.2 FEASIBILITY OF THE SMCR APPROACH

The feasibility of such an approach was demonstrated in a general attempt to synthesize various amino-substituted heterocyclic systems of type **III** (Scheme 3.1) from the parent precursors **I** and **II**

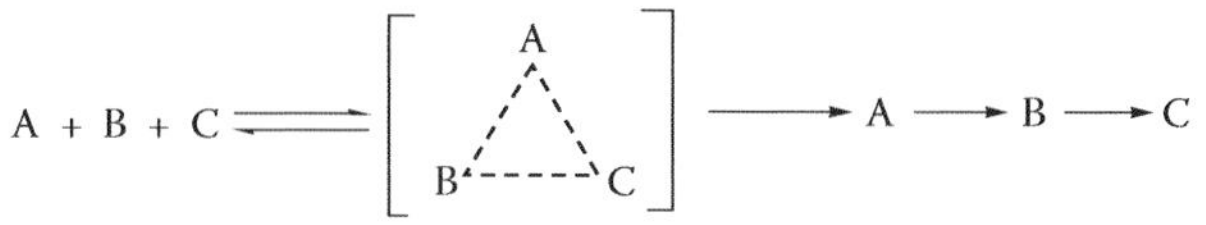

FIGURE 3.3 MCR.

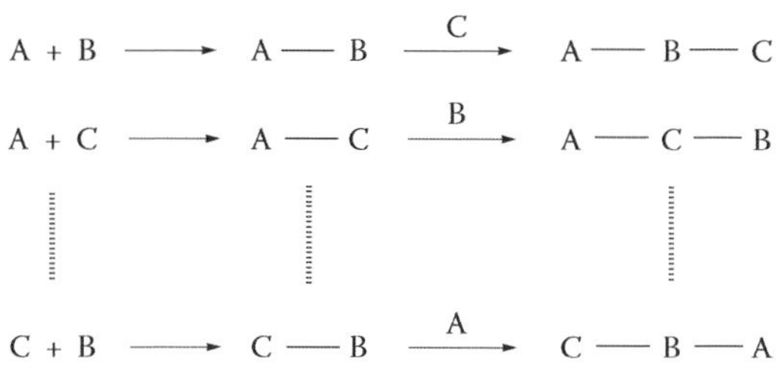

FIGURE 3.4 SMCR.

by nucleophilic substitution with primary amines on solid phase and in solution.[9] Precursors **I** and **II**, in turn, were synthesized using sequential multi-component reactions.

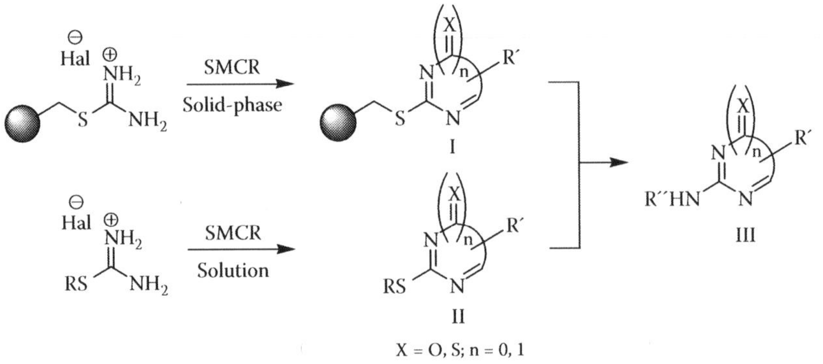

SCHEME 3.1   SMCR aminoheterocycles.

Compounds of type **III** have generated a lot of interest as kinase inhibitors[10] as exemplified by olomoucine[11] and Gleevec® (Imatinib, STI571)[12] (Figure 3.5). Gleevec® was the first kinase inhibitor to enter the market for treatment of chronic myelogenous leukemia (CML).

In this approach, reactive building blocks **1–15** (composed of bis-nucleophiles **1–6**, bis-acceptors **7–11**, acceptor-donors **12** and **13**, electrophiles **14**, and amines and anilines **15**) were used in SMCRs to generate a variety of different heterocycles of type **16** (e.g., **17**), **18** (e.g., **19**), **20** (e.g., **21**), **22** (e.g., **23**), **24** (e.g., **25**), **26** (e.g., **27**), and **28** (e.g., **29**) as summarized in Figure 3.6.

Scheme 3.2 depicts one example in more detail. Condensation of polymer-bound isothiourea with various acetylenic ketones gave sulfur-linked pyrimidines in high yields, which after oxidation with meta-chloroperbenzoic acid and reaction with anilines and amines **15** gave 4,6-disubstituted 2-amino-pyrimidines of type **16**.[13] Compound **17** is a typical example of one of the library members.

By screening various possible SMCRs using the reactive components **1–15** as outlined in Figure 3.6, a novel three-component reaction to 2,4-diamino-1,3-thiazoles of type **28** was found as outlined in Scheme 3.3.[14,15] Reaction of thiouronium salts (**4**) with isothiocyanates (**12**) in the presence of DBU and subsequent addition of α-bromo-methyl ketones (**8**) afforded 2,4-diamino-thiazoles of type **28** via the putative mechanism described in Scheme 3.3. Compound **29** is a typical member of this library. Similar MCR approaches yielding different types of thiazoles were known in the literature.[16]

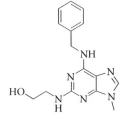

Gleevec® (Imatinib, STI571, Novartis)
First kinase inhibitor on the market for treatment of *chronic myelogenous leukemia* (CML)

Olomoucine
P.G. Schultz et al. *Tetrahedron Lett.* 1997, 38, 1161

FIGURE 3.5   Kinase inhibitors.

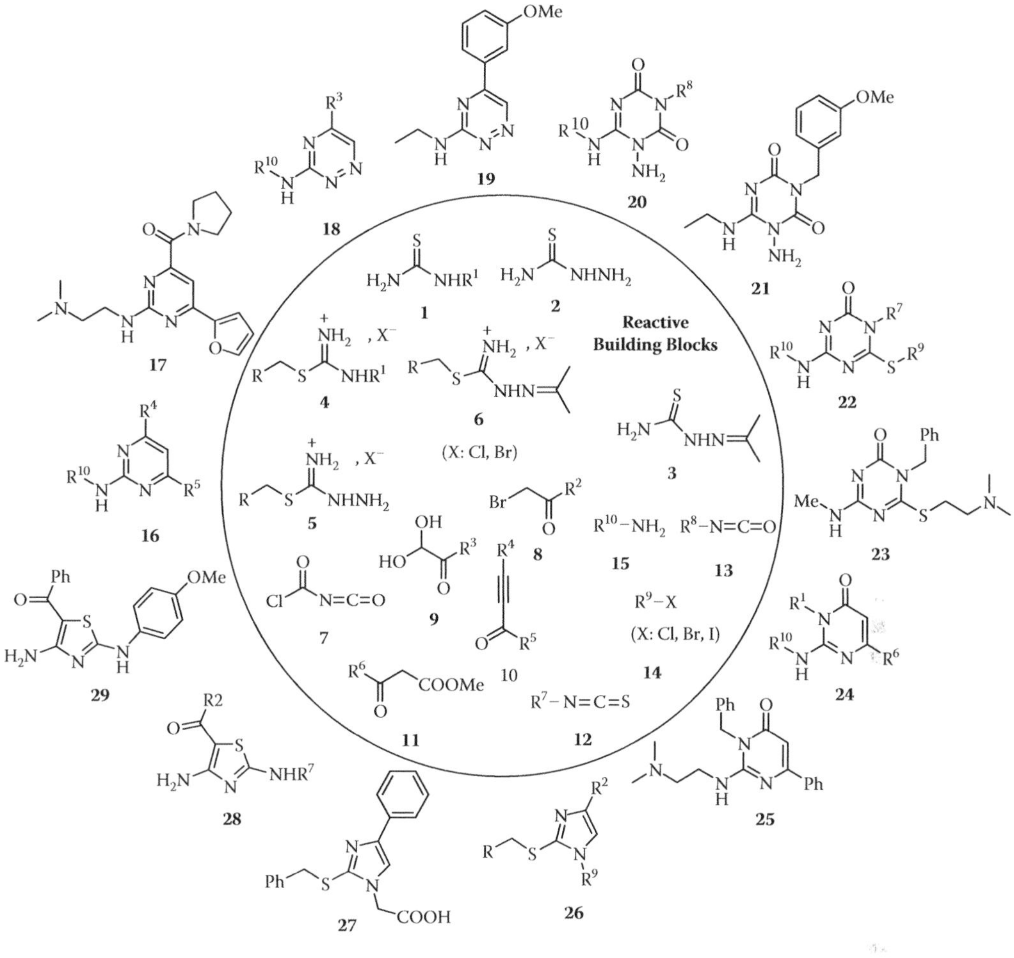

FIGURE 3.6   Reactive building blocks.

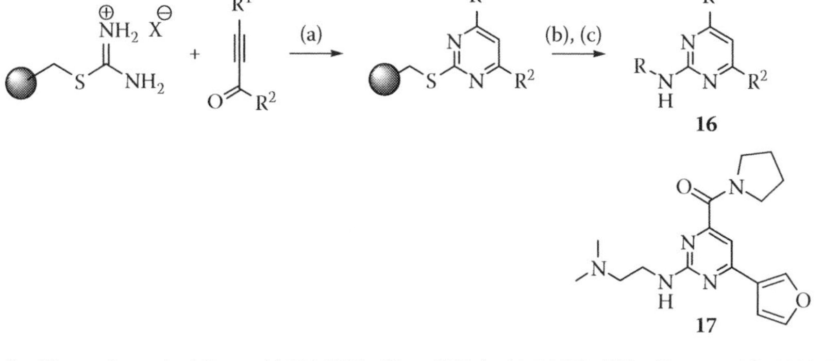

SCHEME 3.2   Example pyrimidines. (a) EtN*i*PR$_2$ (b) *m*CPBA, (c) RNH$_2$ (**15**), dioxane, 80–100°C.

3.2.3   Library Design and Parallel Synthesis

Based on the initial hit compounds **30** and **31** (Figure 3.7) discovered in an insecticidal random screening program at Syngenta AG, two libraries **A** and **B** were designed as a rapid follow-up program for hit validation and optimization.

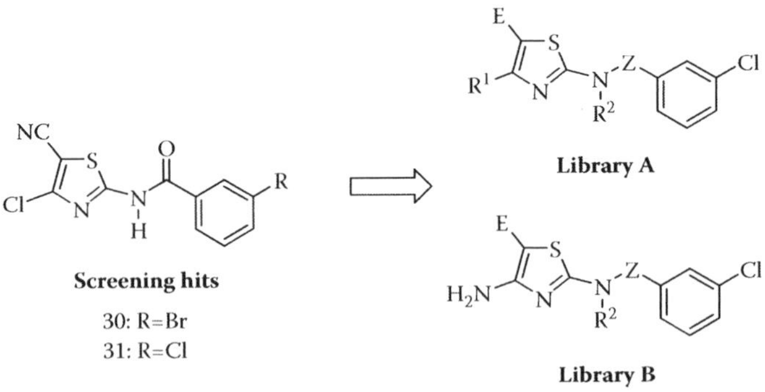

FIGURE 3.7 Initial screening hits (**30, 31**) and design of the thiazole libraries **A** and **B**.
R^1: 4-Cl-C_6H_4, 2-thienyl; CF_3; E: CN; CH_3CO; 2,4-$(Cl)_2$-C_6H_3CO; CF_3CO; 4-NO_2-C_6H_4CO; 4-CF_3-C_6H_4CO;
R^2: H, Me, CH_2OEt; Z: CO, SO_2, direct bond between NR^2 and 3-chlorophenyl group.

The approach took advantage of a three-component SMCR thiazole synthesis described in Scheme 3.3 (*vide supra*).[9,14,15] A parallel synthesis approach seemed an attractive option to speed up early lead discovery.

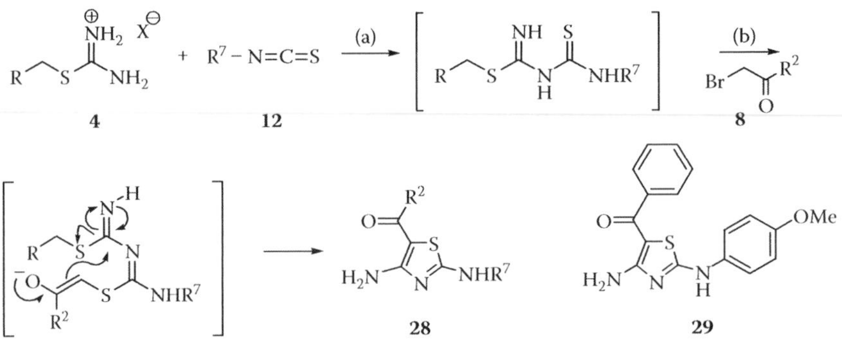

SCHEME 3.3 Thiazoles. (a) DBU, DMF, 0°C, (b) DBU, **8**, rt.

In library **A**, R^1 comprises aromatic and heteroaromatic substituents as well as CF_3; whereas in library **B**, R^1 was chosen as NH_2 (electron-donating group, H-bond donor). In both libraries, E (electron-withdrawing group) was selected from CN, CH_3CO, 2,4-$(Cl)_2$-C_6H_3CO, 4-NO_2-C_6H_4CO, and 4-CF_3-C_6H_4CO. For groups Z, we chose CO, SO_2, and direct connection of the nitrogen to the 3-chloro-phenyl group, and R^2 was chosen from Me and $EtOCH_2$ groups.

The synthesis of thiazoles **36** (library A) and **37** (library B) were performed as described in Scheme 3.4, taking advantage of an SMCR starting from amidines **32** or isothiourea **33**,

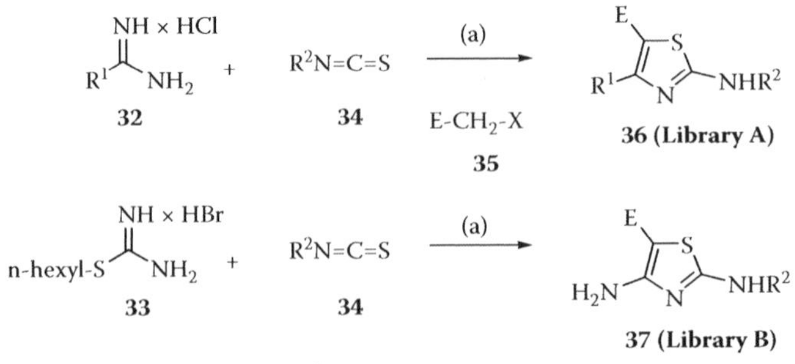

SCHEME 3.4 SMCR thiazoles. (a) DMF, DBU or EtNiPr$_2$; then **35**, DBU or EtNiPr$_2$
R^1: 4-Cl-C_6H_4, thiophen-2-yl, CF_3; R^2: 3-Cl-C_6H_4; 2,4-dimethoxy-benzyl; X: Cl, Br, I; E: CN, CH_3CO,
2,4-$(Cl)_2$-C_6H_3CO, CF_3CO, 4-NO_2-C_6H_4CO, 4-CF_3-C_6H_4CO.

isothiocyanate **34**, and ECH$_2$X (**35**, E: electron-withdrawing group). Yields of the final compounds ranged from 40 to 80%.

Scheme 3.5 describes the three-component reaction of **32** or **33** with 2,4-dimethoxy-benzyl-isothiocyanate **38** to yield intermediate thiazoles **39**. Thiazoles **39** were either treated with 3-chlorobenzenesulfonyl chloride followed by removal of the 2,4-dimethoxybenzyl group with trifluoroacetic acid in CH$_2$Cl$_2$ to yield sulfonamides **40,** or were treated with 3-chlorobenzoyl chloride followed by deprotection of the intermediates with trifluoroacetic acid in CH$_2$Cl$_2$ to yield amides **41**. All reactions were performed in a parallel format and all products were subsequently purified using an automated parallel preparative HPLC/UV/MS-system on either silica gel or reversed phase. The use of the 2,4-dimethoxybenzyl group proved advantageous for the solubility and purification of the intermediates. Overall yields were in the range of 40 to 70%. Subsequent alkylation of **40** and **41** with either MeI or ClCH$_2$OEt in DMF and in the presence of K$_2$CO$_3$ gave products **42** and **43**, which were again purified in parallel on silica gel using the preparative HPLC/UV/MS-system. A library of more than 150 compounds was prepared for biological testing.

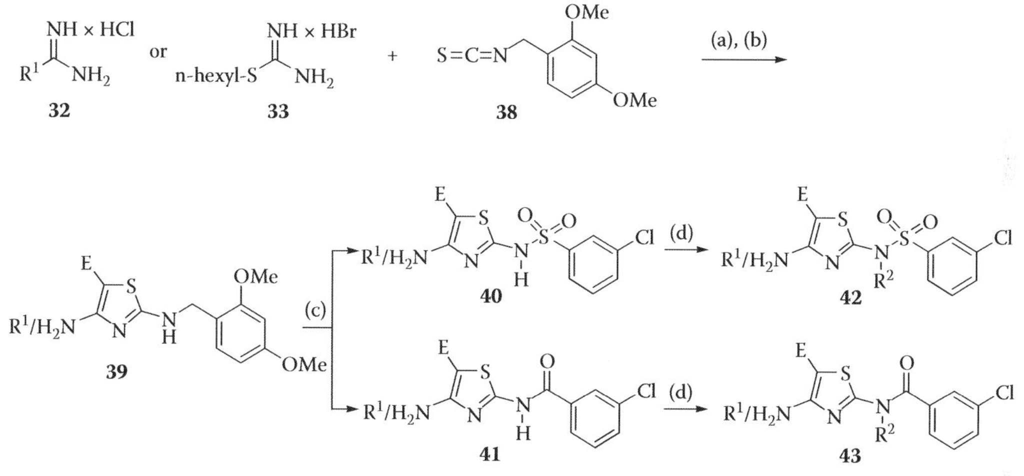

SCHEME 3.5 (a) EtNiPR$_2$ or DBU, DMF; (b) ETNiPR$_2$ or DBU, DMF, ECH$_2$X (**35**); (c) i: 3-Cl-C$_6$H$_4$COCl or 3-Cl-C$_6$H$_4$SO$_2$Cl, CH$_2$Cl$_2$, pyridine, DMAP; ii: TFA, CH$_2$Cl$_2$, H$_2$O; (d) k$_2$CO$_3$, DMF, MeI or ClCH$_2$OEt.

3.2.4 BIOLOGICAL RESULTS

The library was evaluated in a standard HTS for insecticidal, herbicidal, and fungicidal activity. None of these compounds reached the insecticidal activity of the lead structures **30** and **31**. However, some thiazoles from Library **A** displayed interesting fungicidal (nine compounds) and herbicidal (three compounds) activity. The detected fungicidal activity was mainly on one pathogen, *Pyricularia oryzae* (rice blast), whereas herbicidal activity was observed on *Stellaria media* (common chickweed). The most potent compounds were those with R^1 = CF$_3$ and E = CN, 4-NO$_2$-C$_6$H$_4$-CO or 2,4-(Cl)$_2$C$_6$H$_3$CO. Among all compounds prepared, best activity was observed for thiazole **44** (Figure 3.8).

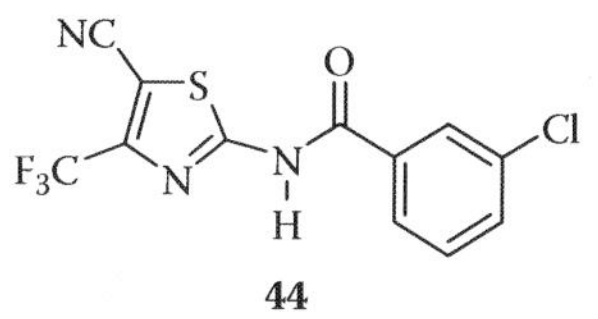

FIGURE 3.8 Most active compound.

3.2.5 SUMMARY AND CONCLUSIONS

Polyphor has developed suitable reaction conditions of an SMCR for the synthesis of new 2-aminothiazoles with a structurally diverse pattern of substituents at C(4) and C(5). The final products were effectively synthesized in a parallel format from easily available building blocks using a one-pot procedure. This approach minimized the number of steps required for the preparation and purification compared to the classical approach and afforded more than 150 final compounds in a few weeks. The major part of the time was used for the development of optimal reaction conditions applicable for parallel synthesis and for the development of a suitable protocol applicable for the parallel purification by HPLC on a 100-mg scale.

This example shows that a relatively small number of 2-aminothiazoles with a high variation of structurally diverse substituents at C(4) and C(5) has led to compounds possessing potent fungicidal and herbicidal activity and serving as new entry points for further optimization programs.

3.3 CASE STUDY II: NEW INSECTICIDAL AND FUNGICIDAL 1,2,4-TRIAZINES APPLYING A MULTI-GENERATION APPROACH[17]

3.3.1 INTRODUCTION: MULTI-GENERATION APPROACH

In contrast to the sequential multi-component approach, in which the formed intermediate(s) are not isolated or isolable, the multi-generation reaction proceeds via intermediates that can be trapped or isolated. Thus, any reaction sequences, in which a transformation generates a next (n)-generation of reactive species suitable for the subsequent transformation to the (n+1)-generation, can be regarded as a multi-generation reaction. The advantage of the multi-generation approach compared to the multi-component approach is that it allows the synthesis of a larger number of core structures, which ultimately lead to a greater structural diversity of the synthesized libraries derived from these cores (Scheme 3.6).

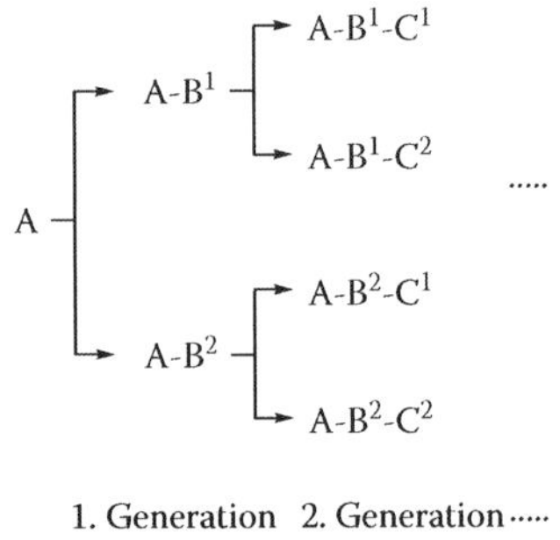

1. Generation   2. Generation·····

SCHEME 3.6   Multi-generation approach.

The following example illustrates this approach.

3.3.2 SCREENING AND BIOLOGICAL ACTIVITY

In a random screening campaign, Syngenta AG identified 1,2,4-triazines as novel lead compounds with activity against species such as *Phytophthora infestans, Pyrenophora teres,* and *Septoria nodum.* The 1,2,4-triazines with an aryl substituent in position-3 showed high potency against a broad range of Lepidoptera and some sucking insects. The mode of action of these compounds was not known.

3.3.3 LIBRARY DESIGN AND SYNTHESIS

The initial screening hits (**1–3**) served as the starting point for the design of a library of 3,6-("linear") and 3,5-("nonlinear") substituted 1,2,4-triazines (Figure 3.9). The residues chosen for position-3 were methylsulfonyl (as found in the initial hits **1** and **2**), thiomethyl, methylsulfoxy, and methoxy,

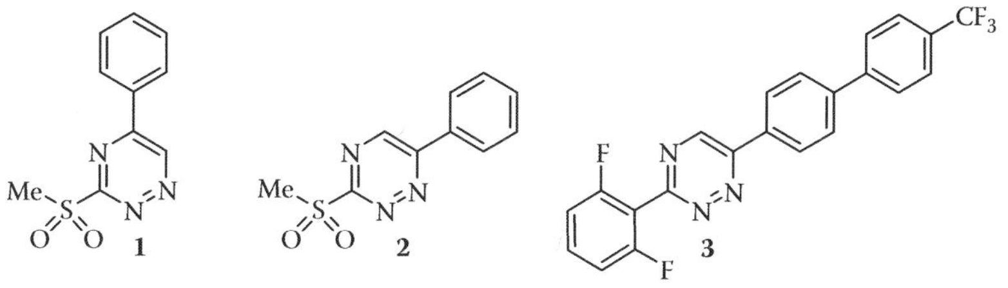

FIGURE 3.9 Initial HTS hits.

as well as 2,6-difluorphenyl and 2,4-dichlorphenyl. A number of substituted phenyl, biphenyl, and heteroaromatic residues were selected for the variation at positions-5 and -6, respectively.

Based on the design, a multi-generation approach was envisaged for the nonaromatic residues selected for position-3 (Scheme 3.7).

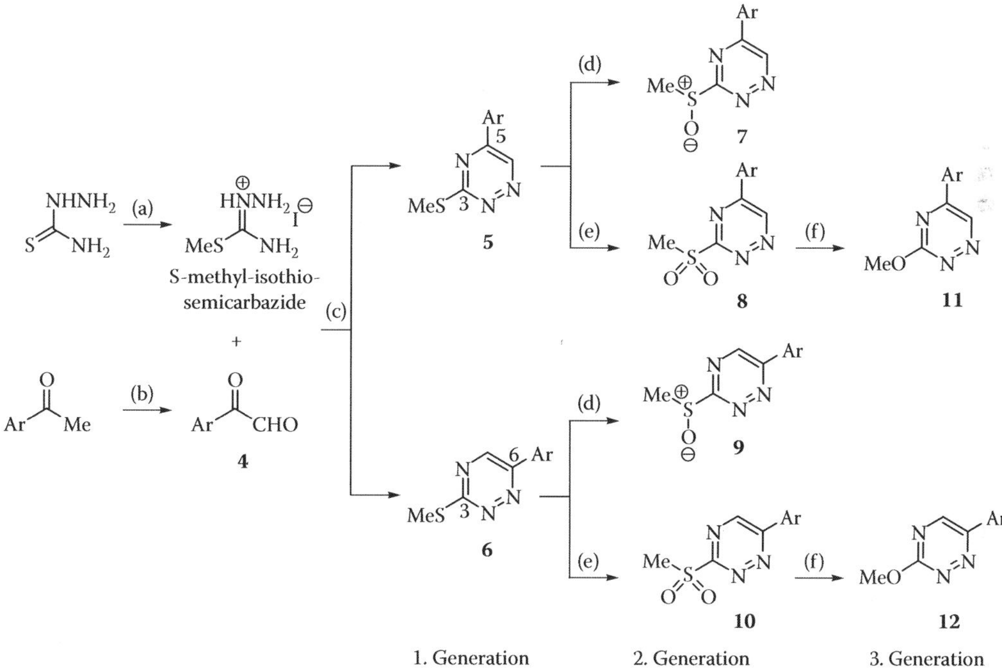

SCHEME 3.7 (a) MeI, dioxane; (b) SeO$_2$, dioxane/H$_2$O, 80°; (c) ETNiPr$_2$, DMF or CH$_2$Cl$_2$; then separation; (d) mCPBA (1.1 eq.) NaOAc, CH$_2$Cl$_2$; (e) mCPBA (2.5 eq.); (f) NaOME, THF.

The first generation of 3-SMe-substituted 1,2,4-triazines (**5** and **6**) was prepared by condensation of S-methyl-isothiosemicarbazide[18] with glyoxals **4** (available by SeO$_2$ oxidation from the corresponding commercially available acetophenones) followed by chromatographic separation of the two regioisomers; the 5-isomers (**5**) were formed predominantly. The following mCPBA oxidations afforded cleanly the second-generation sulfoxides (**5** → **7** and **6** → **9**) or sulfones (**5** → **8** and **6** → **10**) using appropriate equivalents of the oxidation reagent. Treatment of the sulfones with methoxide afforded the third-generation compounds **11** and **12**. Typically, 300 mg of **5** and **6** were required to prepare >20 mg of final compound in each generation.

3.3.4 CHEMICAL OPTIMIZATION OF 6-SUBSTITUTED 1,2,4-TRIAZINES

The low yield of the formation and sometimes difficult isolation of the 6-isomer was unsatisfactory (Scheme 3.8). For example, treatment of glyoxal **13** with S-methyl-isothiosemicarbazide afforded the two regioisomers in good overall yield but in a ratio of approximately 9:1 in favor of the 5-isomer **15**. However, we anticipated using the oxime **14** (obtained from **13** by trans-oximation with acetone oxime in dioxane in the presence of aqueous Na_2HPO_4 solution as mild acid catalyst) might lead to the reversed selectivity.[19] After parallel screening of various reaction conditions, it was found that using n-butanol as solvent in the presence of 4N HCl in dioxane afforded the triazines in equally good yields with almost reversed selectivity (approx. 5:1 in favor of the 6-isomer **16**).

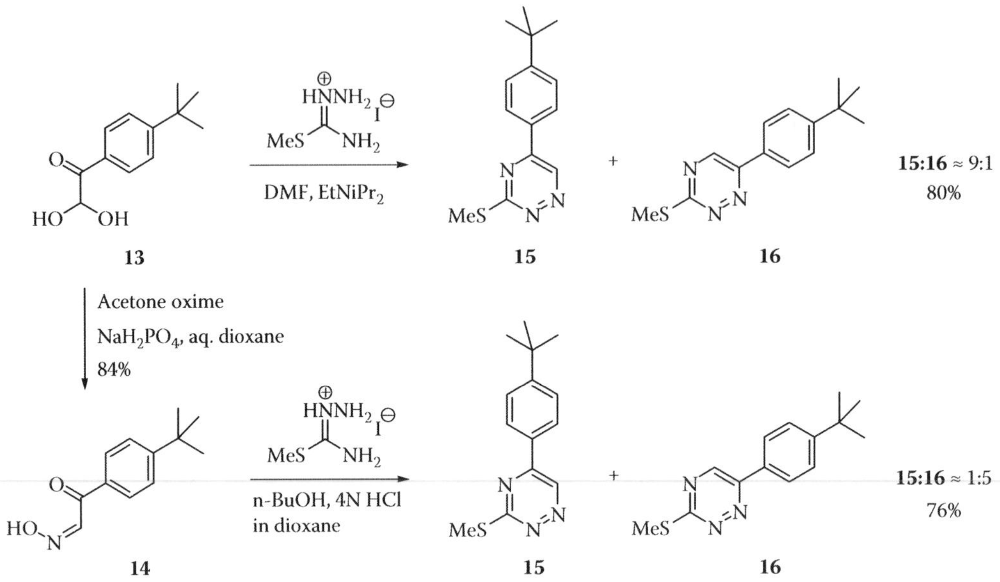

SCHEME 3.8

The synthesis of the 3-aryl-substituted 1,2,4-triazines (e.g., **3**) might be accessible by treating the corresponding sulfoxides with the appropriate aryl-Grignard reagents,[20] however, the synthesis started from the appropriate amidrazone, which was condensed with either the glyoxal or the oxime using the aforementioned conditions. An example (**17** + **18** → **19**, 82%) is given in Scheme 3.9.

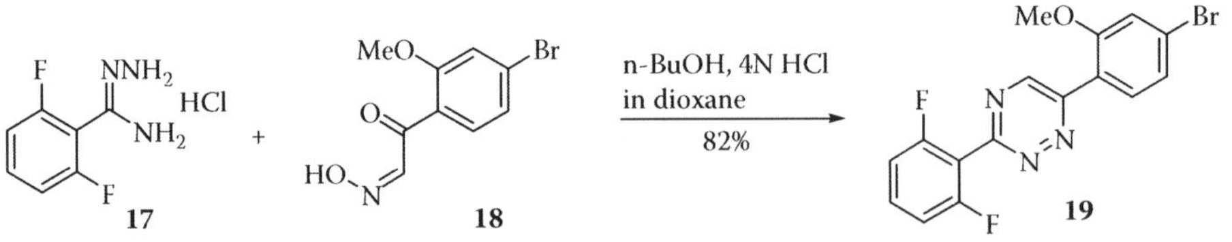

SCHEME 3.9

These conditions were applied to parallel synthesis in solution and a first library of more than 180 5- and 6-substituted 1,2,4-triazines, followed by a second-generation focused library of more than 20 6-substituted 1,2,4-triazines were prepared in 4 months.

3.3.5 BIOLOGICAL RESULTS

The synthesized compounds were screened for fungicidal and insecticidal activities. The compounds possessing thiomethyl, sulfoxymethyl, sulfonylmethyl, and methoxy in position-3 showed interesting fungicidal activities against certain oomycetes such as *Phytophthora infestans* (late blight), *Plasmopora viticola* (grape downy mildew), and *Pythium ultimum* (ashy stem blight), as well as *Pyrenophora teres* (net blotch), and *Septoria nodorum* (glume blotch).

The 3,6-substituted 1,2,4-triazines were highly potent compounds against key chewing insects such as *Spodoptera littoralis* (beet armyworm), *Heliothis virescens* (tobacco budworm), and *Plutella xylostella* (diamondback moth), and also against some sucking pests, at a level comparable to the best commercial standards.

3.3.6 SUMMARY AND CONCLUSION

Polyphor has developed suitable reaction conditions for the selective preparation of either 5- or 6-substituted 1,2,4-triazines using parallel synthesis. The multi-generation approach followed in this project required significantly fewer synthetic steps compared to the classical linear approach and allowed researchers to finish the synthesis of more than 200 compounds in 4 months. Some of the compounds synthesized (structures not shown) reached a level of biological activity comparable to the established marketed standards.

This example has shown that the fast development of optimal reaction conditions applicable for parallel synthesis following the multi-generation approach is the necessary prerequisite to reach the objective. Once this has been successfully established, the parallel synthesis of the targeted compounds can be executed in an efficient and effective way.

3.4. CASE STUDY III: PARALLEL SYNTHESIS OF NEW TETRAMIC ACID TYPE DERIVATIVES AS INFLUENZA ENDONUCLEASE INHIBITORS[21]

3.4.1 INTRODUCTION

The influenza virus infects more than 120 million people every year and is the major cause of mortality worldwide. With the upcoming global spread of the H5N1 virus, the so-called "avian flu," mankind faces the threat of a flu pandemic eventually leading to an estimated 100 to 300 million deaths. Thus far, this virus has spread between humans in only a few suspected cases; however, it might be only a few mutations away to become deadly contagious. As reported recently in *Science*,[22] the influenza virus responsible for the "Spanish flu" crossed the species barrier between birds and humans and killed an estimated 20 to 50 million people.

Vaccination against influenza is the prophylaxis most widely used; however, this is not always completely protective due to the fast mutating nature of the virus. The marketed antiviral drugs Tamiflu® and Relenza®, which target the enzyme neuraminidase, have shown virustatic efficacy (the infection is stalled) if taken within 48 hr after the infection. But there are cases known where mutations in the neuraminidase led to ineffectiveness of the drug. Thus, there is a high medical need for safe and efficacious antiviral drugs.

3.4.2 THE TARGET AND BIOLOGICAL ACTIVITY OF KNOWN INHIBITORS

The targeted enzyme in this case is the endonuclease of the influenza RNA polymerase, a key component of the viral transcription initiation mechanism. This enzyme has no cellular counterpart and offers the opportunity for the discovery of selective and safe drugs for an effective treatment. As complete inhibition of the endonuclease would block the viral transcription machinery, it is expected to have not a virustatic, but rather a virucidal effect. There is strong evidence that the influenza endonuclease belongs to the family of two metal ion groups of phosphate-processing enzymes. This class of enzymes bind inorganic phosphate and two metal ions (particularly Mn^{2+} and Mg^{2+}) in its active site and transfer the phosphate moiety on a nucleophilic serine residue. Several members of this enzyme family have been structurally and mechanistically characterized.[23]

Two lead series were discovered from high throughput screening campaigns. Merck & Co. reported[24] the discovery of 2,4-diketobutanoic acids and N-hydroxy-imides as potent inhibitors of the influenza endonuclease (Figure 3.10).

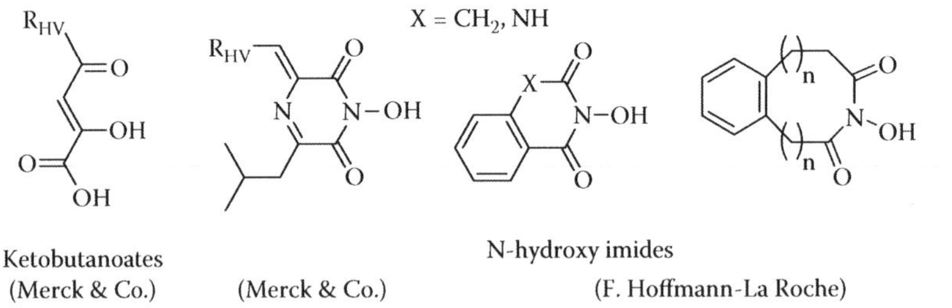

FIGURE 3.10 Structure of known inhibitors.

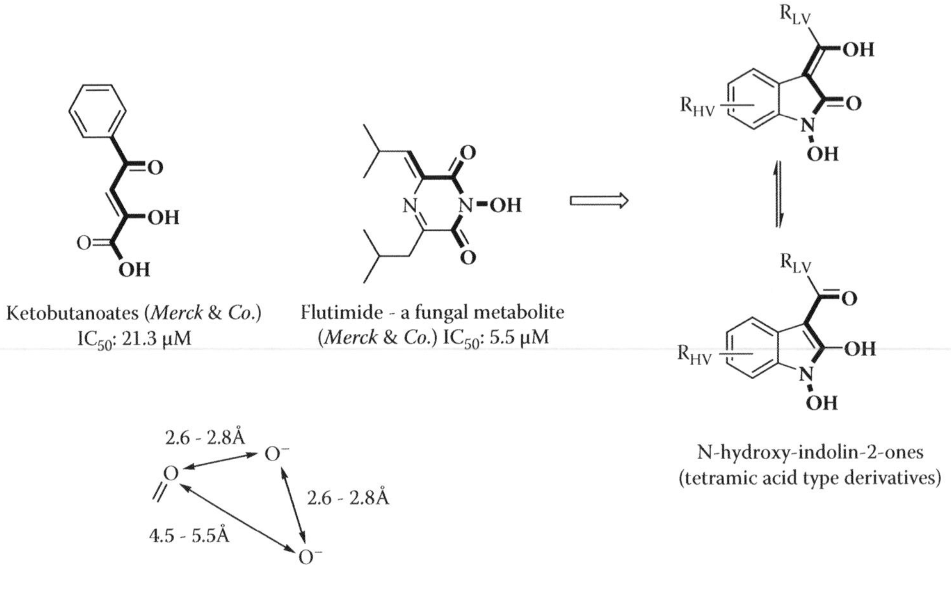

FIGURE 3.11

3.4.3 PHARMACOPHOR MODEL AND INHIBITOR DESIGN

Based on pH-studies of the enzymatic and inhibitory activity and molecular modeling studies, Roche proposed a minimal pharmacophor model for influenza endonuclease inhibitors and suggested a tetramic acid type library to corroborate this hypothesis (Figure 3.11).

This new compound class possesses the appropriate geometry of the three critical oxygen atoms and pKa values in line with the proposed pharmacophor model. A library of 144 compounds was designed having high variation substituents (a diverse set of acyl, carbamoyl, and sulfonyl residues) attached to an anilic nitrogen. The low variation next to the β-keto amide moiety was set to R^{LV} = Me, Et, Ph.

3.4.4 DEVELOPMENT AND SYNTHESIS OF THE KEY INTERMEDIATE

The strategy of the library synthesis was to effectively develop a suitable protected common intermediate with the least number of steps, from which the targeted library is synthesized efficiently using parallel synthesis and purification. Several synthetic challenges were anticipated, as the envisaged N-hydroxytetramic acid type analogs have not been described in the literature and were expected to be polar compounds with physicochemical properties not cooperative for rapid and facile purification. Thus, to find suitable protection groups of the acidic β-keto amide and hydroxamic acid moieties that are stable under the parallel synthesis conditions and can be easily removed to result in final compounds with high purity and yields required considerable experimentation.

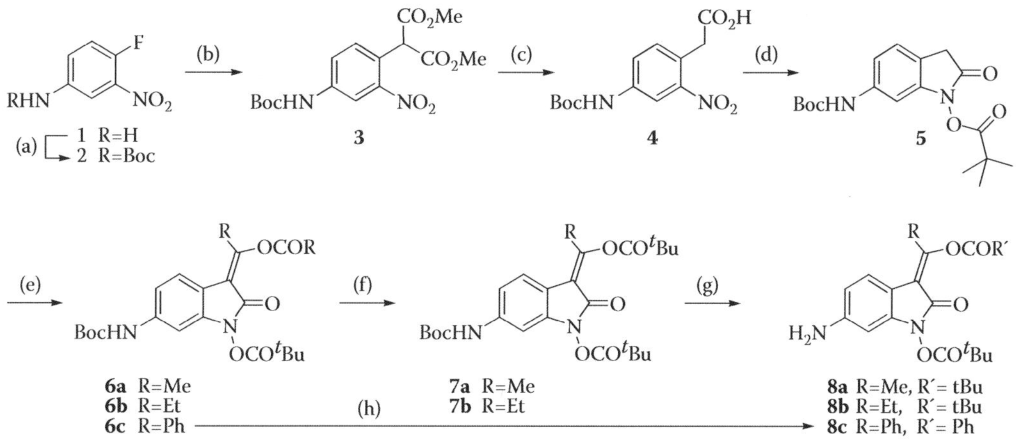

SCHEME 3.10 (a) Boc$_2$O, THF, reflux, 96%; (b) dimethylmalonate, NaH, DMSO, 80°, 93%; (c) aq. NaOH, MeOH, reflux, 78%; (d) i: Pt/C (5%), H$_2$ (1 atm.), EtOH, DMSO, then AcOH; ii: pivaloyl chloride, EtNiPr$_2$, Ch$_2$Cl$_2$, 50%; (e) (RCO)$_2$O (for R = Me) or RCOCl (for R = Et, Ph), EtNiPr$_2$, DMAP; CH$_2$/Cl$_2$/THF, 84% (**6a**), 73% (**6b**), 62% (**6c**); (f) pivaloyl chloride, Bu$_4$NCN or NaCN, pyridine, CH$_2$Cl$_2$, 76% (**7a**), 65% (**7b**); (g) 4N HCl in dioxane, 71% (**8a**), 83% (**8b**); (h) TFA, CH$_2$Cl$_2$, 73% (**8c**).

Treatment of the commercially available aniline **1** with Boc$_2$O in THF afforded the Boc-protected derivative **2** in 96% yield on a 500 g-scale (Scheme 3.10). Substitution of the fluoride by dimethylmalonate in the presence of NaH in DMSO at 80°C (**2** → **3**, 93%), followed by saponification/decarboxylation with aqueous NaOH in MeOH, led to the nitrophenyl acetic acid **4** (87%). The subsequent conversion of **4** into the protected N-hydroxy-indolinon offered two challenges: (1) the partial reduction NO$_2$→NH-OH and (2) the selection of an appropriate protective group that is stable under the conditions of all subsequent transformations but cleaved quantitatively at the end of the sequence. After screening numerous reduction conditions, it was found that a modification of the hydrogenation conditions described by Kende et al.[25] gave best results. Hydrogenation using 5% Pt/C in a solvent mixture of EtOH:DMSO (60:1) at atmospheric pressure, followed by acetic acid-catalyzed cyclization afforded the N-hydroxy-indolinon in approximately 65% yield, together with 5 to 10% of the corresponding lactam as a solid mixture, which was sparsely soluble in common organic solvents and difficult to purify further on a larger scale. Treatment of the crude mixture with pivaloylchloride in the presence of Hünig's base afforded the protected indolinone **5** in 50% overall yields on a 50 g-scale. The subsequent Knoevenagel condensation (**5** → **6a–6c**) to introduce the β-keto amide moiety proved difficult. Following the established procedure using the appropriate carboxylic acid in dichloromethane in the presence of dicyclohexylcarbodiimide (DCC) and N,N-dimethylaminopyridine (DMAP) gave the corresponding tetramic acid type derivatives **6a–6c** in low yields (<10%). Two issues were observed in this reaction: (1) the separation of the products from the starting material (**5**) was very difficult, and (2) the product slowly reverted back to the indolinone **5** under the reaction conditions. It was rationalized that the intermediate tetramic acid type **A** had to be acylated quickly to block the reverse reaction (**A** → **5**; Scheme 3.11).

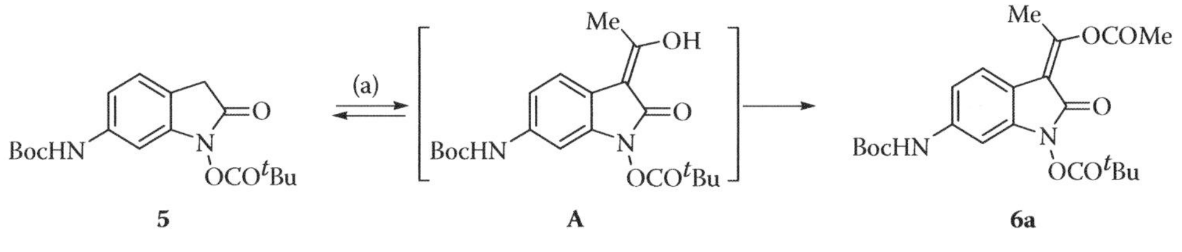

SCHEME 3.11 (a) MeCO$_2$H, DCC, DMAP, CH$_2$Cl$_2$ < 10% yield *or* Ac$_2$O, EtNiPr$_2$, DMAP, THF, 84% yield.

Again, parallel screening of numerous reaction conditions revealed that stronger acylating conditions led to the requested compounds. In our hands, treatment of **5** with the appropriate acid chloride or anhydride in dichloromethane in the presence of Hünig's base and DMAP led to the tetramic acid type derivatives **6a–6c** in good yields (62 to 82%) on a 10 g-scale after chromatographic purification on silica gel. The products were isolated as 1:2 to 1:3 mixtures of the *E*- and *Z*-diastereoisomers. Pure isomeric samples have been obtained by (tedious) chromatographic separation. However, as it turned out that under the conditions of the final deprotection step (*vide infra*, Scheme 3.12), the products are again formed as a diastereoisomeric mixture in a similar ratio, the intermediates were carried forward as diastereoisomeric mixtures.

The methyl- (**6a**) and ethyl- (**6b**) intermediates partially decomposed under the conditions developed for the parallel synthesis. Thus, transesterification using pivaloyl chloride in the presence of NaCN or Bu$_4$NCN afforded the corresponding *tert*-.butyl derivatives **7a**, **7b** in good yields. Final cleavage of the N-Boc protective group with either 4N HCl in dioxane or trifluoroacetic acid in dichloromethane led to the three key intermediates **8a–8c**. These building blocks served as starting material for the parallel synthesis and have been synthesized on a 5 g-scale.

3.4.5 PARALLEL SYNTHESIS AND PURIFICATION OF THE LIBRARIES

The requested library consisted of three sublibraries (16x amides, 17x sulfonamides, 15x ureas) from each building block comprising 3x(16+17+15) = 144 final compounds.

The development of the parallel synthesis comprised two steps: derivatization followed by final deprotection. The first step, derivatizations at the anilic nitrogen into a series of anilides (→**9a–9c**),

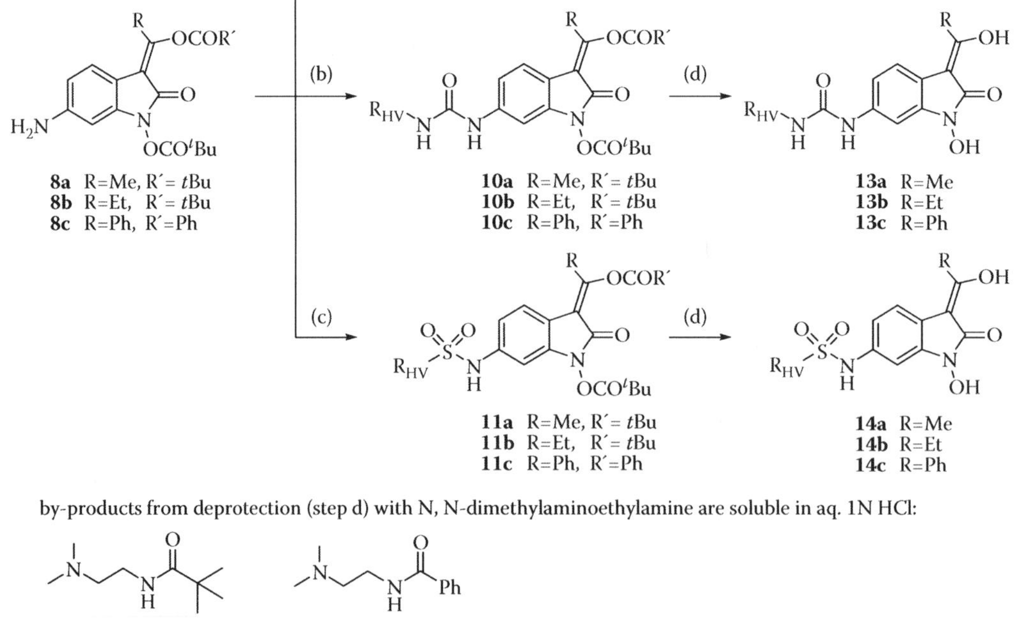

by-products from deprotection (step d) with N,N-dimethylaminoethylamine are soluble in aq. 1N HCl:

SCHEME 3.12

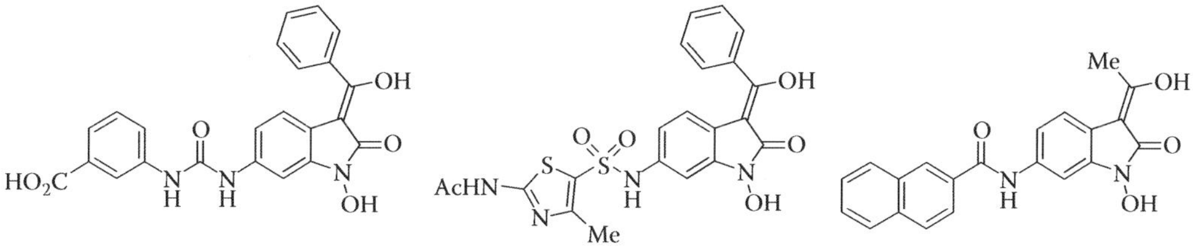

15 (IC_{50} = 3 μM)   **16** (IC_{50} = 32 μM)   **17** (IC_{50} = 9 μM, EC_{50} = 21 μM)

FIGURE 3.12

ureas (→**10a–10c**), and sulfonamides (→**11a–11c**), was performed under standardized conditions using parallel synthesis in solution (Scheme 3.12), followed by automated parallel purification on silica gel. At that stage, typically 100 mg of each intermediate was prepared.

The final deprotection step (**9a–9c → 12a–12c**, **10a–10c → 13a–13c**, **11a–11c → 14a–14c**) was achieved using either lithium hydroxide in aq. THF/MeOH or, more conveniently, by treatment with excess N,N-dimethylaminoethylamine in dichloromethane or chloroform, followed by extraction with aq. 1N HCl. The basic amide by-products (*cf.* Scheme 3.12) remained in the acidic aqueous phase, and the final products were isolated in high purity after removal of the solvent and purified either by precipitation from ethyl acetate/hexane mixtures or preparative reverse-phase HPLC. From the 144 designed final compounds, Polyphor delivered 131 in average quantities of >20 mg and high purity (over 90% of the final compounds had a purity of >90% according to HPLC/UV at 254 nm or ¹H-NMR spectroscopy) to F. Hoffman-La Roche for biological profiling.

3.4.6 BIOLOGICAL RESULTS

As a result of the biological testing, 26 final compounds showed significant potency (IC_{50} < 50 μM) in the endonuclease assay, with compound **15** as the most potent one with an IC_{50} of 3 μM (Figure 3.12). Also members of the other sublibraries like sulfonamide **16** (IC_{50} = 32 μM) and amid **17** (IC_{50} = 9 μM) showed potent inhibitory activity against influenza A endonuclease (Figure 3.3). A subset of six final compounds were tested in cell culture, and compound **17** showed significant antiviral activity (EC_{50} = 21 μM).

3.4.7 SUMMARY AND CONCLUSION

Based on a pharmacophor model rationalizing the key interaction of compounds with the influenza endonuclease active site, Hoffmann-La Roche designed a new class of tetramic acid type derivatives to test this model. Polyphor developed a suitable route for the parallel synthesis of these complex compounds and prepared a total of 131 compounds in 4 months. Biological evaluation revealed a number of potent influenza endonuclease inhibitors (e.g., **15–17**) that might serve as starting points for further lead optimization programs. The g-scale seven-step synthesis of the three building blocks (**8a–8c**) enclosed several synthetic challenges that have been solved through careful observation, meticulous analysis, and rigorous experimentation in a parallel format.

This case demonstrates that highly complex and synthetically challenging compounds are amenable to parallel synthesis. To meet this objective requires skills for both the effective development of new multi-step synthesis routes in g-scale *and* the efficient parallel synthesis and parallel purification of the final compounds.

ACKNOWLEDGMENTS

The syntheses described in these case studies were a collaborative effort of a number of colleagues at Polyphor. We would like to thank, in particular, Philipp Ermert and Jürg Fässler for their excellent contributions. We also thank André Jeanguenat, Salem Farooq, Elke Hillesheim, Thomas Pitterna, and Peter Maienfisch from Syngenta AG, and Kevin E.B. Parkes, Joseph A. Martin, John H. Merrett, and Klaus Klumpp from F. Hoffmann-La Roche (Welwyn, United Kingdom) for the collaboration on the cases described in this chapter, and our scientific advisors, Sir Professor Jack Baldwin (Oxford University) and Professor Andrea Vasella (ETH Zürich), for stimulating discussions and valuable advice.

REFERENCES

1. K.H. Bleicher, H.-J. Böhm, K. Müller, and A. Alanine. Hit and lead generation: beyond high-throughput screening. *Nature Rev. Drug Discov.*, 2003, 2, 369.

2. S.J. Shuttleworth, R.V. Connors, J. Fu, J. Liu, M.E. Lizarzaburu, W. Qiu, R. Sharma, M. Wañska, and A.J. Zhang. Design and synthesis of protein superfamily-targeted chemical libraries for lead identification and optimization. *Curr. Med. Chem.*, 2005, 12, 1239–1281.

3. D. Obrecht and J.-M. Villalgordo. Solid-supported combinatorial and parallel synthesis of synthesis small-molecular-weight compound libraries. J.E. Baldwin, and R.M. Williams, Eds., *Tetrahedron Series*, Vol. 18, Pergamon, 1998.

4. Ch. Hulme and V. Gore. Multi-component reactions: emerging chemistry in drug discovery 'from Xylocain to Crixivan'. *Curr. Med. Chem.*, 2003, 10, 51-80.

5. M. Passerini. Sopra gli Isonitrili (I). Composto del *p*-Isonitrilazobenzolo con Acetone ed Acido Acetico. *Gazz. Chim. Ital.*, 1921, 51, 126; L. Banfi and R. Riva. *The Passerini Reaction*. Organic Reactions, 2005, Vol. 65, L.E. Overman, Ed., John Wiley & Sons.

6. I. Ugi. Neuere Methoden der präparativen organischen Chemie IV Mit Sekundär-Reaktionen gekoppelte α-Additionen von Immonium-Ionen und Anionen an Isonitrile. *Angew. Chem.*, 1962, 74, 9. *Angew. Chem. Int. Ed.* 1962, 1, 8. A. Dömling, I. Ugi. Multikomponentenreaktionen mit Isocyaniden. *Angew. Chem.*, 2000, 112, 3300. *Angew. Chem. Int. Ed.*, 2000, 39, 3168 and refs. cited therein.

7. A. Hantzsch. Die Bildungsweise von Pyrrolderivaten. *Ber. Dtsch. Chem. Ges.*, 1890, 23, 1474.

8. H.T. Bucherer and V.A. Lieb. Über die Bildung substituierter Hydantoine aus Aldehyden und Ketonen. Synthese von Hydantoinen. *J. Prakt. Chem.*, 1934, 141, 5.

9. D. Obrecht and P. Ermert, *5th International Conference on Synthetic Organic Chemistry (ECSOC-5)*, September 1–30, 2001 [B0005]. Applications of sequential three- and four component reactions for parallel synthesis. http://www.mdpi.net/ecsoc/ecsoc-5/Papers/b0005/b0005.htm

10. J. Bridges. Chemical inhibitors of protein kinases. *Chem. Rev.*, 2001, 101, 2541; D. Fabbro and C. Garcia-Echeverria. Targeting protein kinases in cancer therapy. *Curr. Opin. Drug Disc. & Dev.*, 2002, 5, 701; T. Cohen. Protein kinases — the major drug targets for the twenty-first century? *Nature Rev. Drug Discov.*, 2002, 1, 309.

11. Olomucin: N.S. Gray, S. Kwon, and P.G. Schultz. Combinatorial synthesis of 2,9-substituted purines. *Tetrahedron Lett.*, 1997, 38, 1161.

12. Gleevec®: R. Capdeville, E. Buchdunger, J. Zimmermann, and A. Matter. Glivec (STI 571, Imatinib), a rationally developed targeted anti cancer drug. *Nature Rev. Drug Discov.*, 2002, 1, 493; F. de Bree, L.A. Sorbera, R. Fernández, and J. Castañer. Imatinib Mesilate. Treatment of chronic myeloid leukemia, Brc-Abl tyrosine kinase inhibitor. *Drugs of the Future*, 2001, 26, 545; J. Zimmermann, E. Buchdunger, H. Mett, T. Meyer, and N.B. Lydon. Potent and selective inhibitors of the Abl-kinase: phenylaminopyrimidine (PAP) derivatives. *Bioorg. Med. Chem. Lett.*, 1997, 7, 187.

13. D. Obrecht, C. Abrecht, A. Grieder, and J.-M. Villalgordo. A novel and efficient approach for the combinatorial synthesis of structurally divers pyrimidines on solid support. *Helv. Chim. Acta*, 1997, 80, 65; L.M. Gayo and M.J. Suto. Traceless linker: oxidative activation and displacement of a sulfur-based linker. *Tetrahedron Lett.*, 1997, 38, 211.

14. T. Masquelin and D. Obrecht. A new general three component solution-phase synthesis of 2-amino-1,3-thiazole and 2,4-diamino-1,3-thiazole combinatorial libraries. *Tetrahedron*, 2001, 57, 153.

15. M. Altorfer, P. Ermert, J. Fässler, S. Farooq, E. Hillesheim, A. Janguenat, K. Klumpp, P. Maienfisch, J.A. Martin, J.H. Merrett, K.E.B. Parkes, J.-P. Obrecht, T. Pitterna, and D. Obrecht. Applications of parallel synthesis to lead optimization. *Chimia*, 2003, 57, 262.

16. K. Gewald, P. Blauschmidt, and R. Mayer. 4-Amino-thiazole. *J. Prakt. Chem.*, 1967, 35, 97; K.N. Rajasekharan, P.K. Nair, and G.C. Jenardanan. Studies on the synthesis of 5-acyl-2,4-diaminothiazoles from amidinothioureas. *Synthesis*, 1986, 5, 353.

17. Part of this study has been published in C. Abrecht, P. Ermert, P. Jeger, D. Obrecht, A. Jeanguenat, and S. Farooq, *10th IUPAC Int. Congress on the Chemistry of Crop Protection*, Basel 2002, Poster 6 (abstract 3a.06).

18. Synthesized from thiosemicarbazide, *cf.* W.W. Paudler and T.-K. Chen. 1,2,4-Triazines. III. A convenient synthesis of 1,2,4-triazines and their covalent hydration. *J. Heterocycl. Chem.,* 1970, 7, 767.

19. M. Sagi, M. Amano, S. Konno, and H. Yamanaka. Studies on *as*-triazine derivatives. XIII. A facile synthesis of fusaric acid from thienyl-as-triazine derivatives. *Heterocycles*, 1989, 29, 2249.

20. T. Shibutani, H. Fujihara, and N. Furukawa. A new approach for preparation of atropisomers of arylpyridyls via cross-coupling reactions of aryl 2-(3-substituted)pyridyl sulfoxides with Grignard and organolithium reagents. *Tetrahedron Lett.*, 1991, 32, 2943.

21. Details of this study have been published in K.E.B. Parkes, P. Ermert, J. Fässler, J. Ives, J.A. Martin, J.H. Merrett, D. Obrecht, G. Williams, and K. Klumpp, *J. Med. Chem.*, 2003, 46, 1153.

22. T.M. Humpey, C.F. Basler, P.V. Aguilar, H. Zeng, A. Solórzano, D.E. Swayne, N.J. Cox, J.M. Katz, J.K. Taubenberger, P. Palese, and A. Garcia-Sastre. Characterization of the reconstructed 1918 Spanish influenza pandemic virus. *Science,* 2005, 310, 77.

23. For the structure and mechanism of alkaline phosphatase from *E.coli,* see E.E. Kim and H.W. Wyckoff. Reaction mechanism of alkaline phosphatase based on crystal structures: two-metal ion catalysis. *J. Mol. Biol.,* 1991, 218, 449.

24. J. Tomassini, H. Selnick, M.E. Davies, M.E. Armstrong, J. Baldwin, M. Bourgeois, J. Hastings, D. Hazuda, J. Lewis, and W. McClements. Inhibition of cap (m7GpppXm)-dependent endonuclease of influenza virus by 4-substituted 2,4-dioxobutanoic acid compounds. *Antimicrob. Agents Chemother.,* 1994, 38, 2827; and J.C. Hastings, H. Selnick, B. Wolanski, and J.E. Tomassini. Anti-influenza virus activities of 4-substituted 2,4-dioxobutanoic acid inhibitors. *Antimicrob. Agents Chemother.,* 1996, 40, 1304.

25. S. Kende and J. Thurston. Synthesis of 1-hydroxyoxindoles. *Synth. Commun.,* 1990, 20, 2133.

4 A Successful Application of Parallel Synthesis to Computer-Assisted Structural Optimization of New Leads Targeting Human Immunodeficiency Virus-1 Reverse Transcriptase

The Case of Acylthiocarbamates and Thiocarbamates

Angelo Ranise, Andrea Spallarossa, and Sara Cesarini

CONTENTS

4.1 INTRODUCTION

4.1.1 HIV-1 AND AIDS

Human immunodeficency virus (HIV), a member of the lentivirus group of retroviruses, is the causative agent of acquired immune deficiency syndrome (AIDS).[1,2] The disease results in an alteration of immune functions and exposes the infected individuals to a wide range of opportunistic infections.

4.1.2 KEY EVENTS IN THE HIV-1 LIFE CYCLE

The HIV-1 life cycle[3,4] starts with the interaction between the viral gp120 glycoprotein and the immune system cell's CD4 receptor. The viral RNA is then released into the cell and converted in a double-stranded DNA (proviral DNA) by reverse transcriptase (RT), an RNA/DNA-dependent DNA polymerase (see Section 4.2). Integrase (IN) inserts the proviral DNA in the host-cell genome. The integrated DNA is designated as provirus. HIV-1 gene expression and transcription is another essential step, tuned by various factors such as nuclear factor NF-κB, and Tat and Rev proteins.[5] The cellular transcription nuclear factor NF-κB binds to the long terminal repeat (LTR) promotor and strongly induces HIV-1 gene expression; the viral Tat protein seems to play a pivotal role in sustaining a high level of HIV-1 replication by transactivation of the LRT promotor; and the viral regulatory protein Rev conducts the nuclear export control of viral mRNA. The proviral DNA transcription leads to a productive infection generating viral mRNAs that are then translated into viral proteins. The viral RNA and viral proteins assemble at the cell membrane into a new virus. Protease (PR) processes HIV proteins into their functional forms. Following assembly at the cell surface, the virus then buds forth from the cell and is released to infect another cell.

The fast HIV-1 turnover, combined with a high mutation rate, gives rise to a large pool of viral variants (referred to as a "quasispecies").[6] Even if many of these variants are relatively capable of carrying out normal HIV functions, in the presence of drugs that suppress the replication of wild-type HIV-1, drug-resistant variants are selected and eventually these mutants dominate the viral population.[7,8]

4.1.3 Biological Targets of HIV-1

Knowledge of the HIV-1 replication cycle has allowed the identification of different pharmacological targets to be hit in anti-AIDS therapy. The processes catalyzed by viral-encoded RT, IN, and PR (see Section 4.1.2) are essential for viral replication and hence these enzymes are validated targets for the design of inhibitors.[9] At present, several RT and PR inhibitors are marketed but no IN-targeting agent has been approved yet.[10] In addition to these inhibitors, various agents can be considered for chemotherapeutic intervention toward targets involved in other key events (1 to 5) of the HIV replicative cycle:[11]

1. *Viral entry into the cell*: CD4-targeting agents (cellular CD4 receptor down-modulators), CXCR4 and CCR5 antagonists (in particular, the inhibition of the recognition process, that is, the gp120/CD4 interaction, was intensively studied to prevent the entry of HIV into the host cell and therefore to develop vaccines).[12]
2. *Viral adsorption:* agents binding to the viral envelope glycoprotein gp120.
3. *Virus-cell fusion*: agents binding to the viral envelope glycoprotein gp41.
4. *Viral assembly and disassembly*: NCp7 zinc finger-targeting agents.
5. *Proviral DNA expression*: transcription (transactivation) inhibitors.

As far as the two last points are concerned, nucleocapside protein NCp7, Tat and Rev proteins, and NF-κB are particularly attractive targets for the development of new anti-HIV-1 agents.[5,13–16] At present, despite the identification of various HIV-1 targets useful for AIDS therapy, the most important results have been achieved with RT and PR inhibitors.

4.1.4 Highly Active Antiretroviral Therapy (HAART)

The highly active antiretroviral therapy (HAART) is a multi-drug approach that consists of a triple drug combination of RT and PR inhibitors.[17] HAART has dramatically changed the prognosis of patients infected with HIV (decrease in the incidence of deaths, opportunistic infections, and of the need for hospitalization).[18–20] Even if both life span and quality of the infected people are increased, HAART does not succeed in defeating HIV infection (even in patients with undetectable plasma HIV RNA loads, low-grade replication still occurs).[21] Thereby, HIV infection is now envisioned as a chronic disease necessitating continuous, possibly life-long treatment. Other severe drawbacks include drug cross-resistance development due to RT and PR mutation, drug compatibilities, and adverse side effects (bone marrow depression, anemia, peripheral neuropathy). Therefore, the need for more selective and potent drugs able to inhibit HIV mutated forms is still a challenge.

4.2 REVERSE TRANSCRIPTASE (RT) AND RT TARGETING AGENTS

4.2.1 Structure and Functional Role of HIV-1 RT

Functional HIV-1 RT is a heterodimer containing two subunits of 66 kDa (p66) and 51 kDa (p51) (Figure 4.1). The p66 subunit is formed by an N-terminal polymerase domain (further subdivided into fingers, palm, thumb, and connection subdomains) and a C-terminal RNAse H domain (120 residues). The p66 N-terminal portion hosts the catalytic aspartate triad (namely Asp110, 185, 186) primarily involved in the formation of the phosphamidic bond between the triphosphorylated nucleotides and the growing DNA filament.[22,23] The p51 subunit is processed by proteolytic cleavage of p66 and corresponds to the polymerase domain of the p66 subunit.

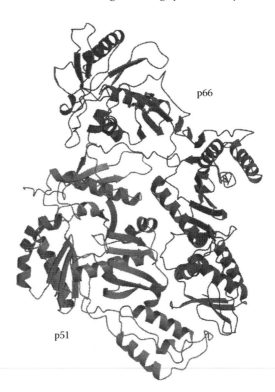

FIGURE 4.1 Overall view of HIV-1 RT: p51 and p66 subunits are shown in dark and light grey, respectively. The polymerase active site residues (Asp110, 185, 186) are represented in dark orange balls and sticks.

RT is a multifunctional enzyme that catalyzes the conversion of the viral RNA into a double-stranded DNA (dsDNA) through three subsequent steps: (1) polymerization of a single-stranded DNA, copy of the viral RNA (RNA-dependent polymerase activity); (2) synthesis of a dsDNA starting from the single-stranded filament (DNA-dependent polymerase activity); and (3) degradation of the original RNA (RNAse H activity).

4.2.2 RT Targeting Agents: HIV-1 Nucleoside and Non-Nucleoside Reverse Transcriptase Inhibitors

RT targeting agents are classified into two classes: nucleoside and non-nucleoside inhibitors (NRTIs and NNRTIs, respectively).

4.2.2.1 Nucleoside Reverse Transcriptase Inhibitors (NRTIs)

NRTIs are nucleoside analogs that compete with the biological substrate in binding to the RT polymerase site. They display a common chemical scaffold with a nitrogen base-like portion linked to a sugar moiety lacking the 3′-hydroxyl group. After intracellular phosphorylation, NRTIs bind to the RT polymerase site. When incorporated in the growing template/primer, the lack of a 3′-OH prevents further elongation of the primer with other desoxyribonucleoside triphosphates, resulting in chain termination.

4.2.2.2 Non-Nucleoside Reverse Transcriptase Inhibitors (NNRTIs)

NNRTIs are lipophilic, highly selective, allosteric, noncompetitive HIV-1 RT modulators that do not need to be metabolized by cellular enzymes. They specifically interact with an allosteric non-substrate binding site (NNIBS), consisting of a hydrophobic pocket largely contained within the RT p66 subunit, some 10 Å from the polymerase active site (Figure 4.2).[24] The NNIBS lies between two

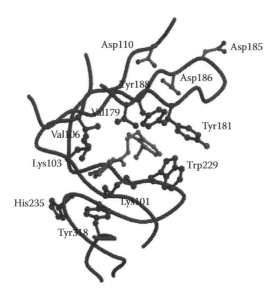

FIGURE 4.2   The NNRTI binding site with bound PETT-1 (Chart 4.1). Amino acids lining the NNRTI binding pocket Asp110, 185, 186 are shown in balls and sticks and PETT-1 is shown in light grey.

β sheets, one (β9-β10) containing the aspartates (*vide supra*) and the other (β12-β13-β14) containing the primer grip. This cavity is generated upon binding of NNRTIs through torsional rotations of the Tyr181 and Tyr188 sidechains and repositioning of the second β sheet, containing Phe227 and Trep229.[25] Therefore, RT-NNRTI interaction locks the enzyme in an inactive form by causing conformational changes of the polymerase active site (short-range distortion) and of the p66 thumb domain (long-range distortion). According to their activity against RT mutated forms (see Section 4.2.4), nowadays NNRTIs are divided into first- and second-generation inhibitors. The former are particularly sensitive to the development of drug resistance, while the latter result in minor losses of activity against variants carrying either single or double mutations (Chart 4.1).

4.2.3   THE BIOACTIVE CONFORMATION OF FIRST- AND SECOND-GENERATION NNRTIS

Most first-generation NNRTIs, despite their chemical diversity,[11,26–44] assume a common butterfly-like conformation inside the NNIBS (Figure 4.2).[45, 46]

However, some recent crystal structures of HIV-1 RT in complexes with second-generation NNRTIs (e.g., imidoylthiourea (ITU), diaryltriazine (DATA), and diarylpyrimidine (DAPY) compounds Chart 4.7) revealed alternative and multiple binding modes. The conformational flexibility and positional adaptability[38,47] of ITU derivative R100943, DAPY analogs Dapivirine (TMC-20-R147681) and R185545, and the majority of DATA molecules lead to a "U" or "horseshoe" bioactive conformation.[47] In addition, the crystal structure of the HIV-1 RT/R120393 complex showed that this DATA analog binds unexpectedly with a more extended conformation (i.e., a "sea-horse" conformation).

4.2.4   CLINICALLY RELEVANT NNRTI-RESISTANCE MUTATIONS

The therapeutic advantage of NNRTIs is strictly limited by the emergence of resistant HIV mutants. The mutation of one or more residues lining the NNIBS and strongly interacting with the inhibitors (e.g., Leu100, Val106, Tyr101) causes a dramatic decrease in the affinity of the inhibitor for RT. Inhibitor/RT affinity is also compromised by mutations involving solvent-exposed residues (e.g., Lys103). These amino acids weakly interact with the NNRTIs but their modification induces significant conformational rearrangements within the NNIBS.[48] Commonly observed

CHART 4.1

Selected Structures of (a) First- and (b) Second-Generation NNRTIs

(a)

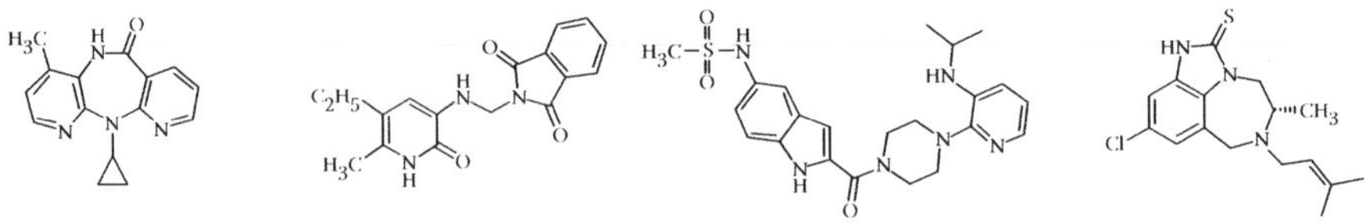

Nevirapine

L-345, 516
(2-pyridinone derivative)

Delavirdine

9-Cl-TIBO

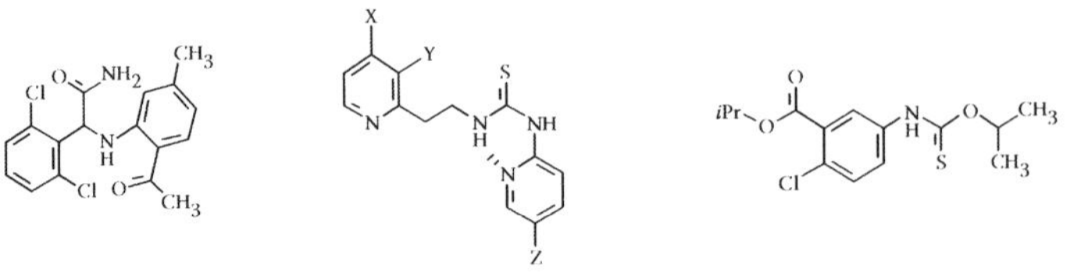

α-APA derivative

Trovirdine (X,Y = H; Z = Br)
PETT-1 (X = OC$_2$H$_5$; Y = F; Z = CN)

UC-38

(b)

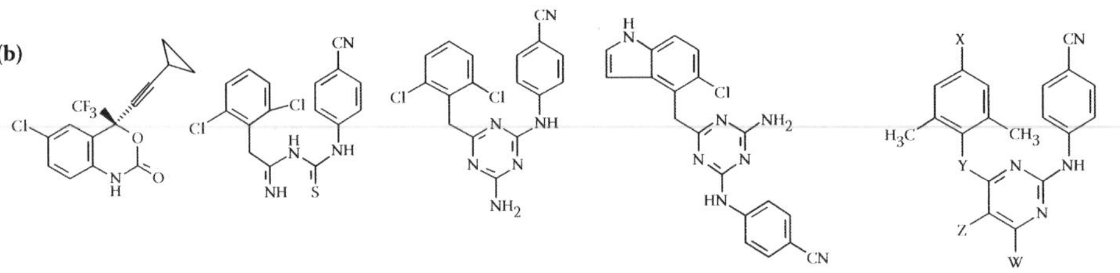

Efavirenz

R100943
(ITU)

R106168
(DATA)

R120393
(DATA)

DAPY
Dapivirine (X = CH$_3$;
Y = NH; Z,W = H)
R185545 (X = CN;
Y = O; Z = Br; W = CH$_2$OH)

single NNRTI-resistance mutations include Leu100Ile, Lys103Asn, Tyr181Cys, Tyr188Leu, and Gly190Ala.[25,49,50]

4.3 THE ROLE OF PARALLEL SYNTHESIS AND RATIONAL DRUG DESIGN IN DRUG DISCOVERY

Parallel synthesis methodology and related multiple synthesis technologies for the preparation of small-molecular-weight drug-like compound libraries have been widely used in the lead generation and lead optimization stages. In comparison with solid-phase synthesis (SPS), parallel solution-phase synthesis has some advantages in that all products (and their precursors when isolable) are generated separately and in sufficient amounts for full characterization and biological screening without the need for laborious, time-consuming deconvolution procedures.

During the 1970s, a more detailed appreciation of enzyme mechanism heralded the advent of rational drug design.[51–53] The structural information derived from crystallographic and NMR studies of biological targets, combined with modeling and computational methods, has further increased the sophistication of the rational design approach. These techniques allow one to filter out redundant compounds in chemical libraries[54,55] and therefore to reduce the number of compounds needed to be made to find and optimize a lead compound.

For these reasons, rational drug design and parallel synthesis have speeded up the process of lead generation and optimization.

4.3.1   IMPORTANCE OF DEVELOPING PARALLEL, CONVERGENT ONE-POT SYNTHETIC STRATEGY FOR LEAD FINDING AND OPTIMIZATION

One of the tasks of our research group is to develop parallel, highly convergent, one-pot, solution-phase procedures, coupled with rapid purification methods to prepare various types of libraries. Our interest in convergent syntheses is due to their advantages in comparison with linear syntheses. In general, a convergent synthesis is accomplished with a lesser number of steps, gives higher overall yields, and is technically simpler and faster than a linear synthesis. Convergent synthesis is especially well suited for solution-phase techniques and would be precluded by SPS, where the combining components are on mutually exclusive solid phases. In general, solution-phase synthesis is more versatile than SPS, in that the former can be adapted for either a linear or convergent synthetic strategy, whereas the latter is necessarily limited to a linear synthetic approach.[56] In solution phase, it is necessary to minimize the number of synthetic steps in which the number of compounds is not multiplied.

One-pot reactions, being characterized by reaction sequences in a same reactor, can be efficiently used in parallel syntheses, because products are prepared with a minimum number of work-up procedures and vessel transfers. In general, this goal is fulfilled by *in situ* generation of versatile, transient, or isolable intermediates ready to undergo the next transformation. Some of the most common examples of one-pot reactions include (1) reductive amination of aldehydes or ketones, followed by acylation of the newly formed secondary amine, (2) double amination of cyclic anhydrides, (3) successive acylation of diamines, etc.[57] When multicomponent reactions (MCRs) are involved, this kind of process allows the introduction of a high degree of molecular diversity in just a single step.[58–63]

4.3.2   TOWARD NEW CLASSES OF THIOCARBAMATE-BASED NNRTIS

Biological activities of thiocarbamate derivatives are fairly wide.[64] Moreover, secondary acylthiocarbamates (the names *primary*, *secondary*, and *tertiary thiocarbamates* refer to the degree of substitution at the thiocarbamic nitrogen atom)[64] can act as elastase inhibitors[65] and exhibit antineoplasic, antiinflammatory and antiarthritic effects.[66] These considerations, along with our pharmacological interest in compound isosteres of the previously described *N,N*-disubstituted-*N'*-acyl-*N'*-aryl-thioureas,[67–69] prompted us to investigate the anti-HIV-1 activity of tertiary acylthiocarbamates (ATCs) and their thiocarbamate precursors. Herein we report the parallel, highly convergent, one-pot solution-phase synthesis of libraries of ATCs and thiocarbamates (TCs) (Chart 4.2), structurally related to *N*-phenethyl-*N'*-thiazolylthiourea (PETT) derivatives (Chart 4.1), a potent class of NNRTIs.[70–72]

Molecular docking simulations were performed on the basis of RT/PETT-1 x-ray structure (PDB code 1DTQ)[73] assisting the ATC and TC SAR strategy. We chose to synthesize libraries of discrete compounds in a spatially separate fashion rather than libraries of compound mixtures to allow a rigorous analytical characterization of the library compounds. The ATC and TC syntheses were carried out by parallel synthesizers such as Carousel 12 Reaction Stations™.

It should be noted that in recent years, Combinatorial Chemistry has firmly contributed to the generation of libraries of anti-HIV-1 agents (an entire issue of *Combinatorial Chemistry & High*

CHART 4.2

General Structure of the Thiocarbamate-Based Libraries (R, R', X: diversity points)

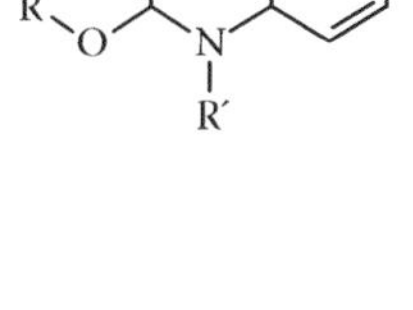

R' = H or Ar-CO or CO-OR''

Throughput Screening was devoted to the publication of timely reviews on the combinatorial synthesis of anti-HIV drugs and their computer-aided design).[74]

4.4 ACYLTHIOCARBAMATES (ATCs)

4.4.1 ATC LIBRARY DESIGN

To the best of our knowledge, ATCs represent the first example of tertiary TCs in which one of the substituents on the nitrogen atom is an acyl group. The ATC highly convergent synthetic process[75] was carried out through a three-step sequence by covalently combining three different building blocks: (1) alcohols, (2) isothiocyanates, and (3) acyl chlorides (for further details, *vide infra*). In the first phase of our studies, we designed, prepared, and tested a number of ATCs[76] sharing with PETT derivatives two aromatic rings joined by a five-atom linker embodying the thiocarbamate backbone instead of the thiourea moiety. The *N*-acyl motif was introduced in the TC molecule with the aim to make additional interactions with the amino acid residues surrounding the NNIBS, as suggested by docking simulations (*vide infra*). The primary screen for evaluating the ATC derivatives was potency in inhibiting replication of wild-type HIV-1. Also, a Selectivity Index (S.I.), expressed as a ratio of toxic concentration (CC_{50}) to inhibitory concentration (EC_{50}) was determined. Within the exploratory ATC series, O-[2-(phenoxy)]ethyl(benzoyl)phenylthiocarbamate **17c** (Chart 4.3) proved to be a selective micromolar anti-HIV-1 agent (CC_{50} = 122 μM, EC_{50} = 8 μM, S.I. = 15.2) in MT-4 cell-based assay. On this basis, we prepared two analog libraries, assuming **17c** as an initial lead. Considering the building blocks used for the library preparation, the chemical structure of **17c** can be retrosynthetically divided in the three molecular portions 1, 2, and 3 (Chart 4.3). Owing to the highly modular character of the synthesis, single or multiple modifications of these portions could be carried out at the same time.

The **17c** structure was modified as follows. (1) Alteration of the *N*-acyl moiety (analogs **17a–d, g, j, k, m–o, q, r**) and the *N*-phenyl ring (saturation, introduction of a fluorine at the 4-position, analogs **16c** and **18c**, respectively). (2) With the aim to enhance hydrophobic contacts with NNIBS amino acid residues, a methyl group was introduced on the carbon adjacent to the thiocarbamic oxygen (ATCs **19a, c, h, p–r**; in the case of **20r, 21r**, and **22r**, electron-withdrawing groups were also introduced on the *N*-phenyl ring at positions *meta-* (Cl) and *para-* (Cl, NO_2)). (3) To vary the conformational freedom of the lead molecule, it was thought either to diminish progressively the number of rotatable bonds by reducing the length of the oxyethyl spacer (**13q, 14c**, and **15**); or (4) to change the 2-phenoxyethyl moiety with the cage-like adamantane nucleus (**12c,q**), being that this framework is employed in the synthesis of potential HIV-1 agents.[77–79] (5) To assess the involvement of the 2-phenyl ring in hydrophobic interactions with RT, we planned compounds **11c,q** devoid of this moiety. (6) As concerns isosteric replacements, the ether oxygen was replaced by sulfur (**23c**) and the aroyl moiety with alkoxy- or phenoxy-carbonyl groups (**24u,v,w**). To gain a better understanding of the RT/**17c** interactions, we derived a hypothetical model for docking **17c** in the

---

CHART 4.3

Structure of Lead Compound 17c Segmented in Three Molecular Portions

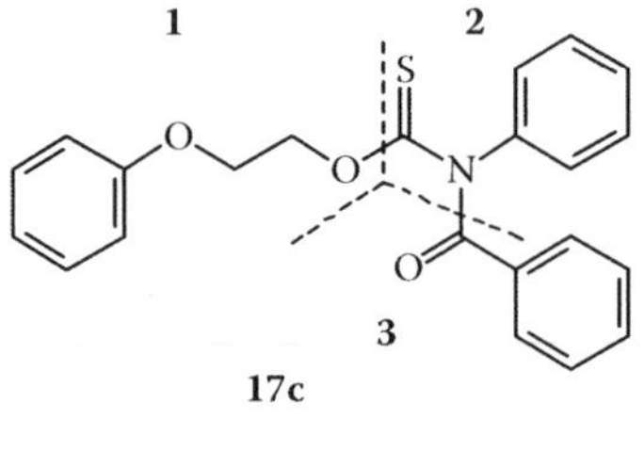

17c

NNIBS (Figure 4.3A). In the RT/**17c** complex, in addition to the hydrophobic contacts established between the phenyl ring of portion 3 and Lys101, Glu138, and Ile135 (from the p51 subunit), there is a hydrogen bond involving the thiocarbonyl sulfur with the side chain of Lys103. An additional polar interaction would occur between the ether oxygen and the Glu138 (from the p51 subunit) side chain, resembling an H-bond. This suggested that we should select 2-phthalimidoethanol **9** (Chart 4.4) as an alcohol building block incorporating two proper hydrogen-bond acceptor groups adjacent to the phenyl ring. On the other hand, the phthalimide moiety is present in other NNRTIs, such as the 2-pyridinone compound L-345,516[80] (Chart 4.1) and some PETT derivatives.[72, 81] Thus, we planned the synthesis of lead analog **25c** also because computational simulations indicated that the RT/**25c** complex (Figure 4.3B) is stabilized by two hydrogen bonds: the former involving the imidic carbonyl oxygen and the Lys103 ε-amino group, the latter between the thione sulfur and the Lys101 backbone amide. Interestingly, **25c** revealed a reverse binding mode as compared with **17c**, despite the fact that they belong to the same chemical class. As this compound actually turned out to be more active ($EC_{50} = 0.4$ μM; $CC_{50} > 200$ μM) than **17c**, it was selected as a new lead for a computer-assisted optimization strategy focused on the variation of the *N*-phenyl ring substituent and/or the modification of the acyl moiety (**26r–46t**). In ATCs **26r–38r**, **40r**, **41r**, **43r**, **44r**, and **45r**,

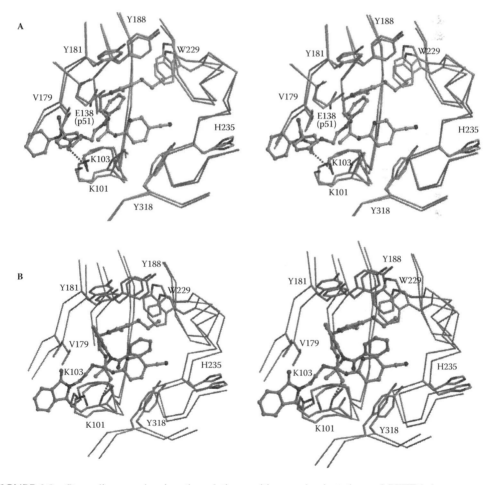

FIGURE 4.3 Stereodiagram showing the relative positions and orientations of PETT-1 (x-ray structure model), **17c (A),** and **25c (B)** (model structures) in the NNIBS of HIV-1 RT. The H-bonds are indicated by dotted lines.

the 2-thenoyl was kept constant to better evaluate the effects determined by the exclusive modification of the *N*-phenyl ring.

4.4.2 Parallel Synthesis of Focused ATC Libraries

ATCs **11c–23c and 25c–46t** were resynthesized by parallelizing the general procedure previously described (Scheme 4.1).[75,76,82]

Owing to their well-established low reactivity toward isothiocyanates,[64] starting alcohols A_{1-9} (Chart 4.4a) were transformed into their corresponding salts (A^-_{1-9}) in the presence of 60% sodium hydride dispersion in mineral oil in anhydrous aprotic solvents (toluene, THF, DMF, pyridine).

CHART 4.4

Building Blocks Used for Preparation of ATC Libraries 1 and 2

(a) Alcohls A_{1-9}

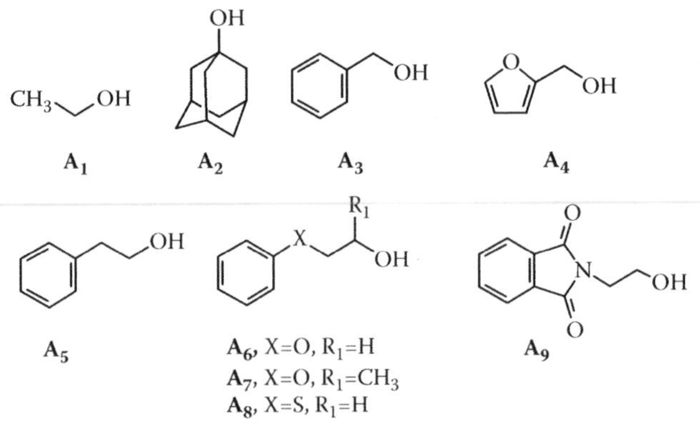

(b) Isothiocyanates I_{1-23}

	Ar-NCS
I_1	cyclohexylisothiocyanate
I_2	phenylisothiocyanate
I_3	*o*-tolylisothiocyanate
I_4	*m*-tolylisothiocyanate
I_5	*p*-tolylisothiocyanate
I_6	*o*-ethylphenylisothiocyanate
I_7	*o*-ethylphenylisothiocyanate
I_8	*m*-trifluoromethylphenylisothiocyanate
I_9	*m*-acetylphenylisothiocyanate
I_{10}	*m*-fluorophenylisothiocyanate
I_{11}	*p*-fluorophenylisothiocyanate
I_{12}	*o*-chlorophenylisothiocyanate
I_{13}	*m*-chlorophenylisothiocyanate
I_{14}	*p*-chlorophenylisothiocyanate
I_{15}	*m*-bromophenylisothiocyanate
I_{16}	*p*-bromophenylisothiocyanate
I_{17}	*p*-iodophenylisothiocyanate
I_{18}	*p*-diethylaminophenylisothiocyanate
I_{19}	*m*-nitrophenylisothiocyanate
I_{20}	*p*-nitrophenylisothiocyanate
I_{21}	*o*-methoxyphenylisothiocyanate
I_{22}	*m*-methoxyphenylisothiocyanate
I_{23}	*p*-ethoxyphenylisothiocyanate

(c) Acyl chlorides AC_{a-t}

	Ar_1COCl
AC_a	phenoxyacetyl chloride
AC_b	*trans*-cinnamoyl chloride
AC_c	benzoyl chloride
AC_d	4-toluoyl chloride
AC_e	4-phenylbenzoyl chloride
AC_f	4-fluorobenzoyl chloride
AC_g	4-chlorobenzoyl chloride
AC_h	3-nitrobenzoyl chloride
AC_i	4-anisoyl chloride
AC_j	2-acetoxybenzoyl chloride
AC_k	2,4-dichlorobenzoyl chloride
AC_l	3,4-dichlorobenzoyl chloride
AC_m	3,5-dichlorobenzoyl chloride
AC_n	4-chloro-3-nitrobenzoyl chloride
AC_o	3,4,5-trimethoxybenzoyl chloride
AC_p	1-naphthoyl chloride
AC_q	2-furoyl chloride
AC_r	2-thenoyl chloride
AC_s	2-chloronicotinoyl chloride
AC_t	6-chloronicotinoyl chloride

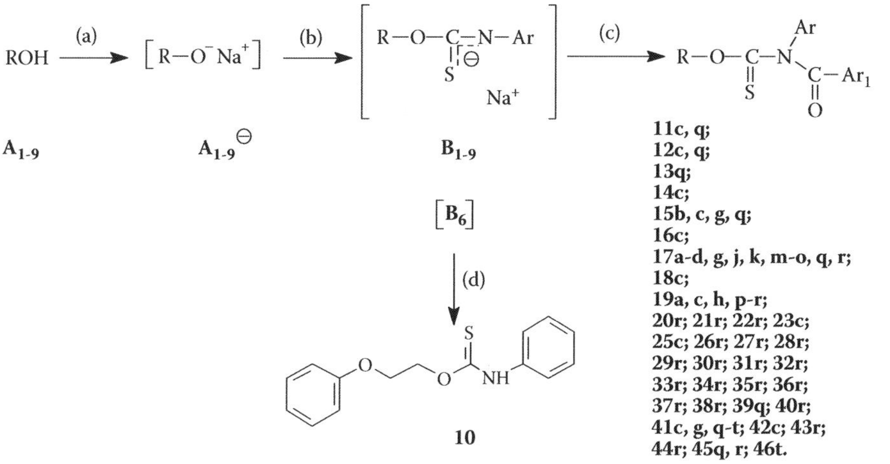

SCHEME 4.1[a] Parallel, convergent, one-pot, solution-phase synthesis of ATCs **11c–23c** and **25c–46t** (Library 1 and 2).

[a] Reaction conditions: **Procedure P$_1$**: (a) NaH, anhydrous THF (**A$_{1,5-8}$**) or dry toluene (**A$_3$, A$_4$**), r.t.; (b) Isothiocyanates, 15 min.; (c) anhydrous pyridine, acyl chlorides, 6 hr. **Procedure P$_2$**: (a) **A$_9$**, Isothiocyanates, anhydrous pyridine, 0–5°C; (b) NaH, 4 hr; (c) TMEDA, acyl chlorides, r.t., 6 hr. **Procedure P$_3$** (only for **12c, q**): (a) **A$_2$**, NaH, dry DMF, 90–95°C; (b) isothiocyanates, 90–95°C, 30 min; and (c) acyl chlorides, anhydrous pyridine, r.t., 6 hr, then 55°C, 1 hr.

Library 1 (Procedure P$_1$)[a]		Library 2 (Procedure P$_2$)[a]	
11c, (A$_1$, I$_2$, AC$_c$)	**17m**, (A$_6$, I$_2$, AC$_m$)	**25c**, (A$_9$, I$_2$, AC$_c$)	**39q**, (A$_9$, I$_7$, AC$_q$)
11q, (A$_1$, I$_2$, AC$_q$)	**17n**, (A$_6$, I$_2$, AC$_n$)	**26r**, (A$_9$, I$_3$, AC$_r$)	**40r**, (A$_9$, I$_{11}$, AC$_r$)
13q, (A$_3$, I$_2$, AC$_q$)	**17o**, (A$_6$, I$_2$, AC$_o$)	**27r**, (A$_9$, I$_6$, AC$_r$)	**41c**, (A$_9$, I$_{14}$, AC$_c$)
14c, (A$_4$, I$_2$, AC$_c$)	**17q**, (A$_6$, I$_2$, AC$_q$)	**28r**, (A$_9$, I$_{12}$, AC$_r$)	**41g**, (A$_9$, I$_{14}$, AC$_g$)
15b, (A$_5$, I$_2$, AC$_b$)	**17r**, (A$_6$, I$_2$, AC$_r$)	**29r**, (A$_9$, I$_{21}$, AC$_r$)	**41q**, (A$_9$, I$_{14}$, AC$_q$)
15c, (A$_5$, I$_2$, AC$_c$)	**18c**, (A$_6$, I$_{11}$, AC$_c$)	**30r**, (A$_9$, I$_4$, AC$_r$)	**41r**, (A$_9$, I$_{14}$, AC$_r$)
15g, (A$_5$, I$_2$, AC$_g$)	**19a**, (A$_7$, I$_2$, AC$_a$)	**31r**, (A$_9$, I$_8$, AC$_r$)	**41s**, (A$_9$, I$_{14}$, AC$_s$)
15q, (A$_5$, I$_2$, AC$_q$)	**19c**, (A$_7$, I$_2$, AC$_c$)	**32r**, (A$_9$, I$_9$, AC$_r$)	**41t**, (A$_9$, I$_{14}$, AC$_t$)
16c, (A$_6$, I$_1$, AC$_c$)	**19h**, (A$_7$, I$_2$, AC$_h$)	**33r**, (A$_9$, I$_{10}$, AC$_r$)	**42c**, (A$_9$, I$_{16}$, AC$_c$)
17a, (A$_6$, I$_2$, AC$_a$)	**19p**, (A$_7$, I$_2$, AC$_p$)	**34r**, (A$_9$, I$_{13}$, AC$_r$)	**43r**, (A$_9$, I$_{17}$, AC$_r$)
17b, (A$_6$, I$_2$, AC$_b$)	**19q**, (A$_7$, I$_2$, AC$_q$)	**35r**, (A$_9$, I$_{15}$, AC$_r$)	**44r**, (A$_9$, I$_{18}$, AC$_r$)
17c, (A$_6$, I$_2$, AC$_c$)	**19r**, (A$_7$, I$_2$, AC$_r$)	**36r**, (A$_9$, I$_{19}$, AC$_r$)	**45q**, (A$_9$, I$_{20}$, AC$_q$)
17d, (A$_6$, I$_2$, AC$_d$)	**20r**, (A$_7$, I$_{13}$, AC$_r$)	**37r**, (A$_9$, I$_{22}$, AC$_r$)	**45r**, (A$_9$, I$_{20}$, AC$_r$)
17g, (A$_6$, I$_2$, AC$_g$)	**21r**, (A$_7$, I$_{14}$, AC$_r$)	**38r**, (A$_9$, I$_5$, AC$_r$)	**46t**, (A$_9$, I$_{23}$, AC$_t$)
17j, (A$_6$, I$_2$, AC$_j$)	**22r**, (A$_7$, I$_{20}$, AC$_r$)		
17k, (A$_6$, I$_2$, AC$_k$)	**23c**, (A$_8$, I$_2$, AC$_c$)		

[a] In parentheses, *A*, *I*, and *AC* represent the alcohol, isothiocyanate, and acyl chloride building blocks, respectively, Chart 4.3 used in sequence for the ATC analog synthesis.

Then, alcoholates **A$^-_{1-9}$** were condensed *in situ* with isothiocyanates (**I$_{1-23}$**) at different temperatures to give saline adducts **B$_{1-9}$**, which were subsequently coupled with acyl chlorides (**AC$_{a-s}$**) to afford the desired products (for the isothiocyanate and acyl chloride building blocks employed, see Chart 4.4b, c). To obviate the dissimilar reactivity of **A$^-_{1-9}$** toward isothiocyanates and/or **B$_{1-9}$** versus acylating reagents, different reaction conditions were adopted (procedures P$_{1-3}$; see the experimental section).

Library 1 (ATCs **11c, q, 13q, 14c, 15b, c, g, q, 16c, 17a–d, g, j, k, m–o, q, r, 18c, 19a, c, h, p–r, 20r, 21r, 22r, 23c**) was prepared according to parallel procedure P$_1$.

In parallel procedure P_2 (**9**, dry pyridine, 0 to 5°C - r.t.), to facilitate the acylation step and increase the overall yield and purity of the ATCs (Library 2: **25c, 26r–38r, 39q, 40r, 41c, g, q–t, 42c, 43r, 44r, 45q, r, 46t**), each reaction mixture was treated with TMEDA (N,N,N',N'-tetramethylethylenediamine) before adding the proper acyl chloride. In the preparation of *ortho*-derivatives **26r–29r**, the reaction mixture was heated to 55–60°C for 3 hr before the work-up. Compounds **12c, q** were obtained by procedure P_3 (**2**, dry DMF:pyridine). The reaction mixture needed to be heated to 90–95°C, owing to the poor reactivity of 1-adamantanol **2** and the corresponding alcoholate.

In procedure P_4 (synthetic variant), $\mathbf{B_7}$ was reacted with proyl-, butyl-, and phenyl-chloroformates in dry THF in the absence of pyridine to give N-alkoxycarbonyl-N-phenyl thiocarbamates **24u,v,w** (Scheme 4.2).

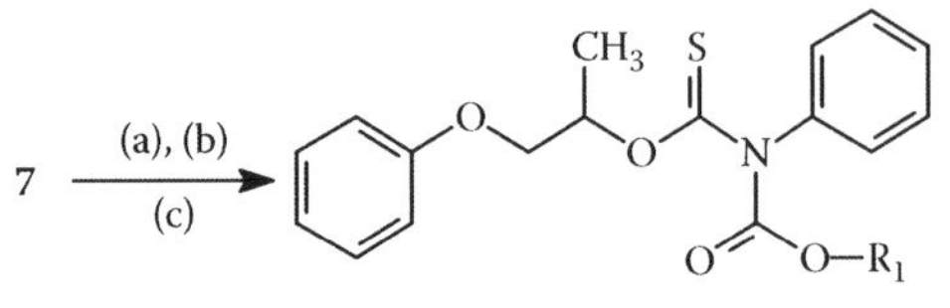

24u, v, w

SCHEME 4.2[a]   Parallel one-pot solution-phase synthesis of O-substituted N-(propoxy- or butoxy- or phenoxy-carbonyl)-N-phenylthiocarbamates **24u, v, w**.

[a]   Reaction conditions: (a) NaH, dry THF, room temperature; (b) C_6H_5-N=C=S; (c) R_1-O-CO-Cl, heat.

4.4.2.1   General Considerations on the Parallel ATC Synthesis

The entire procedure deserves comment. The highly convergent one-pot methodology is featured by high atom economy (only one molecule of HCl is formally lost in the whole process), and all the chemical features of the building blocks selected for the SAR strategy are incorporated into the products. This makes it particularly useful for lead optimization enabling rapid SAR profiling through the generation of analog libraries. Furthermore, no intermediate needs to be isolated and, in general, the overall yields are good to high. Owing to the variety of the chemically accessible building blocks (the ones employed are commercially available or easy to prepare), maximum molecular diversity can be obtained in the smallest number of steps and under the mildest conditions possible. For these last reasons, the above procedures could be easy adaptable also to combinatorial synthesis of primary screening ATC libraries for hit/lead discovery. A great variety of (hetero)aromatic acyl chlorides could be employed, whereas surprisingly the aliphatic ones were, in general, scarcely reactive (only 2-phenoxyacetyl chloride could successfully be used). Probably, the excessive basicity of adducts **B** and the presence of pyridine or TMEDA favored the elimination of hydrogen chloride from α-hydrogen-containing acyl chlorides to afford ketenes[83] rather than the nucleophilic displacement of chloride anion. Furthermore, the absence of reactivity of **B** toward *in situ* generated ketenes (well-known acylating agents) could be explained according to the HSAB principle.[84,85] From this point of view, adducts **B** (owing to the presence of the sulfur atom) would behave as soft bases unable to react with ketenes, due to the hardness of the latter reagents. The ATC synthesis correlates with the one-pot method developed for the preparation of N,N-disubstituted-N'-acyl-N'-phenylthioureas starting from weakly basic amines.[67,68]

Mechanistically and analogously to the acylated thioureas,[86–94] the acylation step would involve the attack of the acyl chloride on the sulfur atom of thiocarbamate sodium salts (**B**) under kinetic control (Scheme 4.3). Then, the transient S-acylisothiocarbamate regioisomers would undergo a 1,3-sulfur-nitrogen acylic four-center Chapman-like transposition[95] under basic catalysis (pyridine, TMEDA) to give the thermodynamically more stable N-acylthiocarbamate regioisomers.

In parallel procedure P_2, the addition of TMEDA improves the purity and the yields of the final products. The better results obtained with TMEDA compared to those obtained with other tertiary bases used, such as Et_3N, DMAP, DABCO, and the TMEDA lower homolog N,N,N',N'-

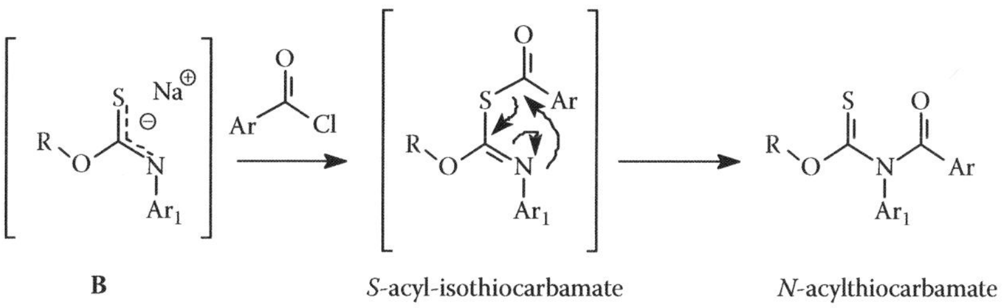

SCHEME 4.3   Proposed mechanism for the acylation step.

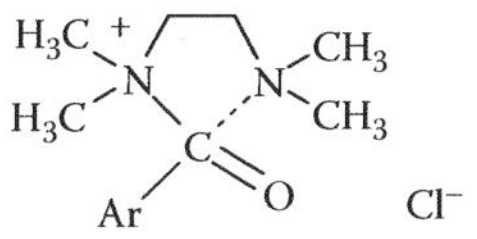

FIGURE 4.4   Hypothetical cyclic complexes between TMEDA and (hetero)aroyl chlorides.

tetramethylmethylenediamine, (unpublished results) are probably due to the formation of a hypothetical cyclic complex[96] (Figure 4.4) with better acylating properties.

As for N-acylated thioureas,[97] to unambiguously assign the position of the acyl group and to confirm the nitrogen regioisomer, we obtained ^{13}C NMR spectra of lead compound **17c** and of its parent thiocarbamate **10**, which exhibited the thione carbon signals at δ 192.13 and 188.62, respectively. It is apparent that the thione carbon signal for **17c** cannot be consistent with the S-acylisothiocarbamate structure.

4.4.3   EXPERIMENTAL SECTION

4.4.3.1   General Procedure P₁ for the Parallel Synthesis of *Library 1* (ATCs 11c,q; 13q; 14c; 15b,c,g,q; 16c; 17a-d,g,j,k,m,n,o,q,r; 18c; 19a,c,h,p,q,r; 20r; 21r; 22r; 23c)

A sodium hydride dispersion (60%) in mineral oil (0.44 g, ~10 mmol) was added in a single portion at r.t. to each numbered reaction tube of a Carousel Reaction Station, containing a stirred solution (15 mL) of the starting alcohol in anhydrous toluene (**1, 5, 6, 7, 8**; 10 mmol) or anhydrous THF (**3** and **4**). As soon as the hydrogen evolution ceased, the proper isothiocyanate (phenyl-, cyclohexyl-, or *p*-fluorophenyl-isothiocyanate, 10 mmol) and anhydrous pyridine (5 mL) were added (after 5 and 15 min, respectively) to each reaction mixture. After 15 min, the proper, freshly distilled or recrystallized acyl chloride (11 mmol) was added in a single portion to each reaction tube, and the reaction was stirred for 6 hr. Each reaction was then transferred into a set of separating funnels, diluted with water (200 mL) and extracted with diethyl ether. The combined extracts were washed with water (30 mL × 2), 2 M HCl (30 mL × 2), 1 M NaHCO₃ (30 mL), and dried over anhydrous Na₂SO₄. After parallel filtration through pads of Florisil (diameter 5 × 2 cm) by an in-house device, evaporating *in vacuo* gave residues that were purified by crystallization.

Isolation of O-(2-Phenoxyethyl) Phenylthiocarbamate Intermediate 10. A sodium hydride dispersion (60%) in mineral oil (0.44 g, ˜10 mmol) was poured in a single portion into a stirred solution of **6** (1.38 g, 10 mmol) in anhydrous THF (15 mL). As soon as the hydrogen evolution ceased, phenylisothiocyanate (1.35 g, 10 mmol) was added to the mixture at r.t. and the reaction was stirred for 30 min. Evaporation *in vacuo* gave a residue that was treated with water (20 mL) and 1 M HCl (10 mL). Then, the mixture was extracted with diethyl ether and the combined organic layers were washed with water (20 mL × 4) and 5% NaHCO₃ (20 mL × 2). The dried organic layer was evaporated under reduced pressure, and the oily residue was crystallized from methanol-diethylether to afford **10** (2.05 g, 75%).

4.4.3.2   General Procedure P₂ for the Parallel Synthesis of *Library 2*
              (ATCs 25c; 26r; 27r; 28r; 29r; 30r; 31r; 32r; 33r; 34r; 35r; 36r;
              37r; 38r; 39q; 40r; 41c,g,q–t; 42c; 43r; 44r; 45q,r; 46t)

A sodium hydride dispersion (60%) in mineral oil (0.44 g, ~10 mmol) was added in a single portion to each numbered reaction tube of a Carousel Reaction Station, containing a stirred, ice-cooled solution of **9** (1.91 g, 10 mmol) and of the proper isothiocyanate (10 mmol) in anhydrous pyridine (25 mL). Each reaction mixture was allowed to react under stirring for 4 hr; then, TMEDA (1.74 g, 15 mmol) and the proper acyl chloride (12 mmol) were added and the reaction was stirred at r.t. for 6 hr. Then, each reaction was transferred into a set of separating funnels, diluted with water (150 mL), and extracted with dichloromethane (in the preparation of *ortho*-derivatives **26r–29r**, the reaction mixture was heated to 55–60°C for 3 hr before the work-up). The organic layer was washed with water (four times) and 1 M HCl (20 mL × 4) and dried. After parallel filtration of each extract through a plug of Florisil by an in-house device, evaporation under reduced pressure gave residues that were purified by crystallization.

4.4.3.3   General Procedure P₃ for the Parallel Synthesis of ATCs 12c,q

A sodium hydride dispersion (60%) in mineral oil (0.44 g, ~10 mmol) was added at r.t. to two numbered reaction tubes containing a stirred, dry DMF (20 mL) solution of 1-adamantanol **2** (1.53 g, 10 mmol). After heating to 90–95°C for 20 min, a dry DMF (5 mL) solution of phenylisothiocyanate (1.35 g, 10 mmol) was added to the reaction mixture, prolonging the heating and stirring for 30 min. Then, anhydrous pyridine (5 mL) and the proper acyl chloride (12 mmol) were added in a single portion to each reaction mixture cooled to r.t. Each reaction mixture was stirred for 6 hr at r.t. and heated to 55°C for 1 hr. Each reaction was transferred into a set of separating funnels and diluted with water (200 mL). Following extraction with dichloromethane, each organic layer was washed with water (30 mL × 5), dried, filtered in parallel through a pad of Florisil (diameter 5 × 2 cm) by an in-house device, and evaporated *in vacuo* to yield a residue that was purified by crystallization.

4.4.3.4   General Procedure P₄ for the Parallel Synthesis of Compounds 24u,v,w

A sodium hydride dispersion (60%) in mineral oil (0.50 g, ~12.5 mmol) was added in a single portion to each numbered reaction tube of a Carousel Reaction Station, containing a stirred, ice-cooled solution of **7** (1.52 g, 10 mmol) in anhydrous THF (15 mL). As soon as hydrogen evolution subsided, neat phenylisothiocyanate (1.35 g, 10 mmol) was added to the mixtures at r.t.. After 20 min, n-propyl- (1.35 g, 11 mmol), n-butyl- (1.50 g, 11 mmol), and phenyl- (1.72 g, 11 mmol) chloroformate were added. Each reaction mixture was heated at 60°C for 3 hr, and, after removing THF by evaporation under reduced pressure, each residue was treated with water (100 mL) and extracted with diethyl ether. The organic layer was dried, filtered in parallel through a plug of silica gel, and evaporated under reduced pressure to give an oily residue that was purified by crystallization.

Refer to Reference 82 for melting points (°C), crystallization solvents, yields, spectroscopic (IR and ¹H NMR), and microanalytical data for all synthesized compounds, and experimental procedures for molecular modeling and biological evaluation.

4.4.4   Biological Results of ATC Screening and Discussion

The ATCs and related isosteres were evaluated for their cytotoxicity and anti-HIV-1 activity in MT-4 cells (Tables 4.1 and 4.2) using Trovirdine as the reference compound. Derivatives **41q, r, s** tested in enzyme assays against the HIV-1 virionic RT (vRT) proved to target RT. The most potent derivatives were also tested against some NNRTI resistant strains[98,99] carrying clinically relevant mutations (Lys103Arg, Tyr181Cys and Lys103Asm plus Try181Cys).[82]

TABLE 4.1.[a]

Cytotoxicity and Anti-HIV-1 Activity of Lead Compound 17c, of its Analogs and (Bio)isosteres

Cpd	Ar	R	Ar$_2$	G-CO	CC$_{50}$[b]	EC$_{50}$[c]	S.I.[d]
13q	phenyl	H	C$_6$H$_5$	2-furoyl	111	58	1.9
14c	2-furyl	H	C$_6$H$_5$	benzoyl	38.3	>38.3	—
15b	benzyl	H	C$_6$H$_5$	trans-cinnamoyl	40.4	4.5	9
15c	benzyl	H	C$_6$H$_5$	benzoyl	91	4.2	22
15g	benzyl	H	C$_6$H$_5$	4-chlorobenzoyl	43	4	10.7
15q	benzyl	H	C$_6$H$_5$	2-furoyl	102	4.3	24
16c	phenoxymethyl	H	C$_6$H$_{11}$	benzoyl	>200	200	—
17a	phenoxymethyl	H	C$_6$H$_5$	phenoxyacetyl	103	6	17.1
17b	phenoxymethyl	H	C$_6$H$_5$	trans-cinnamoyl	66.6	7.7	8.65
17c	phenoxymethyl	H	C$_6$H$_5$	benzoyl	122	8	15.2
17d	phenoxymethyl	H	C$_6$H$_5$	4-toluoyl	>200	9.9	>20.2
17g	phenoxymethyl	H	C$_6$H$_5$	4-chlorobenzoyl	133	10.3	12.9
17j	phenoxymethyl	H	C$_6$H$_5$	2-acetoxybenzoyl	>200	9.5	>21
17k	phenoxymethyl	H	C$_6$H$_5$	2,4-dichlorobenzoyl	43	11.6	3.7
17m	phenoxymethyl	H	C$_6$H$_5$	3,5-dichlorobenzoyl	43	8.8	4.8
17n	phenoxymethyl	H	C$_6$H$_5$	4-chloro-3-nitrobenzoyl	80	7.6	10.5
17o	phenoxymethyl	H	C$_6$H$_5$	3,4,5-trimethoxybenzoyl	>200	9.6	>20.8
17q	phenoxymethyl	H	C$_6$H$_5$	2-furoyl	82	8.4	9.7
17r	phenoxymethyl	H	C$_6$H$_5$	2-thenoyl	125.2	8.6	14.5
18c	phenoxymethyl	H	4-F-C$_6$H$_4$	benzoyl	70.7	4	17.6
19a[e]	phenoxymethyl	CH$_3$	C$_6$H$_5$	phenoxyacetyl	122.4	1.4	87.4
19c[e]	phenoxymethyl	CH$_3$	C$_6$H$_5$	benzoyl	>200	1.3	>153.8
19h[e]	phenoxymethyl	CH$_3$	C$_6$H$_5$	3-nitrobenzoyl	>200	3.6	>55.5
19p[e]	phenoxymethyl	CH$_3$	C$_6$H$_5$	1-naphthoyl	>200	6	>33.3
19q[e]	phenoxymethyl	CH$_3$	C$_6$H$_5$	2-furoyl	51.5	1.3	39.6
19r[e]	phenoxymethyl	CH$_3$	C$_6$H$_5$	2-thenoyl	>200	2	>100
20r[e]	phenoxymethyl	CH$_3$	3-Cl-C$_6$H$_4$	2-thenoyl	>100	11	>9
21r[e]	phenoxymethyl	CH$_3$	4-Cl-C$_6$H$_4$	2-thenoyl	>100	11	>9
22r[e]	phenoxymethyl	CH$_3$	4-NO$_2$-C$_6$H$_4$	2-thenoyl	63	6	10.5
23c	phenylthiomethyl	H	C$_6$H$_5$	benzoyl	44	>44	—
24u[e]	phenoxymethyl	CH$_3$	C$_6$H$_5$	n-propoxycarbonyl	>200	131	>1.5
24v[e]	phenoxymethyl	CH$_3$	C$_6$H$_5$	n-butoxycarbonyl	>200	47	>4
24w[e]	phenoxymethyl	CH$_3$	C$_6$H$_5$	phenoxycarbonyl	>200	>200	—
26r	phthalimidomethyl	H	2-CH$_3$-C$_6$H$_4$	2-thenoyl	>100	11	>9
27r	phthalimidomethyl	H	2-C$_2$H$_5$-C$_6$H$_4$	2-thenoyl	≥100	11	≥9
28r	phthalimidomethyl	H	2-Cl-C$_6$H$_4$	2-thenoyl	>100	11	9
29r	phthalimidomethyl	H	2-OCH$_3$-C$_6$H$_4$	2-thenoyl	>100	11	>9
30r	phthalimidomethyl	H	3-CH$_3$-C$_6$H$_4$	2-thenoyl	>100	10	>10
31r	phthalimidomethyl	H	3-CF$_3$-C$_6$H$_4$	2-thenoyl	>100	11	>9
32r	phthalimidomethyl	H	3-COCH$_3$-C$_6$H$_4$	2-thenoyl	>100	6	>17
33r	phthalimidomethyl	H	3-F-C$_6$H$_4$	2-thenoyl	>100	11	>9
34r	phthalimidomethyl	H	3-Cl-C$_6$H$_4$	2-thenoyl	>100	6	>17
35r	phthalimidomethyl	H	3-Br-C$_6$H$_4$	2-thenoyl	53	1.2	44
37r	phthalimidomethyl	H	3-OCH$_3$	2-thenoyl	>100	3.5	>28
44r	phthalimidomethyl	H	4-N(C$_2$H$_5$)$_2$	2-thenoyl	>100	11	9
Trovirdine					60	0.02	3000

[a] Data represent mean values for three separate experiments. Variation among triplicate samples was less than 10%. [b] Compound concentration [μM] required to reduce the viability of mock-infected cells by 50%, as determined by the MTT method. [c] Compound concentration [μM] required to achieve 50% protection of MT-4 cell from the HIV-1 induced cytopathogenicity, as determined by the MTT method. [d] Selectivity index: CC$_{50}$/EC$_{50}$ ratio. [e] Tested as racemic mixture.

TABLE 4.2ᵃ

Cytotoxicity and HIV-1 Activity of *O*-(2-Phthalimidoethyl) ATCs

Cpd	R	Acyl	CC_{50}^{b}	EC_{50}^{c}	S.I.ᵈ
25c	H	benzoyl	>200	0.4	>500
36r	3-NO$_2$	2-thenoyl	100	0.38	263
38r	4-CH$_3$	2-thenoyl	>100	0.4	>250
39q	4-C$_2$H$_5$	2-furoyl	>100	0.03	>3,333
40r	4-F	2-thenoyl	>100	0.1	>1,000
41c	4-Cl	benzoyl	≥200	0.1	≥2,000
41g	4-Cl	4-chlorobenzoyl	>100	0.025	>4,000
41q	4-Cl	2-furoyl	41	0.007	5,857
41r	4-Cl	2-thenoyl	>200	0.008	>25,000
41s	4-Cl	2-chloronicotinoyl	>200	0.005	>40,000
41t	4-Cl	6-chloronicotinoyl	>200	0.006	>33,333
42c	4-Br	benzoyl	>200	0.035	>5,714
43r	4-I	2-thenoyl	18	0.01	1,800
45q	4-NO$_2$	2-furoyl	18	0.008	2,250
45r	4-NO$_2$	2-thenoyl	168	0.01	16,800
46t	4-OC$_2$H$_5$	6-chloronicotinoyl	>100	0.5	>200
Trovirdine			60	0.02	3,000

ᵃ Data represent mean values for three separate experiments. Variation among triplicate samples was less than 10%. ᵇ Compound concentration [μM] required to reduce the viability of mock-infected cells by 50%, as determined by the MTT method. ᶜ Compound concentration [μM] required to achieve 50% protection of MT-4 cell from the HIV-1 induced cytopathogenicity, as determined by the MTT method. ᵈ Selectivity index: CC_{50}/EC_{50} ratio.

Table 4.1 lists the ATCs active at micromolar concentrations (EC$_{50}$ range: 1.3 to 200 μM). *O*-ethyl derivatives **11c, q** as well as *O*-(1-adamantyl) compounds **12c, q** were inactive (data not shown). Methyl-branched ATCs **19a, c, h, p–r**, (tested as racemic mixtures) led to a four- to sixfold increase in potency with respect to the unbranched analogs (compare **17a** with **19a**, **17c** with **19c**, **17q** with **19q**, and **17r** with **19r**). Removal of the ether oxygen (2-phenethyl derivatives **15b, c, g, q**) led to a two-fold increase in potency (compare **15b** with **17b**, **15c** with **17c**, **15g** with **17g**, and **15q** with **17q**).

Table 4.2 displays that the ATCs endowed with submicromolar or nanomolar activity (EC$_{50}$ values range from 0.5 to 0.005 μM) share the 2-phthalimidoethyl portion. The ATCs bearing sterically demanding electron-withdrawing substituents at position *para* of the *N*-phenyl ring and the *para*-chlorobenzoyl/heteroaroyl moieties are the most potent derivatives. As suggested by docking simulations, these moieties would be in proximity to amino acid residues suitable for additional interactions.

The most active derivatives were also tested in cell-based assays against the clinically relevant Lys103Arg, Tyr181Cys and Lys103Asn+Tyr181C HIV-1 mutated strains.[82] Derivatives **41g, q–t** and **45q, r** inhibited Tyr181Cys mutant, while Lys103Arg and Lys103Asn+Tyr181Cys mutant strains were unsusceptible to ATC compounds.[82] The potency of ATCs against this mutant was in the same micromolar range as that of Trovirdine and correlates with the presence of the *N*-heteroaroyl moieties.

4.5   PETT ISOSTER THIOCARBAMATES (TCs)

4.5.1   TC LIBRARY DESIGN

As previously described (Section 4.4), the *O*-phthalimidoethyl-ATC NNRTIs proved selective and potent against wild-type HIV-1 and moderately active against the Y181C mutant. Because screening of synthesis intermediates represents one of the strategies in the search for new lead compounds,[100] the TCs (i.e., the precursors of the ATCs) were investigated, also considering their isosteric relationship with PETT derivatives. The TCs are characterized by a greater conformational flexibility as compared with PETT compounds, not having any intramolecular H-bond. In this regard, it was speculated that a greater conformational freedom of inhibitors might be a useful design feature for reducing drug resistance.[47,101–103] Furthermore, our interest in the TC field was stimulated by thiocarbamate UC-38 (NSC 629243) (Chart 4.1), which was selected as an anti-HIV-1 agent in early 1990 for preclinical development.[104]

To evaluate the ability of TCs to target RT, a docking simulation was performed for TC **12**, the precursor of ATC **25c**. The RT/**12** (Figure 4.5) complex is stabilized by a hydrogen bond, involving the NH group and the Lys101 main-chain carbonyl. Another polar interaction occurs between one of the imidic oxygen atoms and the carboxylic group of Glu138 (distance: 3.2 Å) from the p51 sub-unit. Moreover, the phthalimide phenyl ring establishes hydrophobic contacts with Val106 and π-π interactions with Tyr181, Tyr188, and Phe227, while the *N*-phenyl interacts with the Leu100, Lys103, Pro236, and Tyr318 side chains. Finally, the ethyl linker is involved in hydrophobic contacts with the Val179 side chain, and the sulfur atom has van der Waals contacts with the Lys101 main chain.

Differently from ATC-**25c**, **12** would assume the butterfly-shaped bioactive conformation, strongly mimicking the RT-bound PETT-1 bioactive conformation and orientation (Figure 4.6).

As anticipated by its docking model, **12** was able to target virion RT (IC$_{50}$ = 0.6 μM) and to selectively inhibit the HIV-1 induced cytopathogenicity in MT-4 cells (CC$_{50}$ >100 μM, EC$_{50}$ = 1.2 μM) at a concentration three fold higher than ATC **25c** (EC$_{50}$: 0.4 μM, Table 4.1).[105]

Therefore, to explore the potential of TCs as NNRTI agents, we designed two small analog libraries of lead **12**. It should be noted that in the field of NNRTIs, structure-based design of Trovirdine analogs is an attractive research area for medicinal chemists, as documented by recent published papers.[80,106–116] In addition, PETT isostere TCs have not been hitherto reported in the literature.

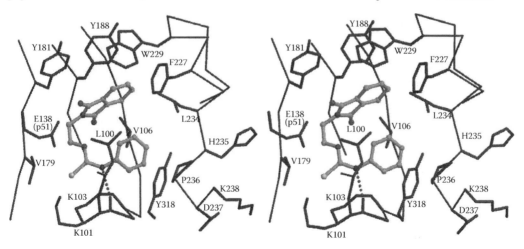

FIGURE 4.5   Stereo view showing the position and orientation of C-TC **12** in the NNIBS. The ligand (ball-and-stick, model structure) is shown along with RT amino acid residues. The hydrogen bond to the Lys103 main chain is depicted as a dotted line.

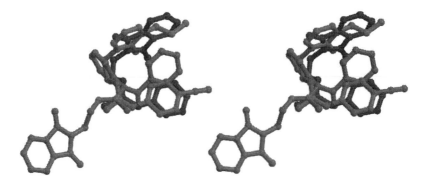

FIGURE 4.6 Stereo diagram showing superposition of TC **12** (model structure), ATC-**25c** (model structure), and PETT-1 (x-ray structure model) within the NNIBS.

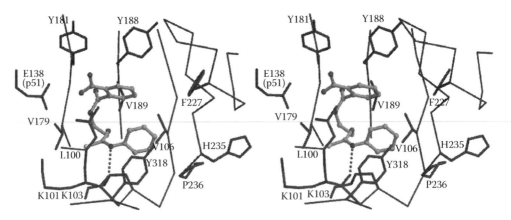

FIGURE 4.7 Stereo view showing the position and orientation of O-TC **51**. The ligand and amino acid residues lining the NNIBS are represented. The hydrogen bond between the NH group and the Lys101 carbonyl is depicted as a dotted line.

With the aim of establishing the TC key structural requirements for anti-HIV-1 activity, our lead computer-assisted optimization strategy focused on (1) substituent variation on the *N*-phenyl ring; (2) modification of the phthalimide moiety; (3) shortening, lengthening, and branching of the ethyl spacer; (4) modification of the thiocarbamate scaffold by isosteric replacements (S/O and NH/O); and (5) opening of the imide ring. Indeed, the last modification was originally planned for future SAR strategies, but several ring-opened analogs (O-TCs) were concomitantly obtained along with their corresponding ring-closed congeners (C-TCs). The O-TC **51** docking model predicted a bioactive conformation similar to that of PETT compounds (Figure 4.7). The following features of the model are noteworthy. The complex is mainly stabilized by a hydrogen bond involving the NH thiocarbamic group and the Lys101 main chain carbonyl group. The *N*-phenyl ring makes hydrophobic interactions with Val106, His235, Pro236, and Tyr318, while the phthalimide moiety is in contact with Leu100, Tyr181, and Tyr188. The carbamoyl-ethyl linker has a number of van der Waals interactions with Val179, Tyr181, Tyr188, Val189, and Gly190; and the carboxylic group interacts with the Leu100 and Tyr181 side chains. Therefore, all the isolated O-TCs were tested. Finally, some structural variations were performed with the aim not only to expand the SARs, but also to validate the TC docking model (for representative examples, see derivatives **77–81** and **92–95**).

4.5.2 PARALLEL SYNTHESIS OF FOCUSED TC LIBRARIES

The TCs were prepared in parallel according to the ATC procedure by omitting the acylation step. The alcohol and isothiocynates building blocks used to perform our SAR strategy are reported in Chart 4.5.

According to literature procedures,[117] the syntheses of the not commercially available alcohols **3**, **4**, (±)**5**, (±)**6**, **8**, and **9** (Chart 4.5) were carried out by fusion of anhydrides **I–V** and amino-alcohols (Scheme 4.4).

Scheme 4.5 summarizes the general procedure for the preparation of two focused TC libraries (Library 1, solvent = DMF; Library 2, solvent = pyridine). Alcohols **1-7** (Chart 4.5) were converted

---

CHART 4.5

Alcohols 1–10 and Isothiocyanates (Ar$_1$-NCS) Used as Building Blocks for the Parallel Synthesis of TC Libraries

(a)

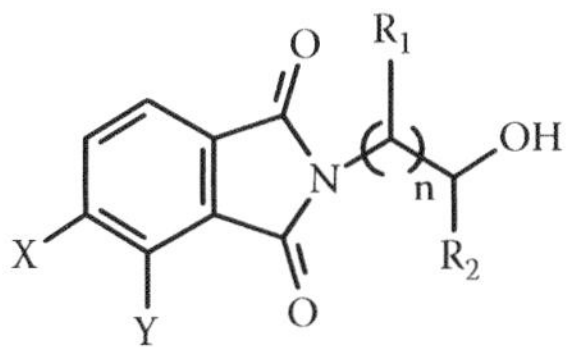

	n	X	Y	R$_1$	R$_2$
1	0	H	H	-	H
2	1	H	H	H	H
3	1	H	CH$_3$	H	H
4	1	CH$_3$	H	H	H
5	1	H	H	CH$_3$	H
6	1	H	H	H	CH$_3$
7	2	H	H	H	H

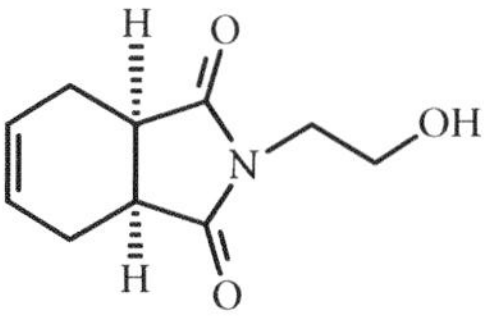

8 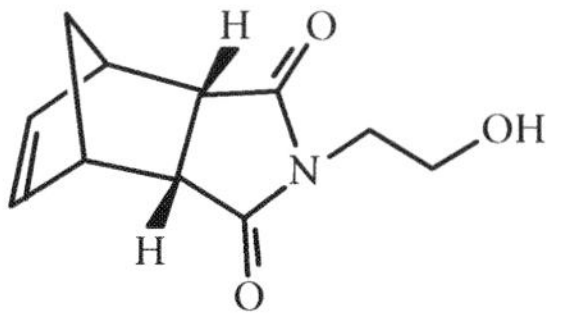   9 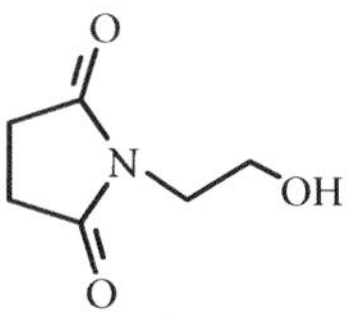   10

(b)

	Ar$_1$-NCS
I$_1$	cyclohexylisothiocyanate
I$_2$	phenylisothiocyanate
I$_3$	2-tolylisothiocyanate
I$_4$	3-tolylisothiocyanate
I$_5$	4-tolylisothiocyanate
I$_6$	4-ethylphenylisothiocyanate
I$_7$	2-isopropylphenylisothiocyanate
I$_8$	4-isopropylphenylisothiocyanate
I$_9$	2-trifluoromethylphenylisothiocyanate
I$_{10}$	3-trifluoromethylphenylisothiocyanate
I$_{11}$	4-trifluoromethylphenylisothiocyanate
I$_{12}$	3-acetylphenylisothiocyanate
I$_{13}$	4-acetylphenylisothiocyanate
I$_{14}$	2-metoxycarbonylphenylisothiocyanate
I$_{15}$	3-metoxycarbonylphenylisothiocyanate
I$_{16}$	4-metoxycarbonylphenylisothiocyanate
I$_{17}$	4-ethoxycarbonylphenylisothiocyanate
I$_{18}$	4-cyanophenylisothiocyanate
I$_{19}$	2-fluorophenylisothiocyanate

(c)

	Ar$_1$-NCS
I$_{20}$	4-fluorophenylisothiocyanate
I$_{21}$	2-chlorophenylisothiocyanate
I$_{22}$	3-chlorophenylisothiocyanate
I$_{23}$	4-chlorophenylisothiocyanate
I$_{24}$	2-bromophenylisothiocyanate
I$_{25}$	4-bromophenylisothiocyanate
I$_{26}$	4-iodophenylisothiocyanate
I$_{27}$	4-dimethylaminophenylisothiocyanate
I$_{28}$	4-diethylaminophenylisothiocyanate
I$_{29}$	2-nitrophenylisothiocyanate
I$_{30}$	3-nitrophenylisothiocyanate
I$_{31}$	4-nitrophenylisothiocyanate
I$_{32}$	2-methoxyphenylisothiocyanate
I$_{33}$	3-methoxyphenylisothiocyanate
I$_{34}$	4-methoxyphenyisothiocyanate
I$_{35}$	4-ethoxyphenylisothiocyanate
I$_{36}$	4-benzyloxyphenylisothiocyanate
I$_{37}$	3-methylsulfanylphenylisothiocyanate
I$_{38}$	1-naphthylisothiocyanate

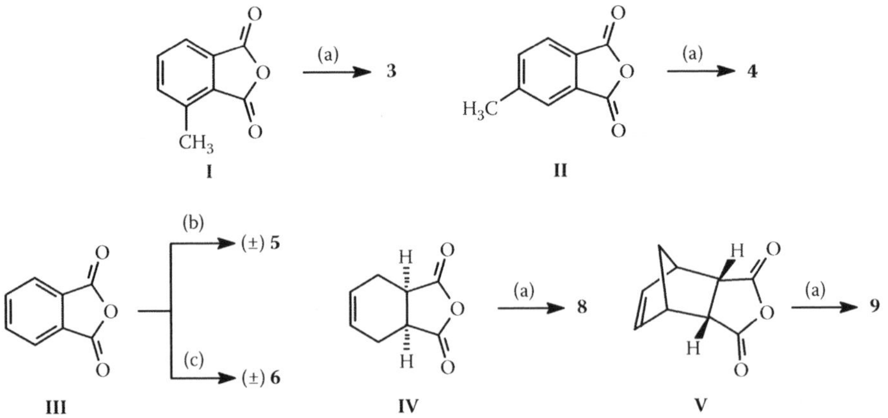

SCHEME 4.4 Syntheses of alcohol building blocks **3**, **4**, (±)**5**, (±)**6**, **8**, and **9**.[a]

[a] Reaction conditions: (a) 2-aminoethanol, 175–180°C, 2 hr; (b) (±) 2-amino-1-propanol, 140–150°C, 2 hr; and (c) (±)1-amino-2-propanol, 160–175°C, 2 hr.

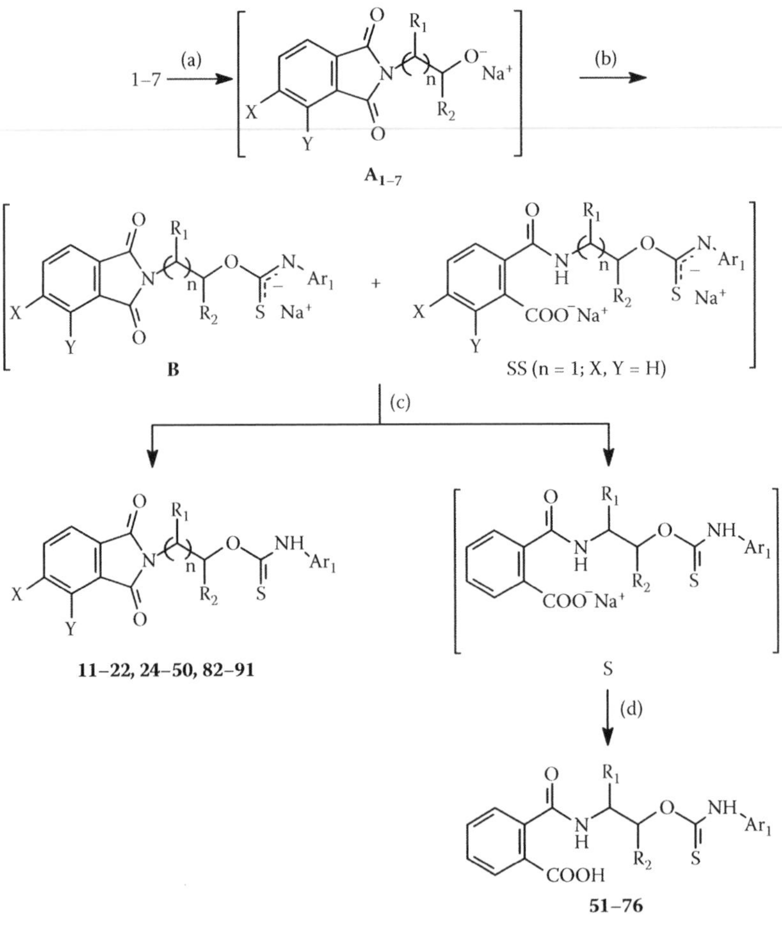

SCHEME 4.5 General procedure for preparation of two focused solution-phase TC libraries by parallel analog synthesis (Libraries 1 and 2).[a]

[a] Reaction conditions: Library 1 (a), (b) isothiocyanate, NaH, dried DMF, 0–5°C, then r.t. for 24 hr. Library 2 (a), (b) isothiocyanate, NaH, anhydrous pyridine, r.t. for 15 min, then 60–65°C for 3 hr; (c) H$_2$O, NH$_4$Cl; (d) 2 N HCl.

TC Library 1[a]

C-TCs					O-TCs	
11, (A_2, I_1)	29, (A_2, I_{30})	39, (A_2, I_{23})	48, (A_3, I_{31})	87, (A_6, I_{23})	51, (A_2, I_2)	67, (A_2, I_{23})
12, (A_2, I_2)	31, (A_2, I_5)	40, (A_2, I_{25})	49, (A_4, I_{23})	88, (A_6, I_{25})	55, (A_2, I_{29})	68, (A_2, I_{25})
13, (A_2, I_{38})	32, (A_2, I_6)	41, (A_2, I_{26})	50, (A_4, I_{31})	89, (A_6, I_{31})	58, (A_2, I_{15})	69, (A_2, I_{26})
21, (A_2, I_{29})	34, (A_2, I_{11})	42, (A_2, I_{27})	82, (A_1, I_{31})	90, (A_7, I_2)	61, (A_2, I_{37})	70, (A_2, I_{31})
24, (A_2, I_{10})	35, (A_2, I_{17})	43, (A_2, I_{28})	83, (A_5, I_{23})	91, (A_7, I_{31})	62, (A_2, I_5)	72, (A_2, I_{35})
25, (A_2, I_{12})	36, (A_2, I_{13})	44, (A_2, I_{31})	84, (A_5, I_{31})		63, (A_2, I_6)	74, (A_6, I_5)
26, (A_2, I_{15})	37, (A_2, I_{18})	46, (A_2, I_{35})	85, (A_6, I_5)		65, (A_2, I_{18})	75, (A_5, I_{23})
28, (A_2, I_{37})	38, (A_2, I_{20})	47, (A_3, I_{23})	86, (A_6, I_{18})		66, (A_2, I_{20})	76, (A_5, I_{31})

TC Library 2

C-TCs		O-TCs	
14, (A_2, I_3)	20, (A_2, I_{24})	52, (A_2, I_3)	59, (A_2, I_{22})
15, (A_2, I_7)	22, (A_2, I_{32})	53, (A_2, I_{14})	60, (A_2, I_{33})
16, (A_2, I_9)	27, (A_2, I_{22})	54, (A_2, I_{19})	64, (A_2, I_8)
17, (A_2, I_{14})	30, (A_2, I_{33})	56, (A_2, I_{32})	71, (A_2, I_{34})
18, (A_2, I_{19})	33, (A_2, I_8)	57, (A_2, I_4)	73, (A_2, I_{36})
19, (A_2, I_{21})	45, (A_2, I_{34})		

[a] In parentheses, *A* and *I* represent the alcohol and isothiocyanate building blocks (Chart 4.5), respectively, used for the TC analog synthesis.

into their corresponding alcoholates A_{1-7} by means of 60% NaH dispersion in mineral oil. A_{1-7} were then condensed *in situ* with appropriate isothiocyanates (Chart 4.5B) to afford adducts **B** (sodium salts of C-TCs) and/or **SS** (disodium salts of O-TCs). Upon treatment of the reaction medium with water and ammonium chloride, C-TCs **11–22, 24–50**, and **82–91** precipitated, while ring-opened double salts **SS** were transformed into **S** (monosodium salts of the O-TCs). The subsequent acidification of the reaction mother liquors with 2 M HCl caused separation of O-TCs **51-76** as solids. Notably, in the synthesis of C-TCs **11, 13, 24, 25, 29, 34-36, 42, 43, 47-50, 82, 86–89, 90, 91** (Library 1) and C-TCs **15, 16, 19, 20** (Library 2), it was impossible to isolate workable amounts of the corresponding O-TCs (yields < 0.5 to 1%, data not shown). In contrast, reaction of alcohol **2** with 3-tolylisothiocyanate and 4-benzyloxyphenylisothiocyanate afforded only O-TCs **53** and **73** (Library 2), respectively. It should be noted that this anomalous ring opening of the phthalimide moiety was also found in hydrochlorides of basically substituted phthalimidyl derivatives, which unexpectedly gave the corresponding ring-opened products on careful neutralization with alkali in cold water.[118]

In addition, in the synthesis of C-TCs **47–50** (Library 1), we isolated their corresponding O-TCs, but these derivatives have not been fully characterized yet (the ring opening of the asymmetric 3-methyl- and 4-methyl-phthalimmide moieties might give a mixture of two regioisomers). Purification of the TCs was carried out by crystallization.

The dehydrative cyclization of O-TC **57** in the presence of excess P_2O_5 in DMF (Scheme 4.6) afforded C-TC **23**, not directly obtainable with the general procedure.

TCs **77–81** were synthesized in parallel by reacting **10, 8** in DMF, or **9** in anhydrous THF with the proper isothiocyanate in the presence of sodium hydride (Scheme 4.7).

Scheme 4.8 illustrates the synthesis of compounds **92–95** by reacting **2** with 4-tolyl- (**92**) and 4-chlorophenyl- (**93**) thionoformate, 4-nitrophenyl isocyanate (**94**), and thiocarbonyldiimidazole (**95**). Refer to Reference 105 for melting points (°C), crystallization solvents, yields, spectroscopic

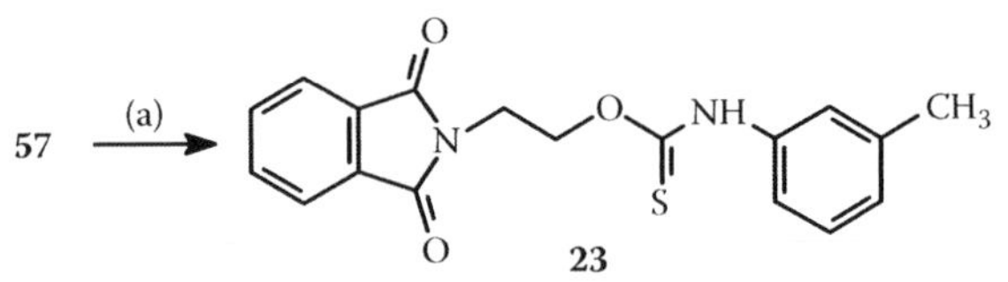

SCHEME 4.6   Synthesis of C-TCs **23**.[a]

[a]   Reaction conditions: (a) dried DMF, P_2O_5, 65°C for 5 hr, Na_2CO_3.

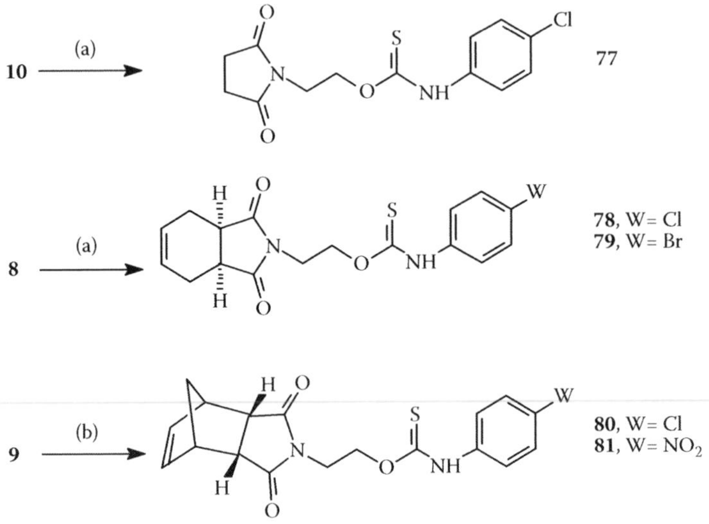

SCHEME 4.7   Parallel synthesis of TCs **77–81**.[a]

[a]   Reaction conditions: (a) dried DMF, NaH, isothiocyanate, r.t. for 20 hr; (b) anh. THF, NaH, isothiocyanate, r.t. 20 hr.

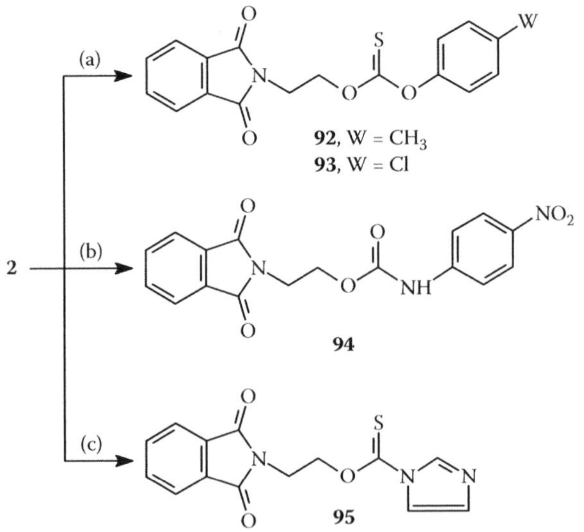

SCHEME 4.8   Non-parallel synthesis of thiocarbonates **92, 93**, carbamate **94**, and phthalimidoethyl imidazole-1-carbothioate **95**.[a]

[a]   *Reaction conditions:* (a) Acetonitrile, DMAP, chlorothionoformate, r.t.; for 24 hr;[119] (b) DMF/Py (1:1), *p*-nitrophenylisocyanate, 140°C for 5 hr; and (c) mixture of **2** and thiocarbonyldiimidazole ground in a mortar for 15 min.[120]

(IR and ^{1}H NMR) and microanalytical data for all synthesized compounds, and experimental procedures for molecular modeling and biological evaluation

4.5.3 General Considerations on the Parallel TC Synthesis

The formation of O-TCs and their separation from C-TCs deserve further comment. Scheme 4.9a illustrates that the double **SS** salts can be generated either directly from **B** (route b) or indirectly from ring-opened intermediates **2′**, **5′**, **6′**, and $A_{2',5',6'}$ according to route c. Clearly, these species,

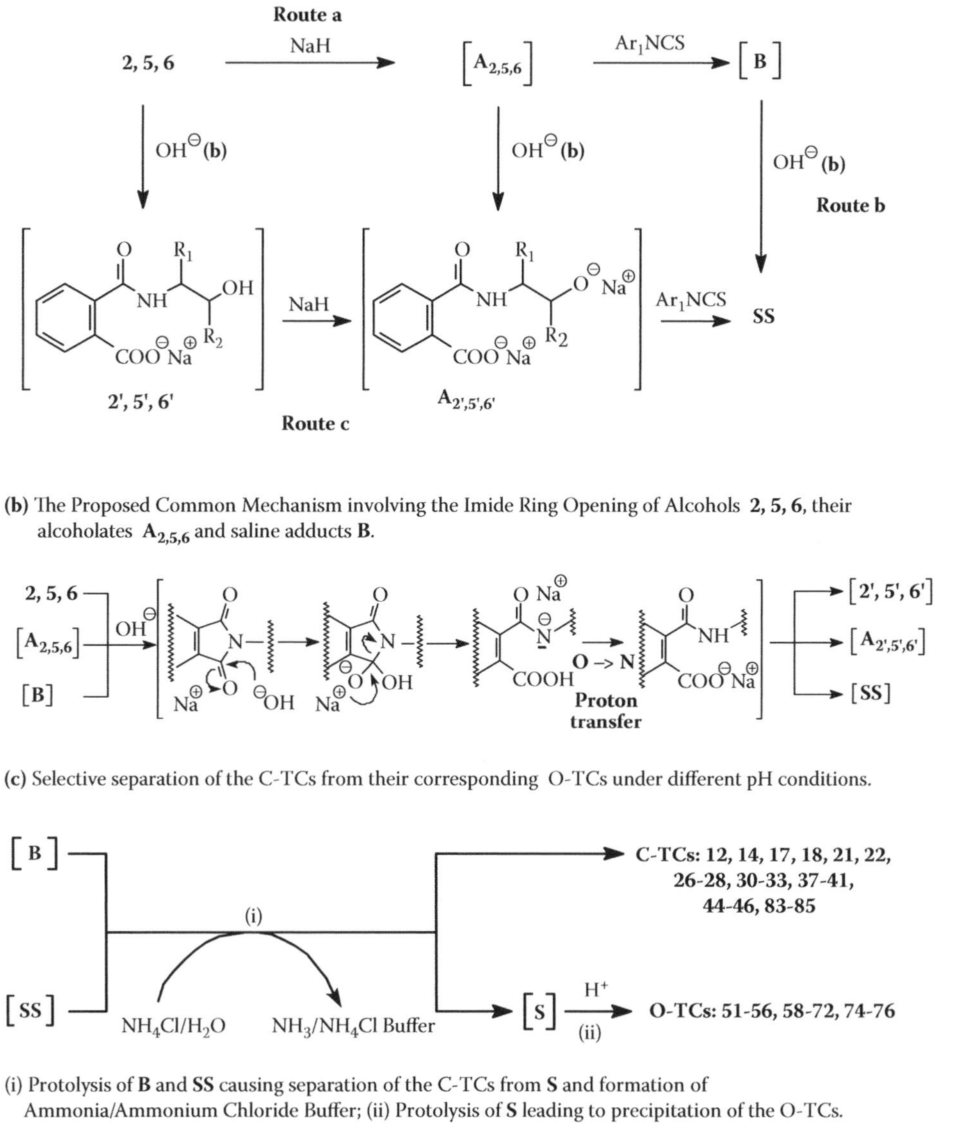

(b) The Proposed Common Mechanism involving the Imide Ring Opening of Alcohols **2, 5, 6**, their alcoholates $A_{2,5,6}$ and saline adducts **B**.

(c) Selective separation of the C-TCs from their corresponding O-TCs under different pH conditions.

(i) Protolysis of **B** and **SS** causing separation of the C-TCs from **S** and formation of Ammonia/Ammonium Chloride Buffer; (ii) Protolysis of **S** leading to precipitation of the O-TCs.

SCHEME 4.9   **(a):** The hypothesized formation of **SS** (routes **b** and **c**).

(b) The proposed common mechanism involving the imide ring opening of alcohols **2, 5, 6**, their alcoholates $A_{2,5,6}$ and saline adducts **B**.

(c) Selective Separation of the C-TCs from their corresponding O-TCs under Different pH Conditions: (i) protolysis of **B** and **SS** causing separation of the C-TCs from **S** and formation of Ammonia/Ammonium Chloride Buffer; and (ii) protolysis of **S** leading to precipitation of the O-TCs.

originated from **2, 5, 6,** and **A$_{2,5,6}$**, are expected to react like the corresponding ring-closed counterparts (see route a). Mechanistically, the attack of the hydroxide ion on the phthalimide carbonyls would cause imide ring opening of **B, 2, 5, 6,** and **A$_{2,5,6}$** (Scheme 4.9b). The easy separation of the C-TCs from the O-TCs is mainly due to protolysis of basic salts **B, SS,** and **S** under different pH conditions (Scheme 4.9c). Thus, upon treatment of the reaction mixture with weak acid (water and ammonium chloride), the thiocarbamic moieties of **B** and **SS** are protonated with concomitant formation of NH_3/NH_4Cl buffer, as indicated by measuring the pH of the final mixture. As a consequence, C-TCs and **S** are generated, that is, the respective conjugate acids of **B** and **SS**. Therefore, C-TCs **12, 14, 17, 18, 21, 22, 26–28, 30–33, 37–41, 44–46, 83–85** precipitate, while salts **S** are kept in solution (indeed, it is not ruled out that the ammonium chloride salting-out effect contributes to the complete separation of the C-TCs from **S**). For protonation of the carboxylate group of **S** to occur, the pH is adjusted at low acidic values by acidifying the reaction mother liquors with 2 M HCl. This treatment causes separation of O-TCs **51–56, 58–72,** and **74–76**.

4.5.4 Biological Results of TC Screening and Discussion

The antiretroviral activity of TCs **11–91** (Tables 4.3 and 4.4), thionocarbonates **92** and **93**, carbamate **94**, and imidazole-1-carbothioate **95** were evaluated in cell-based assays by assessing the reduction of the HIV-1-induced cytopathogenicity in MT-4 cells.[105] Trovirdine was used as the reference compound.

Table 4.3 lists the TC derivatives showing micromolar activity (EC$_{50}$ valve range: 1.2–68 μM). None of them was more active than lead **12**.

Table 4.4 summarizes the TCs endowed with sub-micromolar or nanomolar potencies. The *para*-substituted C-TCs **31–34, 36–41, 44, 45** proved active at low nanomolar concentrations (EC$_{50}$ valve range: 20–140 nM). The electronic properties of the *para*-substituents did not seem to affect activity, while potency correlated with the presence of more sterically demanding groups (iodo **41** > bromo **40** > chloro **39** = trifluoromethyl **34** > cyano **37** > fluoro **38**). Methyl branched C-TCs **83** and **84** were 1.2-fold and 15-fold less potent than unbranched 4-chloro **39** and 4-nitro **44**, respectively. The introduction of a methyl group at position-3 of the phthalimide scaffold slightly diminished the activity (compare **47** with **39** and **48** with **44**). Conversely, positional isomers **49** and **50**, carrying the methyl group at position-4 of the phthalimide moiety, were 1.6- and 7-fold more potent than **39** and **44**, respectively. Congener **50** was the most active C-TC (EC$_{50}$: 10 nM), being twofold more potent than Trovirdine. The *para*-substituted O-TCs **62, 63, 65–71,** and **73** were active at concentrations within the range 0.1 to 0.5 μM with reduced potency in comparison with their corresponding C-TCs. Sub-micromolar activity was also retained by 2-fluorophenyl O-TC **54**.

TABLE 4.3
TCs Active in the Micromolar Concentration Range[a]

Cpd	B$_{-1/2}$	n	R$_1$	R$_2$	Ar$_1$	CC$_{50}$[b]	EC$_{50}$[c]	S.I.[d]
12	B$_{-1}$	1	H	H	phenyl	>100	1.2	>83.3
13	B$_{-1}$	1	H	H	1-naphthyl	>100	6.5	>15.4
14	B$_{-1}$	1	H	H	2-tolyl	>100	3.7	>27
15	B$_{-1}$	1	H	H	2-isopropylphenyl	>100	43	>2.3
16	B$_{-1}$	1	H	H	2-trifluoromethylphenyl	>100	65	>1.5
17	B$_{-1}$	1	H	H	2-methoxycarbonylphenyl	45	>45	—

TABLE 4.3

TCs Active in the Micromolar Concentration Range[a] (continued)

Cpd	$B_{-1/2}$	n	R_1	R_2	Ar_1	CC_{50}[b]	EC_{50}[c]	S.I.[d]
18	B_{-1}	1	H	H	2-fluorophenyl	>100	3.0	>33
19	B_{-1}	1	H	H	2-chlorophenyl	>100	17	>5.9
20	B_{-1}	1	H	H	2-bromophenyl	>100	14	>7.1
21	B_{-1}	1	H	H	2-nitrophenyl	>100	>100	—
22	B_{-1}	1	H	H	2-methoxyphenyl	>100	5.6	>17.9
23	B_{-1}	1	H	H	3-tolyl	>100	1.5	>66.7
24	B_{-1}	1	H	H	3-trifluoromethylphenyl	>100	13	>7.7
25	B_{-1}	1	H	H	3-acetylphenyl	>100	4.3	>23
26	B_{-1}	1	H	H	3-methoxycarbonylphenyl	>100	21	>4.8
27	B_{-1}	1	H	H	3-chlorophenyl	>100	2.0	>50
28	B_{-1}	1	H	H	3-methylsulfanylphenyl	>100	7.0	>14.3
30	B_{-1}	1	H	H	3-methoxyphenyl	>100	4.0	>25
35	B_{-1}	1	H	H	4-ethoxycarbonylphenyl	>100	23	>4.3
43	B_{-1}	1	H	H	4-diethylaminophenyl	>100	1.40	>71
46	B_{-1}	1	H	H	4-ethoxyphenyl	>100	7.0	>14.3
51	B_{-2}	1	H	H	phenyl	>100	3	>33.3
52	B_{-2}	1	H	H	2-tolyl	>100	27	>3.7
53	B_{-2}	1	H	H	2-methoxycarbonylphenyl	>100	>100	—
55	B_{-2}	1	H	H	2-nitrophenyl	>100	>100	—
56	B_{-2}	1	H	H	2-methoxyphenyl	>100	65	>1.5
57	B_{-2}	1	H	H	3-tolyl	>100	5.0	>20
58	B_{-2}	1	H	H	3-methoxycarbonylphenyl	>100	>100	—
59	B_{-2}	1	H	H	3-chlorophenyl	>100	3.0	>33.3
60	B_{-2}	1	H	H	3-methoxyphenyl	>100	2.6	>38.5
61	B_{-2}	1	H	H	3-methylsulfanylphenyl	>100	11	>9.1
64	B_{-2}	1	H	H	4-isopropylphenyl	>100	3.7	>27
72	B_{-2}	1	H	H	4-ethoxyphenyl	100	2.0	50
77	B_{-5}	1	H	H	4-chlorophenyl	>100	2.4	>42
78	B_{-6}	1	H	H	4-chlorophenyl	>100	9.4	>10.6
79	B_{-6}	1	H	H	4-bromophenyl	>100	2.3	>43
80	B_{-7}	1	H	H	4-chlorophenyl	>100	68	>1.5
81	B_{-7}	1	H	H	4-nitrophenyl	>100	>100	—
82	B_{-1}	0	—	H	4-nitrophenyl	23	>23	—
85[e]	B_{-1}	1	H	CH_3	4-tolyl	>100	48	>2.1
86[e]	B_{-1}	1	H	CH_3	4-cyanophenyl	>100	36	>2.7
87[e]	B_{-1}	1	H	CH_3	4-chlorophenyl	>100	1.7	>58.8
88[e]	B_{-1}	1	H	CH_3	4-bromophenyl	>100	2.3	>43.5
89[e]	B_{-1}	1	H	CH_3	4-nitrophenyl	>100	8.6	>11.6
90	B_{-1}	2	H	H	phenyl	>100	>100	—
91	B_{-1}	2	H	H	4-nitrophenyl	≥100	19	≥5.3
74[e]	B_{-2}	1	H	CH_3	4-tolyl	>100	41	>2.4
75[e]	B_{-2}	1	CH_3	H	4-chlorophenyl	>100	5.0	>20
76[e]	B_{-2}	1	CH_3	H	4-nitrophenyl	>100	13	>7.7
Trovirdine						60	0.02	3,000

[a] Data represent mean values for three separate experiments. Variation among triplicate samples was less than 10%. [b] Compound concentration [μM] required to reduce the viability of mock-infected cells by 50%, as determined by the MTT method. [c] Compound concentration [μM] required to achieve 50% protection of MT-4 cell from the HIV-1 induced cytopathogenicity, as determined by the MTT method. [d] Selectivity index: CC_{50}/EC_{50} ratio. [e] Tested as racemic mixture.

TABLE 4.4

TCs Active at Nanomolar Concentrations[a]

B_1: (phthalimide structure); B_2: (2-carboxamide-benzoic acid structure, NH, COOH); B_3: (3-methylphthalimide, CH_3); B_4: (methyl-substituted phthalimide, H_3C)

General structure: $B_{1/4}$—with R_1, R_2, n, O, NH—Ar_1, S

Cpd	$B_{1/2}$	n	R_1	R_2	Ar_1	CC_{50}[b]	EC_{50}[c]	S.I.[d]
11	B_{-1}	1	H	H	cyclohexyl	>100	0.3	>333
29	B_{-1}	1	H	H	3-nitrophenyl	>100	0.6	>167
31	B_{-1}	1	H	H	4-tolyl	>100	0.02	>5,000
32	B_{-1}	1	H	H	4-ethylphenyl	>100	0.08	>1,250
33	B_{-1}	1	H	H	4-isopropylphenyl	94	0.04	2,350
34	B_{-1}	1	H	H	4-trifluoromethylphenyl	>100	0.04	>2,500
36	B_{-1}	1	H	H	4-acetylphenyl	>100	0.14	>714
37	B_{-1}	1	H	H	4-cyanophenyl	>100	0.07	>1,429
38	B_{-1}	1	H	H	4-fluorophenyl	>100	0.10	>1,000
39	B_{-1}	1	H	H	4-chlorophenyl	>100	0.04	>2,500
40	B_{-1}	1	H	H	4-bromophenyl	>100	0.03	>3,333
41	B_{-1}	1	H	H	4-iodophenyl	>100	0.02	>5,000
42	B_{-1}	1	H	H	4-dimethylaminophenyl	>100	0.60	>167
44	B_{-1}	1	H	H	4-nitrophenyl	65	0.04	1,625
45	B_{-1}	1	H	H	4-methoxyphenyl	>100	0.03	>3,333
47	B_{-3}	1	H	H	4-chlorophenyl	>100	0.05	>2,000
48	B_{-3}	1	H	H	4-nitrophenyl	71	0.07	1,014
49	B_{-4}	1	H	H	4-chlorophenyl	>100	0.025	>4,000
50	B_{-4}	1	H	H	4-nitrophenyl	80	0.01	8,000
54	B_{-2}	1	H	H	2-fluorophenyl	>100	0.5	>200
62	B_{-2}	1	H	H	4-tolyl	>100	0.4	>250
63	B_{-2}	1	H	H	4-ethylphenyl	>100	0.3	>333
65	B_{-2}	1	H	H	4-cyanophenyl	>100	0.1	>1,000
66	B_{-2}	1	H	H	4-fluorophenyl	>100	0.25	>400
67	B_{-2}	1	H	H	4-chlorophenyl	>100	0.1	>1,000
68	B_{-2}	1	H	H	4-bromophenyl	>100	0.1	>1,000
69	B_{-2}	1	H	H	4-iodophenyl	>100	0.13	>769
70	B_{-2}	1	H	H	4-nitrophenyl	>100	0.30	>333
71	B_{-2}	1	H	H	4-methoxyphenyl	>100	0.2	>500
73	B_{-2}	1	H	H	4-benzyloxyphenyl	>100	0.5	>200
83[e]	B_{-1}	1	CH_3	H	4-chlorophenyl	44	0.05	880
84[e]	B_{-1}	1	CH_3	H	4-nitrophenyl	>100	0.6	>167
Trovirdine						60	0.02	3,000

[a] Data represent mean values for three separate experiments. Variation among triplicate samples was less than 10%. [b] Compound concentration [μM] required to reduce the viability of mock-infected cells by 50%, as determined by the MTT method. [c] Compound concentration [μM] required to achieve 50% protection of MT-4 cell from the HIV-1 induced cytopathogenicity, as determined by the MTT method. [d] Selectivity index: CC_{50}/EC_{50} ratio. [e] Tested as racemic mixture.

The isosteric replacement of the NH group or thione sulfur of the thiocarbamic moiety with oxygen (**92**, **93**, and **94**) and the incorporation of the thiocarbamic nitrogen into an imidazole nucleus (**95**) was not beneficial for activity (data not shown), and this is consistent with the H-bond donor role of the NH group to the Lys101 carbonyl (Figure 4.5).

In cell-based assays, C-TCs **31**, **37**, **39**, **40**, and **44** significantly inhibited the Tyr181Cys mutant, while analog **41** was as active as Efavirenz against the K103R mutant (IC_{50}: 2.3 μM)[105] in enzyme assays. Conversely, Lys103Arg and Lys103Asn+Tyr181Cys mutated strains turned out generally unsusceptible to the test compounds in cell based assays.[105]

4.6 CONCLUDING REMARKS AND OUTLOOK

This chapter describes a successful application of parallel synthesis in the NNRTI field. Our unprecedented, highly convergent solution-phase procedure for preparation of ATC and TC libraries, along with structure-based drug design approaches (suggestions from the different binding modes between the ATCs and the TCs proved useful for selecting building blocks for their synthesis), led to structural optimization of the ATC and TC lead compounds, structurally related to PETT derivatives. In particular, ATC solution-phase parallel synthesis allows the incorporation of three sources of diversity, due to the sequential assembly of alcohols, isothiocyanates, and acyl chlorides or chloroformates. An inherent limitation to the diversity of ligands that can be prepared in a parallel way is the third step where (hetero)aroyl chlorides can prevalently be used. However, the modular nature of the ATCs (in the structure of type $R-OCSN(Ar_1)COAr_2$, the R, Ar_1, and Ar_2 molecular portions can individually or concurrently be tuned) gives the opportunity for a high degree of library diversity owing to the great number of available or easily accessible alcohols, isothiocyanates, and acyl chlorides/chloroformates. Of course, a much lower diversity degree is achievable for the TCs, because their convergent synthesis was carried out by assembling two types of building blocks (alcohols and isothiocyanates). One of the main advantages of our approach was the simplicity of work-up and purification procedures that did not need the development of new technologies, thus allowing high-speed and efficient purification of intermediates and/ or final products in a massive parallel way. In conclusion, due to its versatility, our diversity-driven parallel procedure may be a valuable tool in the design and synthesis of new ATCs and TCs with an improved resistance profile toward clinically relevant RT mutations.

REFERENCES

1. Gallo, R.C.; Sarin, P.S.; Gelmann, EP; Robert-Guroff, M.; Richardson, E.; Kalyanaraman, V.S.; Mann, D.; Sidhu, G.D.; Stahl, R.E.; Zolla-Pazner, S.; Leibowitch, J.; and Popovic, M. Isolation of human T-cell leukemia virus in acquired immune deficiency syndrome (AIDS). *Science,* 1983, 220, 865–867.
2. Barre-Sinoussi, F.; Chermann, J.C.; Rey, F.; Nugeyre, M.T.; Chamaret, S.; Gruest, J.; Dauguet, C.; Axler-Blin, C.; Vezinet-Brun, F.; Rouzioux, C.; Rozenbaum, W.; and Montagnier, L. Isolation of a T-lymphotropic retrovirus from a patient at risk for acquired immune deficiency syndrome (AIDS). *Science,* 1983, 220, 868–871.
3. Gomez, C; and Hope, T.J. The ins and outs of HIV replication. *Cell Microbiol.,* 2005, 7, 621–626.
4. Bukrinskaya, A.G. HIV-1 assembly and maturation. *Arch. Virol.,* 2004, 149, 1067–1082.
5. Baba, M. Inhibitors of HIV-1 gene expression and transcription. *Curr. Top. Med. Chem.,* 2004, 4, 871–82.
6. Coffin, J.M. HIV population dynamics *in vivo*: implications for genetic variation, pathogenesis, and therapy. *Science,* 1995, 267, 483–489.
7. Ho, D.D.; Neumann, A.U.; Perelson, A.S.; Chen, W.; Leonard, J.M.; and Markowitz, M. Rapid turnover of plasma virions and CD4 lymphocytes in HIV-1 infection. *Nature,* 1995, 373, 123–126.
8. Wei, X.; Ghosh, S.K.; Taylor, M.E.; Johnson, V.A.; Emini, E.A.; Deutsch, P.; Lifson, J.D.; Bonhoeffer, S.; Nowak, M.A.; Hahn, B.H.; et al. Viral dynamics in human immunodeficiency virus type 1 infection. *Nature,* 1995, 373, 117–122.
9. Coffin, J.C.; Hughes, S.H.; and Varmus, H.E. *Retroviruses.* Cold Spring Harbor Laboratory Press: Plainview, NY, 1999.
10. Neamati, N. Patented small molecule inhibitors of HIV-1 integrase: a 10-year saga. *Expert Opin. Ther. Patents,* 2002, 12, 709–724.
11. De Clercq, E. New approaches toward Anti-HIV chemotherapy, *J. Med. Chem.,* 2005, 48, 1297–1313.

12. Vermeire, K.; and Schols, D. Anti-HIV agents targeting the interaction of gp120 with the cellular CD4 receptor. *Expert Opin. Investig. Drugs,* 2005, 14, 1199–1212.

13. Giacca, M. The HIV-1 Tat protein: a multifaceted target for novel therapeutic opportunities. *Curr. Drug. Targets Immune. Endocr. Metabol. Disord.,* 2004, 4, 277–85.

14. Chang, H.K.; Gallo, R.C.; and Ensoli, B. Regulation of cellular gene expression and function by the human immunodeficiency virus type 1 Tat protein. *J. Biomed. Sci.,* 1995, 3, 189–202.

15. Felber, B.K.; Hadzopoulou-Cladaras, M.; Cladaras, C.; Copeland, T.; and Pavlakis, G.N. Rev protein of human immunodeficiency virus type 1 affects the stability and transport of viral mRNA. *Proc. Natl. Acad. Sci U.S.A.,* 1989, 86, 1495–1499.

16. Goel, A.; Mazur, S.J.; Fattah, R.J.; Hartman, T.L.; Turpin, J.A.; Huang, M.; Rice, W.G.; Appella, E.; and Inman, J.K. Benzamide-based thiolcarbamates: a new class of HIV-1 NCp7 inhibitors. *Bioorg. Med. Chem. Lett.,* 2002, 12, 767–770, and references cited therein.

17. Finzi, D.; Hermankova, M.; Pierson, T.; Carruth, L.M.; Buck, C.; Chaisson, R.E.; Quinn, T.C.; Chadwick, K.; Margolick, J.; Brookmeyer, R.; Gallant, J.; Markowitz, M.; Ho, D.D.; Richman, D.D.; and Siciliano, R.F. Identification of a reservoir for HIV-1 in patients on highly active antiretroviral therapy. *Science,* 1997, 278, 1295–1300.

18. Viard, J.P. Treatment strategy issues for chronic HIV-1 infection in adults: the dilemma of life-long antiretroviral treatment. *Curr. Med. Chem. — Anti-Infective Agents,* 2005, 4, 29–36.

19. Palella, F.J.; Delaney, K.M.; Moorman, A.C.; Loveless, M.O.; Fuhrer, J.; Satten, G.A.; and Aschman, D.J. *N. Engl. J. Med.,* 1998, 338, 853.

20. Mocroft, A.; Vella, S.; Benfield, T.L.; Chiesi, A.; Miller, V.; Gargalianos, P.; d'Arminio Monforte, A.; Yust, I.; Bruun, J.N.; Phillips, A.N.; and Lundgren, J.D. Changing patterns of mortality across Europe in patients infected with HIV-1. *Lancet,* 1998, 352, 1725–1730.

21. Blankson, J.N.; Persaud, D.; and Siciliano, R.F. The challenge of viral reservoirs in HIV-1 infection. *Annu. Rev. Med.,* 2002, 53, 557–593.

22. Hsiou, Y.; Ding, J.; Das, K.; Clark, A.D., Jr; Hughes, S.H.; and Arnold, E. Structure of unliganded HIV-1 reverse transcriptase at 2.7 Å resolution: implications of conformational changes for polymerisation and inhibition mechanism. *Structure,* 1996, 4, 853–860.

23. Patel, P.H.; Jacobo-Molina, A.; Ding, J.; Tantillo, C.; Clark A.D.; Raag, R.; Nanni, R.G.; Hughes, S.H.; and Arnold, E. Insight into polymerization mechanisms from structure and function analysis of HIV-1 reverse transcriptase. *Biochemistry,* 1995, 34, 5351–5363.

24. Kohlstaedt, L.A.; Wang, J.; Fiedman, J.M.; Rice, P.A.; and Steitz, T.A. A crystal structure at 3.5 Å resolution of HIV-1 reverse transcriptase complexed with an inhibitor. *Science,* 1992, 256, 1783–1790.

25. Sarafianos, S.G.; Das, K.; Hughes, S.H.; and Arnold, E. Taking aim at a moving target: designing drugs to inhibit drug-resistant HIV-1 reverse transcriptases. *Curr. Opin. Struct. Biol.,* 2004, 14, 716–730.

26. De Clercq, E. Nonnucleoside reverse transcriptase inhibitors (NNRTIs): Past, present, and future. *Chem. Biodiversity,* 2004, 1, 44–64.

27. Balzarini, J. Current status of the non-nucleoside reverse transcriptase inhibitors of human immunodeficiency virus type 1. *Curr. Top. Med. Chem.,* 2004, 4, 921–944.

28. Pauwels, R. New non-nucleoside reverse transcriptase inhibitors (NNRTIs) in development for the treatment of HIV infections. *Curr. Opin. Pharmacol.,* 2004, 4, 437–446.

29. Fattorusso, C.; Gemma, S.; Butini, S.; Huleatt, P.; Catalanotti, B.; Persico, M.; De Angelis, M.; Fiorini, I.; Nacci, V.; Ramunno, A.; Rodriquez, M.; Greco, G.; Novellino, E.; Bergamini, A.; Marini, S.; Coletta, M.; Maga, G.; Spadari, S.; and Campiani, G. Specific targeting highly conserved residues in the HIV-1 reverse transcriptase primer grip region. Design, synthesis, and biological evaluation of novel, potent, and broad spectrum NNRTIs with antiviral activity. *J. Med. Chem.,* 2005, 48, 7153–7165.

30. Jorgensen, W.L.; Ruiz-Caro, J.; Tirado-Rives, J.; Basavapathruni, A.; Anderson, K.S.; and Hamilton, A.D. Computer-aided design of non-nucleoside inhibitors of HIV-1 reverse transcriptase. *Bioorg. Med. Chem. Lett.,* 2005, 16, 663–667.

31. Camarasa, M.J.; Velazquez, S.; San-Felix, A.; and Perez-Perez, M.J. TSAO derivatives the first non-peptide inhibitors of HIV-1 RT dimerization. *Antivir. Chem. Chemother.,* 2005, 16, 147–153.

32. Andreola, M.L.; Nguyen, C.H.; Ventura, M.; Tarrago-Litvak, L.; and Legraverend, M. Antiviral activity of 4-benzyl pyridinone derivatives as HIV-1 reverse transcriptase inhibitors. *Expert Opin. Emerg. Drugs,* 2001, 6, 225–238.

33. De Martino, G.; La Regina, G.; Di Pasquali, A.; Ragno, R.; Bergamini, A.; Ciaprini, C.; Sinistro, A.; Maga, G.; Crespan, E.; Artico, M.; and Silvestri, R. Novel 1-[2-(diarylmethoxy)ethyl]-2-methyl-5-nitroimidazoles as HIV-1 non-nucleoside reverse transcriptase inhibitors. A structure-activity relationship investigation. *J. Med. Chem.*, 2005, 48, 4378–4388.

34. Di Santo, R.; and Costi, R. 2H-Pyrrolo[3,4-b] [1,5]benzothiazepine derivatives as potential inhibitors of HIV-1 reverse transcriptase. *Farmaco,* 2005, 60, 385–392.

35. Barreca, M.L.; Rao, A.; De Luca, L.; Zappala, M.; Monforte, A.M.; Maga, G.; Pannecouque, C.; Balzarini, J.; De Clercq, E.; Chimirri, A.; and Monforte, P. Computational strategies in discovering novel non-nucleoside inhibitors of HIV-1 RT. *J. Med. Chem.*, 2005, 48, 3433–3437.

36. Balzarini, J.; Auwerx, J.; Rodriguez-Barrios, F.; Chedad, A.; Farkas, V.; Ceccherini-Silberstein, F.; Garcia-Aparicio, C.; Velazquez, S.; De Clercq, E.; Perno, C.F.; Camarasa, M.J.; and Gago, F. The amino acid Asn136 in HIV-1 reverse transcriptase (RT) maintains efficient association of both RT subunits and enables the rational design of novel RT inhibitors. *Mol. Pharmacol.*, 2005, 68, 49–60.

37. Masuda, N.; Yamamoto, O.; Fujii, M.; Ohgami, T.; Fujiyasu, J.; Kontani, T.; Moritomo, A.; Orita, M.; Kurihara, H.; Koga, H.; Kageyama, S.; Ohta, M.; Inoue, H.; Hatta, T.; Shintani, M.; Suzuki, H.; Sudo, K.; Shimizu, Y.; Kodama, E.; Matsuoka, M.; Fujiwara, M.; Yokota, T.; Shigeta, S.; and Baba, M. Studies of non-nucleoside HIV-1 reverse transcriptase inhibitors. 2. Synthesis and structure-activity relationships of 2-cyano and 2-hydroxy thiazolidenebenzenesulfonamide derivatives. *Bioorg. Med. Chem.*, 2005, 13, 949–961.

38. Das, K.; Lewi, P.J.; Hughes, S.H.; and Arnold, E. Crystallography and the design of anti-AIDS drugs: conformational flexibility and positional adaptability are important in the design of non-nucleoside HIV-1 reverse transcriptase inhibitors. *Prog. Biophys. Mol. Biol.*, 2005, 88, 209–231.

39. Ragno, R.; Frasca, S.; Manetti, F.; Brizzi, A.; and Massa, S. HIV-reverse transcriptase inhibition: inclusion of ligand-induced fit by cross-docking studies. *J. Med. Chem.*, 2005, 48, 200–212.

40. Sluis-Cremer, N.; Temiz, N.A.; and Bahar, I. Conformational changes in HIV-1 reverse transcriptase induced by nonnucleoside reverse transcriptase inhibitor binding. *Curr. HIV Res.*, 2004, 2, 323–332.

41. Freeman, G.A.; Andrews, C.W.; Hopkins, A.L.; Lowell, G.S.; Schaller, L.T.; Cowan, J.R.; Gonzales, S.S.; Koszalka, G.W.; Hazen, R.J.; Boone, L.R.; Ferris, R.G.; Creech, K.L.; Roberts, G.B.; Short, S.A.; Weaver, K.; Reynolds, D.J.; Milton, J.; Ren, J.; Stuart, D.I.; Stammers, D.K.; and Chan, J.H. Design of non-nucleoside inhibitors of HIV-1 reverse transcriptase with improved drug resistance properties. 2. *J. Med. Chem.*, 2004, 47, 5923–5936.

42. Hopkins, A.L.; Ren, J.; Milton, J.; Hazen, R.J.; Chan, J.H.; Stuart, D.I.; and Stammers, D.K. Design of non-nucleoside inhibitors of HIV-1 reverse transcriptase with improved drug resistance properties. 1. *J. Med. Chem.*, 2004, 47, 5912–5922.

43. Wang, L.Z.; Kenyon, G.L.; and Johnson, K.A. Novel mechanism of inhibition of HIV-1 reverse transcriptase by a new non-nucleoside analog, KM-1. *J. Biol. Chem.* 2004, 279, 38424–38432.

44. Auwerx, J.; Stevens, M.; Van Rompay, A.R.; Bird, L.E.; Ren, J.; De Clercq, E.; Oberg, B.; Stammers, D.K.; Karlsson, A.; and Balzarini, J. The phenylmethylthiazolylthiourea nonnucleoside reverse transcriptase (RT) inhibitor MSK-076 selects for a resistance mutation in the active site of human immunodeficiency virus type 2 RT. *J. Virol.*, 2004, 78, 7427–7437.

45. Ding, J.; Das, K.; Moereels, H.; Koymans, L.; Andries, K.; Janssen, P.A.; Hughes, S.H.; and Arnold, E. Structure of HIV-1 RT/TIBO R 86183 complex reveals similarity in the binding of diverse non-nucleoside inhibitors. *Nat. Struct. Biol.*, 1995, 5, 407–415.

46. Ding, J.; Das, K.; Tantillo, C.; Zhang, W.; Clark, A.D.J.; Jessen, S.; Lu, V.; Hsiou, Y.; Jacobo-Molina, A.; Andries, K.; Pauwels, R.; Moereels, H.; Koymans, L.; Janssen, P.A.; Smith, R.H.J.; Kroeger Koepke, M.; Michejda, C.J.; Hughes, S.H.; and Arnold, E. Structure of HIV-1 reverse transcriptase in a complex with the non-nucleoside inhibitor alpha-APA R 95845 at 2.8 Å resolution. *Structure,* 1995, 3, 365–379.

47. Das, K.; Clark, A.D.J.; Lewi, P.J.; Heeres, J.; De Jonge, M.R.; Koymans, M.H.; Vinkers, H.M.; Daeyaert, F.; Ludovici, D.W.; Kukla, M.J.; De Corte, B.; Kavash, R.W.; Ho, C.Y.; Ye, H.; Lichtenstein, M.A.; Andries, K.; Pauwels, R.; De Bethune, M.P.; Boyer, P.L.; Clark, P.; Hughes, S.H.; Janssen, P.A.; and Arnold E. Roles of conformational and positional adaptability in structure-based design of TMC125-R165335 (etravirine) and related non-nucleoside reverse transcriptase inhibitors that are highly potent and effective against wild-type and drug-resistant HIV-1 variants. *J. Med. Chem.*, 2004, 47, 2550–2560.

48. Ren, J.; Milton, J.; Weaver, K.L.; Short, S.A.; Stuart, D.I.; and Stammers, D.K. Structural basis for the resilience of efavirenz (DMP-266) to drug resistance mutations in HIV-1 reverse transcriptase. *Structure,* 2000, 8, 1089–1094.

49. Balzarini, J.; Karlsson, A.; Perez-Perez, M.J.; Camarasa, M.J.; Tarpley, W.G.; et al. Treatment of human immunodeficiency virus type 1 (HIV-1)-infected cells with combinations of HIV-1-specific inhibitors results in a different resistance pattern than does treatment with single-drug therapy. *J. Virol.*, 1993, 67, 5353–5359.

50. Bacheler, L.; Jeffrey, S.; Hanna, G.; D'Aquila, R.; Wallace, L.; et al. Genotypic correlates of phenotypic resistance to efavirenz in virus isolates from patients failing non-nucleoside reverse transcriptase inhibitor therapy. *J. Virol.*, 2001, 75, 4999–5008.

51. Hirschmann, R. Medicinal chemistry in the golden age of biology: Lessons from steroid and peptide research. *Angew. Chem. Int. Ed.*, 1991, 30, 1278–1301.

52. Patchett, A.A. Excursions in drug discovery. *J. Med. Chem.*, 1993, 36, 2051–2058.

53. Wiley, R.A.; and Rich, D.H. Peptidomimetics derived from natural products. *Med. Res. Rev.*, 1993, 13, 327–384.

54. Salemme, F.R.; Spurlino, J.; and Bone, R. Serendipity meets precision: the integration of structure-based drug design and combinatorial chemistry for efficient drug discovery. *Structure*, 1997, 5, 319–324.

55. Stanton, R.V.; Mount, J.; and Miller, J.L. Combinatorial library design: maximizing model-fitting compounds within matrix synthesis constraints. *J. Chem. Inf. Comput. Sci.*, 2000, 40, 701–705.

56. Boger, D.L.; Desharnais, J.; and Capps, K. Solution-phase combinatorial libraries: modulating cellular signaling by targeting protein-protein or protein-DNA interactions. *Angew. Chem. Int. Ed.*, 2003, 42, 4138–4176.

57. Austel, V. Solution-phase combinatorial chemistry. In *Combinatorial Chemistry, Synthesis, Analysis, Screening*; G. Jung, Ed.; Wiley-VCH, 1999, p. 103.

58. Ugi, I.; Domling, A.; and Werner, B. Since 1995 the new chemistry of multicomponent reactions and their libraries, including their heterocyclic chemistry. *J. Heterocycl. Chem.*, 2000, 37, 647.

59. Ugi, I.; and Heck, S. The multicomponent reactions and their libraries for natural and preparative chemistry. *Comb. Chem. High Throughput Screening*, 2001, 4, 1–34.

60. Ugi, I. Recent progress in the chemistry of multicomponent reactions. *Pure Appl. Chem.*, 2001, 73, 187–191.

61. Domling, A.; and Ugi, I. Multicomponent reactions with isocyanides. *Angew. Chem. Int. Ed.*, 2000, 39, 3168–3210.

62. Orru, R.V.A.; and de Greef, M. Recent advances in solution-phase multicomponent methodology for the synthesis of heterocyclic compounds. *Synthesis*, 2003, 10, 1471–1499.

63. Hulme, C.; and Gore, V. Multi-component reactions: emerging chemistry in drug discovery from xylocain to crixivan. *Curr. Med. Chem.*, 2003, 10, 51–80.

64. Walter, W.; and Bode, K.D. Syntheses of thiocarbamates. *Angew. Chem. Int. Ed.*, 1967, 6, 281–384, and references therein.

65. Digenis, G.A.; and Rodis, N.P. U.S. Patent 5539123, 1996; *Chem. Abstr.*, 1996, 125, 167993n.

66. Trivedi, B.K. U.S. Patent, 8810816, 1988; *Chem. Abstr.*, 1988, 108, 167428d.

67. Ranise, A.; Bondavalli, F.; Bruno, O.; Schenone, S.; Losasso, C.; Costantino, M.; Cenicola, M.L.; Donnoli, D.; and Marmo, E. 3,3-Disubstituted 1-acyl-1-phenylthioureas with platelet antiaggregating and other activities. *Farmaco*, 1991, 46, 317-338.

68. Ranise, A.; Bondavalli, F.; Bruno, O.; Schenone, S.; Donnoli, D.; Parrillo, C.; Cenicola, M.L.; and Rossi, F. 1-Acyl-, 3-acyl- and 1,3-diacyl-3-furfuryl-1-phenylthioureas with platelet antiaggregating and other activities. *Farmaco*, 1991, 46, 1203–1216.

69. Ranise, A.; Spallarossa, A.; Bruno, O.; Schenone, S.; Fossa, P.; Menozzi, G.; Bondavalli, F.; Mosti, L.; Capuano, A.; Mazzeo, F.; Falcone, G.; and Filippelli, W. Synthesis of N-substituted-N-acylthioureas of 4-substituted piperazines endowed with local anaesthetic, antihyperlipidemic, antiproliferative activities and antiarrythmic, analgesic, antiaggregating actions. *Farmaco*, 2003, 58, 765–780.

70. Ahgren, C.; Backro, K.; Bell, F.W.; Cantrell, A.S.; Clemens, M.; Colacino, J.M.; Deeter, J.B.; Engelhardt, J.A.; Hogberg, M.; Jaskunas, S.R.; Johansson, N.G.; Jordan, C.L.; Kasher, J.S.; Kinnick, M.D.; Lind, P.; Lopez, C.; Morin, J.M., Jr.; Muesing, M.A.; Noreen, R.; Oberg, B.; Paget, C.J.; Palkowitz, J.A.; Parrish, C.A.; Pranc, P.; Rippy, M.K.; Rydergard, C.; Sahlberg, C.; Swanson, S.; Ternansky, R.J.; Unge, T.; Vasileff, R.T.; Vrang, L.; West, S.J.; Zhang, H.; and Zhou, X.-X. The PETT series, a new class of potent nonnucleoside inhibitors of human immunodeficiency virus type 1 reverse transcriptase. *Antimicrob. Agents Chemother.*, 1995, 39, 1329–1335.

71. Bell, F.W.; Cantrell, A.S.; Hogberg, M.; Jaskunas, S.R.; Johansson, N.G.; Jordan, C.L.; Kinnick, M.D.; Lind, P.; Morin, J.M. Jr.; Noreen, R.; Oberg, B.; Palkowitz, J.A.; Parrish, C.A.; Pranc, P.; Sahlberg, C.; Ternansky, R.J.; Vasileff, R.T.; Vrang, L.; West, S.J.; Zhang, H.; and Zhou, X.-X. PETT compounds, a new class of HIV-1 reverse transcriptase inhibitors. 1. Synthesis and basic structure-activity relationship studies of PETT analogs. *J. Med. Chem.*, 1995, 38, 4929–4936.

72. Cantrell, A.S.; Engelhardt, P.; Hogberg, M.; Jaskunas, S.R.; Johansson, N.G.; Jordan, C.L.; Kangasmetsa, J.; Kinnick, M.D.; Lind, P.; Morin, J.M. Jr.; Muesing, M.A.; Noreen, R.; Oberg, B.; Pranc, P.; Sahlberg, C.; Ternansky, R.J.; Vasileff, R.T.; Vrang, L.; West, S.J.; and Zhang, H. Phenethylthiazolylthiourea (PETT) compounds as a new class of HIV-1 reverse transcriptase inhibitors. 2. Synthesis and further structure-activity relationship studies of PETT analogs. *J. Med. Chem.,* 1996, 39, 4261–4274.

73. Ren, J.; Diprose, J.; Warren, J.; Esnouf, R.M.; Bird, L.E.; Ikemizu, S.; Slater, M.; Milton, J.; Balzarini, J.; Stuart, D.I.; and Stammers, D.K. Phenylethylthiazolylthiourea (PETT) non-nucleoside inhibitors of HIV-1 and HIV-2 reverse transcriptases. Structural and biochemical analyses. *J.Biol.Chem.,* 2000, 275, 5633–5639.

74. Combinatorial synthesis and computer-aided design of anti-HIV drugs: A review. *Combinatorial Chemistry & High Throughput Screening,* 2005, 8, 375–443.

75. Ranise, A.; Bruno, O.; Schenone, S.; and Bondavalli, F. One pot synthesis of N-acyl-N-phenylthiocarbamates of α-, β-, γ-pyridylcarbinols and phenoxyethanol with potential hypolipidemic activity. *II Italian-Spanish Joint Meeting of Medicinal Chemistry.* Ferrara, Italy, 1995, Poster 115.

76. Ranise, A.; Spallarossa, A.; Bruno, O.; Schenone, S.; and Bondavalli, F. One pot synthesis of N-acyl-N-phenylthiocarbamate derivatives endowed with anti-HIV and/or antiproliferative acitivities. *Italian-Hungarian-Polish Joint Meeting on Medicinal Chemistry.* Taormina, Italy, 1999, Poster 50.

77. Van Derpoorten, K.; Balzarini, J.; De Clercq, E.; and Poupaert, J.H. Anti-HIV activity of N-1-adamantyl-4-aminophthalimide. *Biomed. Pharmacother.,* 1997, 51, 464–468.

78. Vamecq, J.; Van derpoorten, K.; Poupaert, J.H.; Balzarini, J.; De Clercq, E.; and Stables, J.P. Anticonvulsant phenytoinergic pharmacophores and anti-HIV activity—preliminary evidence for the dual requirement of the 4-aminophthalimide platform and the N-(1-adamantyl) substitution for antiviral properties. *Life Sci.,* 1998, 63, 267–274.

79. Kolocouris, N.; Foscolos, G.B.; Kolocouris, A.; Marakos, P.; Pouli, N.; Fytas, G.; Ikeda, S.; and De Clercq, E. Synthesis and antiviral activity evaluation of some aminoadamantane derivatives. *J. Med. Chem.,* 1994, 37, 2896–2902.

80. Goldman, M.E. Discovery and development of 2-pyridinone HIV-1 reverse transcriptase. In *The Search for Antiviral Drugs. Case Histories from Concept to Clinic*; Adams, J., Merluzzi, V.J., Eds.; Birkhäuser: Boston, 1993; p. 105.

81. Campiani, G.; Fabbrini, M.; Morelli, E.; Nacci, V.; Greco, G.; Novellino, E.; Maga, G.; Spadari, S.; Bergamini, A.; Faggioli, E.; Uccella, I.; Bolacchi, F.; Marini, S.; Coletta, M.; Fracasso, C.; and Caccia, S. Non-nucleoside HIV-1 reverse transcriptase inhibitors: synthesis and biological evaluation of novel quinoxalinylethylpyridylthioureas as potent antiviral agents. *Antivir. Chem. Chemother.,* 2000, 11, 141–155.

82. Ranise, A.; Spallarossa, A.; Schenone, S.; Bruno, O.; Bondavalli, F.; Vargiu, L.; Marceddu, T.; Mura, M.; La Colla, P.; and Pani, A. Design, synthesis, SAR, and molecular modeling studies of acylthiocarbamates: a novel series of potent non-nucleoside HIV-1 reverse transcriptase inhibitors structurally related to phenethylthiazolylthiourea derivatives. *J. Med. Chem.,* 2003, 46, 768–781.

83. Smith, B.M.; and March, J. Elimination. In *March's Advanced Organic Chemistry*, 5th ed.; John Wiley & Sons, Inc., New York, 2001, p. 1338.

84. Pearson, R.G. Hard and soft acids and bases. *J. Am. Chem. Soc.,* 1963, 85, 3533–3539.

85. Pearson, R.G. Chemical hardness and bond dissociation energies. *J. Am. Chem. Soc.,* 1988, 110, 7684–7690.

86. Smith, P.A.S. *The Chemistry of the Open-Chain Nitrogen Compounds*, Vol. 1, W.A. Benjamin, New York, 1965, p. 233.

87. Hegarty, A.F.; and Bruice, T.C. Acyl transfer reactions from and to the ureido functional group. I. Mechanisms of hydrolysis of an O-acylisourea (2-amino-4,5-benzo-6-oxo-1,3-oxazine). *J. Am. Chem. Soc.,* 1970, 92, 6561–6567.

88. Hegarty, A.F.; and Bruice, T.C. Acyl transfer reactions from and to the ureido functional group. II. Mechanisms of aminolysis of an O-acylisourea (2-amino-4,5-benzo-6-oxo-1,3-oxazine). *J. Am. Chem. Soc.,* 1970, 92, 6568–6574.

89. Hegarty, A.F.; and Bruice, T.C. Acyl transfer reactions from and to the ureido functional group. III. Mechanisms of intramolecular nucleophilic attack of the ureido functional group upon acyl groups. *J. Am. Chem. Soc.,* 1970, 92, 6575–6588.

90. Hegarty, A.F.; Pratt, R.F.; Giudici, T.; and Bruice, T.C. Acyl transfer reactions from and to the ureido functional group. IV. Neighboring carboxyl group general acid catalysis in the hydrolysis of an O-acylisourea (2-amino-8-carboxy-4-oxo-3,1,4-benzoxazine). *J. Am. Chem. Soc.,* 1971, 93, 1428–1434.

91. McCarty, C.G. *The Chemistry of the Carbon-Nitrogen Double Bond*, S. Patai, Ed., Interscience, New York, 1970, Chapter 9, p 363, and references cited therein.

92. Bruice, T.C.; and Hegarty, A.F. Biotin-bound CO_2 and the mechanism of enzymatic carboxylation reactions. *Proc. Natl. Acad. Sci. U.S.A.*, 1970, 65, 805–809.

93. Dixon, A.E.; and Taylor, J. The acylation of thiocarbamides. *J. Chem. Soc. Trans.*, 1920, 117, 720–728.

94. Curtin, D.Y.; and Miller, L.L. The isolation and rearrangement of simple isoimides (iminoanhydrides). *Tetrahedron Lett.*, 1965, 6, 1869–1876.

95. Schulenberg, J.W., and Archer, S. The Chapman Rearrangement. In *Organic Reactions,* Vol. 14; A.C. Cope, Ed.; Wiley-VCH, 1965; p. 1, and references cited therein.

96. Sano, T.; Ohashi, K.; and Oriyama, T. Remarkably fast acylation of alcohols with benzoyl chloride promoted by TMEDA. *Synthesis*, 1999, 7, 1141–1144.

97. Jirman, J.; and Lycka, A. ^{15}N, ^{13}C, and ^{1}H NMR spectra of acylated ureas and thioureas, *Collection Czechoslovak Chem. Com.*, 1987, 52, 2474–2481.

98. Tantillo, C.; Ding, J.; Jacobo-Molina, A.; Nanni, R.G.; Boyer, P.L.; Hughes, S.H.; Pauwels, R.; Andries, K.; Janssen, P.A.; and Arnold, E. Locations of anti-AIDS drug binding sites and resistance mutations in the three-dimensional structure of HIV-1 reverse transcriptase. Implications for mechanisms of drug inhibition and resistance. *J. Mol. Biol.*, 1994, 243, 369–387.

99. Maass, G.; Immendoerfer, U.; Koenig, B.; Leser, U.; Mueller, B.; Goody, R.; and Pfaff, E. Viral resistance to the thiazolo-iso-indolinones, a new class of nonnucleoside inhibitors of human immunodeficiency virus type 1 reverse transcriptase. *Antimicrob. Agents Chemother.*, 1993, 37, 2612–2617.

100. Wermuth, C.G. Strategies in the search for new lead compounds or original work hypotheses. In *The Practice of Medicinal Chemistry*, 2nd ed.; Wermuth, C.G., Ed.; Academic Press, 2003, pp. 76–77.

101. Mager, P.P. Evidence of a butterfly-like configuration of structurally diverse allosteric inhibitors of the HIV-1 reverse transcriptase. *Drug Des. Discov.*, 1996, 14, 241–257.

102. Hsiou, Y.; Das, K.; Ding, J.; Clark, A.D.J.; Kleim, J.P.; Rosner, M.; Winkler, I.; Riess, G.; Hughes, S.H.; and Arnold, E. Structures of Tyr188Leu mutant and wild-type HIV-1 reverse transcriptase complexed with the non-nucleoside inhibitor HBY 097: inhibitor flexibility is a useful design feature for reducing drug resistance. *J. Mol. Biol.*, 1998, 284, 313–323.

103. Mao, C.; Sudbeck, E.A.; Venkatachalam, T.K.; and Uckun, F.M. Structure-based drug design of non-nucleoside inhibitors for wild-type and drug-resistant HIV reverse transcriptase. *Biochem. Pharmacol.*, 2000, 60, 1251–1265.

104. McMahon, J.B.; Buckheit, R.W.J.; Gulakowski, R.J.; Currens, M.J.; Vistica, D.T.; Shoemaker, R.H.; Stinson, S.F.; Russell, J.D.; Bader, J.P.; Narayanan, V.L.; Schultz, R.J.; Brouwer, W.G.; Felauer, E.E.; and Boyd, M.R. Biological and biochemical anti-human immunodeficiency virus activity of UC 38, a new non-nucleoside reverse transcriptase inhibitor. *J. Pharmacol. Exp. Ther.*, 1996, 276, 298–305.

105. Ranise, A.; Spallarossa, A.; Cesarini, S.; Bondavalli, F.; Schenone, S.; Bruno, O.; Menozzi, G.; Fossa, P.; Mosti, L.; La Colla, M.; Sanna, G.; Murreddu, M.; Collu, G.; Busonera, B.; Marongiu, M.E.; Pani, A.; La Colla, P.; and Loddo, R. Structure-based design, parallel synthesis, structure-activity relationship and molecular modelling studies of thiocarbamates, new potent non-nucleoside HIV-1 reverse transcriptase inhibitor isosteres of phenethylthiazolylthiourea derivatives. *J. Med. Chem.*, 2005, 48, 3858–3873.

106. Campiani, G.; Morelli, E.; Fabbrini, M.; Nacci, V.; Greco, G.; Novellino, E.; Ramunno, A.; Maga, G.; Spadari, S.; Caliendo, G.; Bergamini, A.; Faggioli, E.; Uccella, I.; Bolacchi, F.; Marini, S.; Coletta, M.; Nacca, A.; and Caccia, S. Pyrrolobenzoxazepinone derivatives as non-nucleoside HIV-1 RT inhibitors: further structure-activity relationship studies and identification of more potent broad-spectrum HIV-1 RT inhibitors with antiviral activity. *J. Med. Chem.*, 1999, 42, 4462–4470.

107. Vig, R.; Mao, C.; Venkatachalam, T.K.; Tuel-Ahlgren, L.; Sudbeck, E.A.; and Uckun, F.M. Rational design and synthesis of phenethyl-5-bromopyridyl thiourea derivatives as potent non-nucleoside inhibitors of HIV reserve transcriptase. *Bioorg. Med. Chem.*, 1998, 6, 1789–1797.

108. Sahlberg, C.; Noreen, R.; Engelhardt, P.; Hogberg, M.; Kangasmetsa, J.; Vrang, L.; and Zhang, H. Synthesis and anti-HIV activities of urea-PETT analogs belonging to a new class of potent non-nucleoside HIV-1 reverse transcriptase inhibitors. *Bioorg. Med. Chem. Lett.*, 1998, 8, 1511–1516.

109. Mao, C.; Vig, R.; Venkatachalam, T.K.; Sudbeck, E.A.; and Uckun, F.M. Structure-based design of N-[2-(1-piperidinylethyl)]-N′-[2-(5-bromopyridyl)]-thiourea and N-[2-(1-piperazinylethyl)]-N′-[2-(5-bromopyridyl)]-thiourea as potent non-nucleoside inhibitors of HIV-1 reverse transcriptase. *Bioorg. Med. Chem. Lett.*, 1998, 8, 2213–2218.

110. Mao, C.; Sudbeck, E.A.; Venkatachalam, T.K.; and Uckun, F.M. Rational design of N-[2-(2,5-dimethoxyphenylethyl)]-N′-[2-(5-bromopyridyl)]-thiourea (HI-236) as a potent non-nucleoside inhibitor of drug-resistant human immunodeficiency virus. *Bioorg. Med. Chem. Lett.,* 1999, 9, 1593–1598.

111. Uckun, F.M.; Mao, C.; Pendergrass, S.; Maher, D.; Zhu, D.; Tuel-Ahlgren, L.; and Venkatachalam, T.K. N-[2-(1-cyclohexenyl)ethyl]-N′-[2-(5-bromopyridyl)]-thiourea and N′-[2-(1-cyclohexenyl)ethyl]-N′-[2-(5-chloropyridyl)]-thiourea as potent inhibitors of multidrug-resistant human immunodeficiency virus-1. *Bioorg. Med. Chem. Lett.,* 1999, 9, 2721–2726.

112. Uckun, F.M.; Pendergrass, S.; Maher, D.; Zhu, D.; Tuel-Ahlgren, L.; Mao, C.; and Venkatachalam, T.K. N′-[2-(2-thiophene)ethyl]-N′-[2-(5-bromopyridyl)] thiourea as a potent inhibitor of NNI-resistant and multidrug-resistant human immunodeficiency virus-1. *Bioorg. Med. Chem. Lett.,* 1999, *9,* 3411–3416.

113. Hogberg, M.; Sahlberg, C.; Engelhardt, P.; Noreen, R.; Kangasmetsa, J.; Johansson, N.G.; Oberg, B.; Vrang, L.; Zhang, H.; Sahlberg, B.L.; Unge, T.; Lovgren, S.; Fridborg, K.; and Backbro, K. Urea-PETT compounds as a new class of HIV-1 reverse transcriptase inhibitors. 3. Synthesis and further structure-activity relationship studies of PETT analogues. *J Med Chem.,* 1999, 42, 4150–4160.

114. Hogberg, M.; Engelhardt, P.; Vrang, L.; and Zhang, H. Bioisosteric modification of PETT-HIV-1 RT-inhibitors: synthesis and biological evaluation. *Bioorg. Med. Chem. Lett.,* 2000, 10, 265–268.

115. Dong, Y.; Venkatachalam, T.K.; Narla, R.K.; Trieu, V.N.; Sudbeck, E.A.; and Uckun, F.M. Antioxidant function of phenethyl-5-bromo-pyridyl thiourea compounds with potent anti-HIV activity. *Bioorg. Med. Chem. Lett.,* 2000, 10, 87–90.

116. Venkatachalam, T.K.; Sudbeck, E.A.; Mao, C.; and Uckun, F.M. Stereochemistry of halopyridyl and thiazolyl thiourea compounds is a major determinant of their potency as nonnucleoside inhibitors of HIV-1 reverse transcriptase. *Bioorg. Med. Chem. Lett.,* 2000, 10, 2071–2074.

117. Da Settimo, A.; Primofiore G.; Ferrarini, P.; Livi, O.; Tellini, N.; and Bianchini P. Synthesis and local anesthetic activity of some N-b-diethylaminoethylphthalimides. *Eur. J. Med. Chem.,* 1981, 16, 59–64, and references cited therein.

118. Peck, R.M. Anomalous hydrolytic behavior of some basically-substituted phthalimides. A novel rearrangement of a 4-aminoquinoline side chain. *J. Org. Chem.,* 1962, 27, 2677–2679.

119. For general solvent-free procedures see: Chamberlain, S.D.; Biron, K.K.; Dornsife, R.E.; Averett, D.R.; Beauchamp, L.; and Koszalka, G.W. An enantiospecific synthesis of the human cytomegalovirus antiviral agent [(R)-3-((2-amino-1,6-dihydro-6-oxo-9H-purin-9-yl)methoxy)-4-hydroxybutyl]phosphonic acid. *J. Med. Chem.,* 1994, 37, 1371–1377.

120. For analagous syntheses see: Tanaka, K.; and Toda, F. Solvent-free organic synthesis. *Chem. Rev.,* 2000, 100, 1025–1074.

Rongshi Li

CONTENTS

5.1   INTRODUCTION

High-throughput medicinal chemistry (HTMC) plays a pivotal role in lead discovery and optimization, especially to improve efficiency in the synthesis of library compounds. It originated in Combinatorial Chemistry in the early 1990s and evolved to date as a powerful tool in drug discovery. This efficient tool promotes the rapid generation of both relevant molecules and potential drug candidates, and is widely used in pharmaceuticals, biotechnology, and other chemistry-related discovery programs. HTMC has resulted in the rapid expansion of compound libraries to keep pace with the demands of high-throughput screening. The ability to do novel chemistry in multistep parallel synthesis of privileged structures in both the solution and solid phase has allowed for the creation of diverse chemical libraries in high quality and purity. Efficient synthesis of well-designed compound libraries requires a knowledge-based strategy of both solution- and solid-phase syntheses with appropriate combinations of both.[1,2] Well-designed privileged scaffolds and lead-like or drug-like compound libraries should contain diverse sets of templates and building blocks that bear sufficient functionality to address and answer critical questions of structure-activity relationships (SARs). In this case, a high success rate of the library is also essential. Screening well-designed small-molecule libraries that utilize drug-relevant building blocks and biologically privileged scaffolds can provide better coverage of drug-like chemically accessible space, which will enhance the probability of successful lead discovery.

5.2 CASE STUDY I: DISCOVERY AND OPTIMIZATION OF NOVEL ALK INHIBITORS

5.2.1 Introduction

It is estimated that approximately 33% of drug discovery programs target protein kinases.[3] There are 518 kinases, which constitute approximately 1.7% of the human genome. In this largest enzyme family, more than 90 members are protein tyrosine kinases, which are further divided into receptor tyrosine and nonreceptor tyrosine kinases, all of which have highly conserved catalytic domains.[4] Despite the challenges to discover selective kinase inhibitors for therapeutic use, it has been proven that small molecule kinase inhibitors are effective drugs after FDA approval of Gleevec (**1**), Iressa (**2**), Tarceva (**3**), Nexavar (**4**), Sutent (**5**), and Sprycel (**6**) (Figure 5.1). It is estimated that there are more than 30 kinase inhibitors currently in clinical development, and many more are in preclinical studies.[5]

Anaplastic lymphoma kinase (ALK) is a promising new target for therapy of certain cancers, such as anaplastic large-cell lymphoma (ALCL) and inflammatory myofibroblastic tumor (IMT).[6–8] It was originally identified by virtue of its involvement in the t(2;5)(p23;q35) chromosomal translocation that occurs in a subset of non-Hodgkin's lymphoma (NHL) known as the anaplastic large-cell

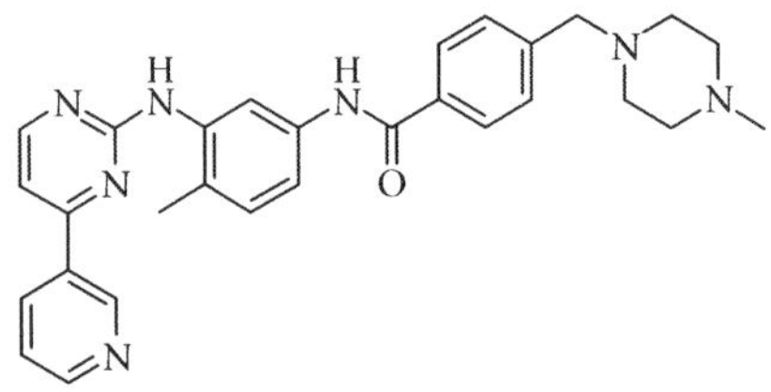

1, Gleevec

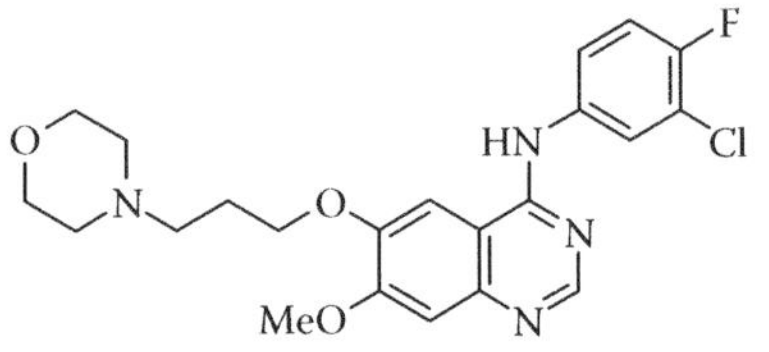

2, Iressa

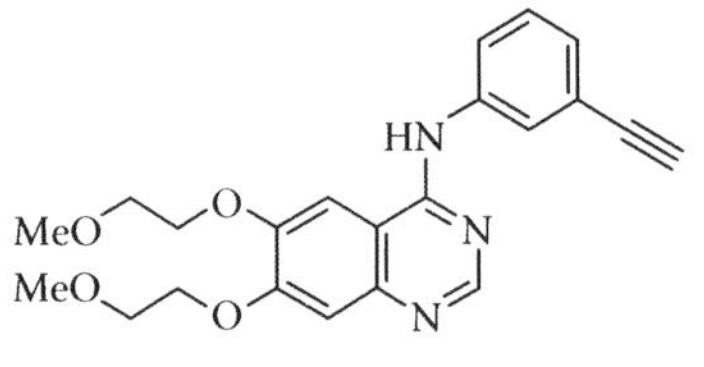

3, Tarceva

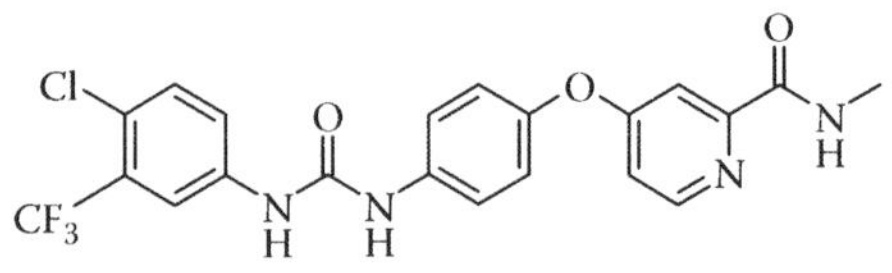

4, Nexavar

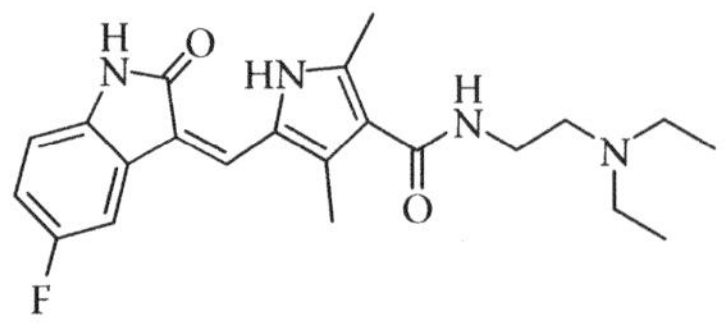

5, Sutent

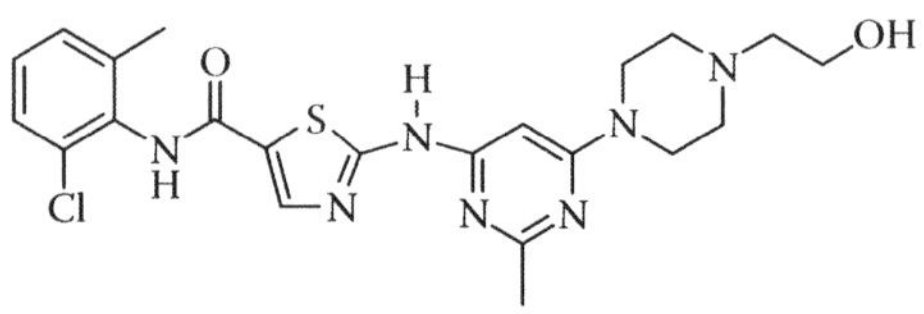

6, Sprycel

FIGURE 5.1 Structures of kinase inhibitors approved by the FDA for cancer therapy.

lymphomas (ALCLs). The positional cloning of the t(2;5) chromosomal rearrangement, identifying the nucleophosmin (NPM)–anaplastic lymphoma kinase fusion gene as reported by Morris.[6] The normal ALK gene is a member of the insulin receptor superfamily and expressed normally in the central and peripheral nervous systems.[6–12] Since ALK as a promising new target for cancer therapy was discovered,[6] there have been many potent inhibitors reported. A thorough and comprehensive review on a single oncogenic protein, ALK, and its small molecule inhibitors can be found in a recent publication.[13]

5.2.2 "EARLY LEAD" IDENTIFICATION ASSISTED BY STRUCTURE-BASED DRUG DESIGN

Our approach to identify a series of novel inhibitors of ALK started with initially screening a set of 2677 compounds from a kinase-targeted library.[14] This screening yielded numerous "hits" from several distinct structural series that showed IC_{50}s < 20 μM in a biochemical ALK inhibition assay. Because no crystal structures of ALK were available, the group chose to use homology modeling to assist in the evaluation of novel scaffolds, rank virtual libraries, and perform docking studies. Since the initial combination of molecular structural determination, together with computation as an important emerging tool for drug development, and its application to acquired immunodeficiency syndrome (AIDS) and bacterial drug resistance issues,[15] docking and scoring technologies have been widely utilized at different stages of the drug discovery process. Among the three main challenges of (1) predicting the binding mode of a known active ligand, (2) predicting the binding affinities of related compounds from a known active series, and (3) identifying new ligands using virtual screening, prediction of the mode of ligand binding in the active site of a protein is the most straightforward and is the application in which most success has been achieved.[16] Templates of the insulin receptor kinase (IRK) and c-ABL crystal structures were used by the ChemBridge Research Laboratories/St. Jude Children's Research Hospital investigators to build the ALK homology model (the amino acid identities of the IRK and c-ABL within their kinase domains to ALK are 45% and 40%, respectively). Employing this homology model, docking studies were performed using an energy function including the internal energy of the ligand based on the ECEPP/3 force field, van der Waals, hydrogen-bonding, electrostatic and hydrophobic ligand/receptor interaction terms.[17–19] A virtual library around a novel template, pyridone, was then designed to follow up and optimize one of the original hits. Docking studies against ALK homology model showed the pyridone virtual library of 724 compounds to yield a 5.5% hit rate, while the average hit rate of a randomly selected reference library was only 1.3%, with a standard deviation of 1.0%; therefore, compounds from a virtual library of the pyridone scaffold had a 4-fold higher likelihood to be a potential ALK ligand than randomly selected compounds.[20]

5.2.3 OPTIMIZATION OF NOVEL PYRIDONE SERIES AS ALK INHIBITORS

To synthesize the first batch of compounds around the pyridone scaffold as shown in Figure 5.2, the development of appropriate chemistry is required. It is well known that 2-hydroxypyridine coexists with the corresponding 2-pyridone through tautomeric equilibrium. 2-Hydroxypyridine can be alkylated on both the oxygen and the nitrogen, unlike 3-hydroxypyridine as a typical phenol.[21]

As postulated in Scheme 5.1, compound **7** underwent Mitsunobu reaction with *p*-methoxybenzyl alcohol to afford one compound **8a** and **8b** in yields of 56% and 4%, respectively. Our attempts to remove the PMB group from **8a** failed under various conditions (DDQ, CAN, $BF_3.Et_2O$, TFA, and hydrogenation), whereas PMB on **8b** is extremely sensitive to TFA treatment. NMR analysis indicated that the major product **8a** exists as a pyridone, the PMB group is attached at *N*, and **8b** is an *O*-alkylated pyridine. Both **8a** and **8b** underwent Suzuki coupling reaction smoothly with 3,4-methylenedioxyphenyl boronic acid to yield **9a** and **9b**, respectively. As expected, compound **7** without a protecting group failed to react with any boronic acid under various Suzuki coupling conditions. These results led us to determine that the solution-phase approach cannot be utilized

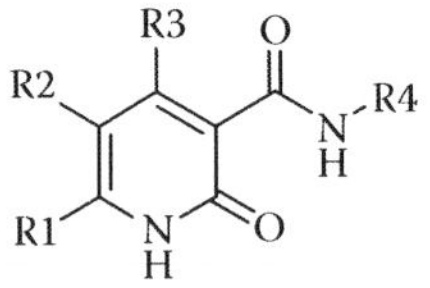

FIGURE 5.2 Pyridone series.

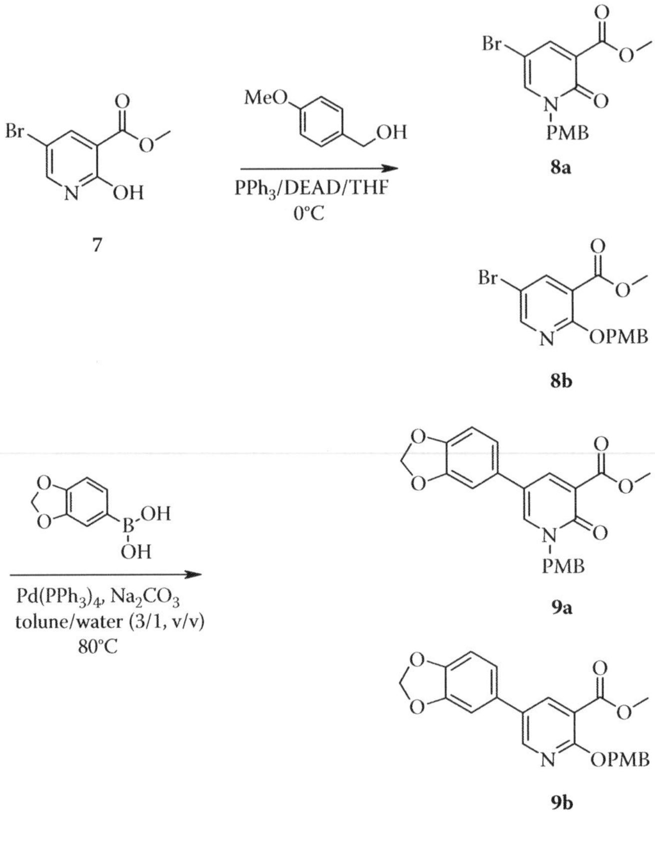

SCHEME 5.1

due to the inertness of the *N*-linked PMB group toward chemical manipulations. Surprisingly, the treatment of resin-bound **7** (hydroxyl group attached to Wang resin) with 10% TFA/DCM for 2 hr released **7** in greater than 50% yield and excellent purity. This surprising difference may have resulted from the predominant formation of acid-sensitive *O*-alkylated product on the solid support. The solid support, in this case, plays a unique role in the synthesis as a "gatekeeper," in that the undesired *N*-alkylated compound cannot be cleaved and has no impact on the purity of the final product. The first dozen compounds (**10 series**) were designed and successfully synthesized using the solid-phase approach as shown in Scheme 5.2.[22] Some of the compounds are confirmed as ALK inhibitors and their enzymatic activities are shown in Table 5.1.

The most potent compounds from the series, **10a–10d**, showed similar or little better enzymatic activity to that of the best compound from the screening set but with fourfold selectivity over IRK. The result is quite encouraging because the series represents a completely novel chemo type as an ALK inhibitor. With the robust chemistry developed in hand, a focused library of 48 compounds was designed and synthesized. The library was designed in a way to keep the left-hand piece of pyridone intact and vary R1 and R2 using a diverse set of aldehydes.[22] However, the screening results

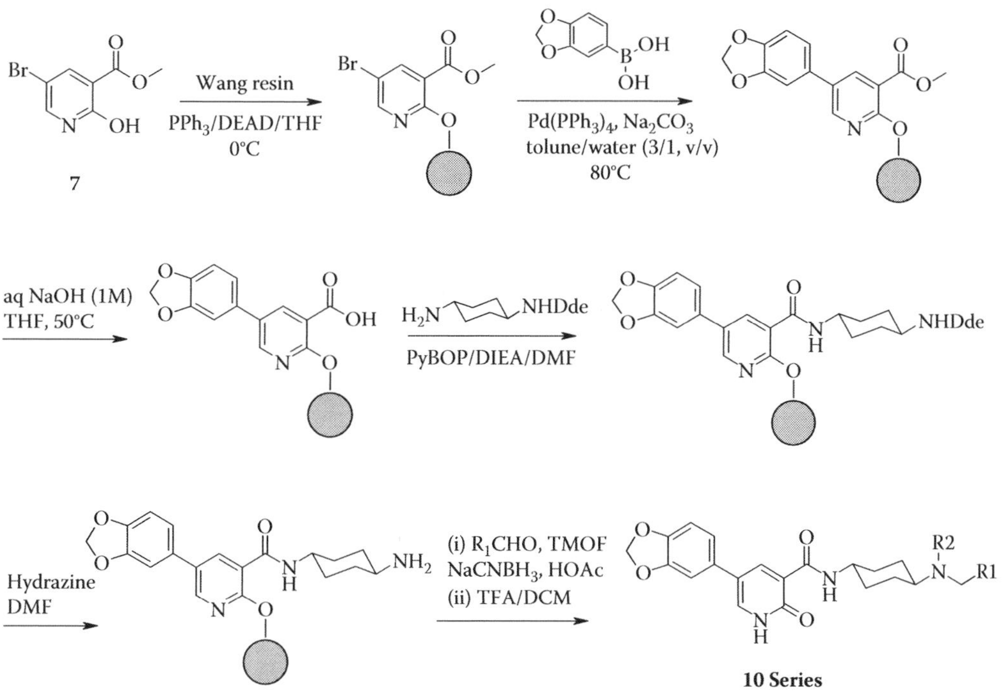

10 Series

SCHEME 5.2

TABLE 5.1
The First Generation of Pyridone Focused Library

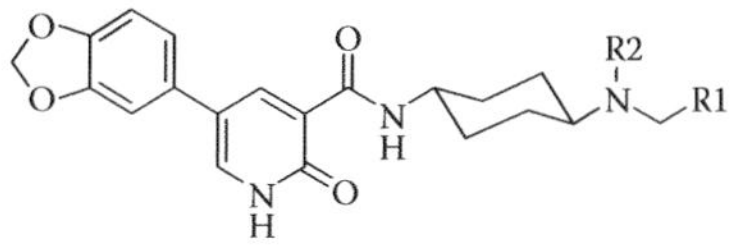

Entry	R1	R2	ALK Inhibition IC_{50} (μM)	IRK Inhibition IC_{50} (μM)
10a	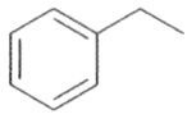	H	4.3	16
10b	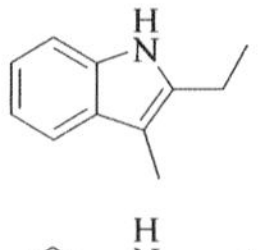	H	4.8	15
10c	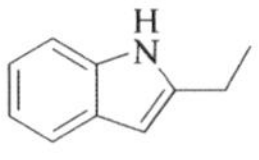	H	6.4	20
10d	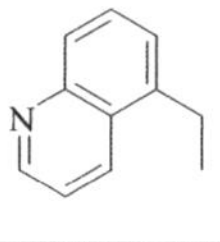	H	6.8	>40

of the library did not yield compounds better than **10a**. A welcome breakthrough was realized by replacing the *trans*-1,4-diaminocyclohexane spacer with an aromatic surrogate. Two compounds (**11a** and **11b**) out of two dozen compounds synthesized exhibited sub-micromolar enzymatic activity (Table 5.2 and Scheme 5.3).

TABLE 5.2

The Second Generation of Pyridone Focused Library

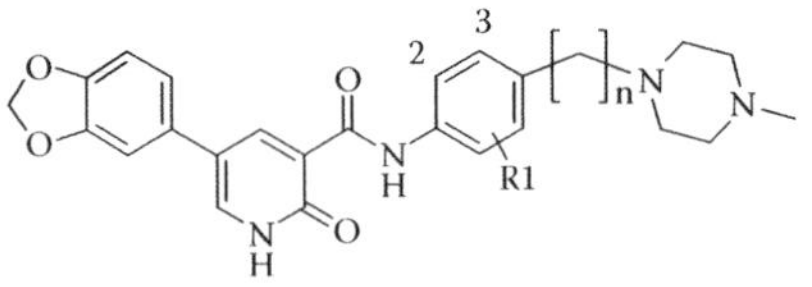

Entry	R1	n	ALK Inhibition IC$_{50}$ (μM)
11a	H	2	0.4
11b	H	0	0.7
11c	H	1	0.9
11d	3-F	0	1.1
11e	H	3	1.3
11f	2-OMe	1	13.8
11g	2-Me	1	14.5

A third iteration of the focused library relies on optimization of the left-hand piece while holding the 4-(4-methylpiperazin-1-ylmethyl)-phenylamine moiety constant. Some 48 boronic acids were selected based on diversity, molecular weight, and other desired calculated properties and subjected to Suzuki coupling reaction (**12 series** in Scheme 5.4).

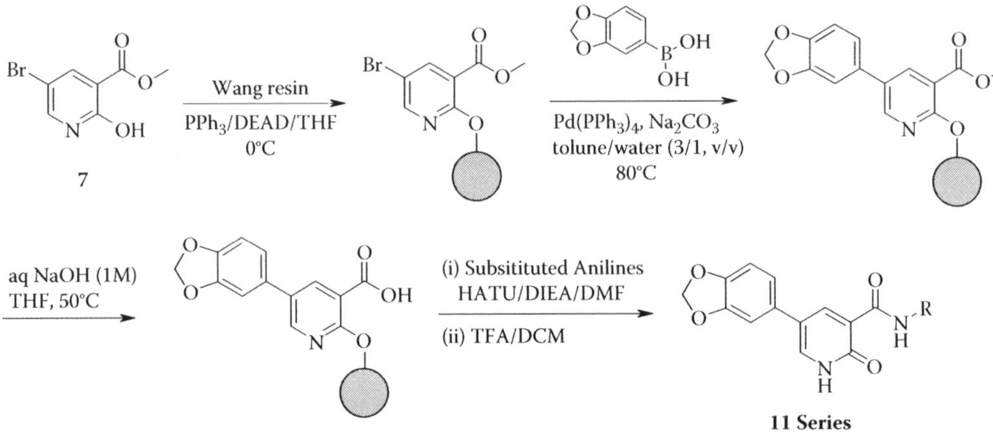

SCHEME 5.3

5.2.4 STRUCTURE-ACTIVITY RELATIONSHIPS

After a thorough analysis, structure-activity relationships revealed the following trends:

1. The compounds with the right-hand piece bearing *trans*-1,4-diaminocyclohexane spacer only exert moderate enzymatic activity, as shown in Table 5.1. This is probably due to the conformation and/or orientation of the spacer being less favored to the enzyme active site.
2. The activity was improved dramatically by replacing the *trans*-1,4-diaminocyclohexane spacer with a more planar aromatic ring such as an IC$_{50}$ of compound **11a** being 0.4 μM, as shown in Table 5.2.

3. Only up to a threefold difference in activity was observed, regardless of the number of carbons between the aromatic spacer and the 4-methylpiperazin-1-yl moiety (**11a–11e**).

4. The activity was reduced at least 15-fold when there was an *ortho*-substitution with a methyl group (**11g** vs. **11c**) or methoxy group (**11f** vs. **11c**) next to the aniline nitrogen. This indicates that the disturbance of co-planarity of the aromatic ring with the amide carbonyl group might be detrimental to the desired activity.

5. The methylenedioxyphenyl group seems to be an optimal functionality as a left-hand piece for desired activity, as shown in Table 5.3.

6. Substitution at the R2 position cannot be tolerated as a fourfold decrease in activity was observed (**12b** vs. **12k** in Table 5.3).

TABLE 5.3

The Third Generation of Pyridone Focused Library

Entry	R1	R2	ALK Inhibition IC$_{50}$ (μM)
11c		H	0.9
12a		H	2.5
12b		H	2.6
12c		H	2.8
12d		H	3.0
12e		H	3.8
12f		H	5.5
12g		H	7.5
12h		H	7.5
12i		H	8.3
12j		H	8.5
12k		Me	10.0

The indication of co-planarity trend (item 4 above) inspired us to perform a thorough investigation of the pyridone core, which resulted in the fourth iteration. Four compounds (**13a**–**13d**) in addition to those from **11 series** were designed and synthesized, and their enzymatic activity is tabulated in Table 5.4. First, the inhibitory activity was reduced fourfold by methyl group substitution on the amide nitrogen (**13a** vs. **11c**, IC_{50} = 3.7 vs. 0.9 μM), presumably due to the loss of a hydrogen bond donor from the amide NH of **11c**. Second, a co-planar geometric arrangement of the pyridone core with the aniline aromatic ring seems to be preferred because a loss of activity by 16-fold was observed with compounds **11f** and **11g** (~14 μM vs. 0.9 μM for **11c**). Obviously, co-planarity was disturbed by the *ortho*-substitution with either a methyl or methoxy group. Third, it appears that the pyridone carbonyl is acting as a hydrogen bond acceptor because a dramatic decrease in activity (20- to 30-fold) was observed in compounds **13b** and **13c**. The latter compounds substitute carbonyl with methoxy and chlorine, respectively (Table 5.4). Finally, the N-H on the pyridone ring seems to be an essential hydrogen bond donor because activity decreased by more than 50-fold when this hydrogen was replaced by a methyl group (**13d**).

TABLE 5.4
Structure-Activity Relationship Data Concerning the Pyridone Core

Entry	Structure	ALK IC_{50} (μM)
11c		0.9 ± 0.4 (n = 6)
13a		3.7
11f		13.8
11g		14.5
13b		20.3
13c		26.8
13d		>40

TABLE 5.5
Kinase Inhibitory Activity of 5-Aryl-pyridone-3-carboxamide Derivatives

Entry	Structure	ALK[a]	IRK[a]	NA/BaF3[c]	BaF3[c]	Karpas299[c]	K562[c]	Jurkat[c]
11a		0.4 0.2±0.1[b]	3.6 >20[b]	17.4±3.0[d]	18.0±2.9	7.5±1.2	9.5±0.2	4.7±0.3
11b		0.7 0.6[b]	4.8 8.8[b]	6.5±2.9	6.5±2.8	3.6±1.1	3.8±1.3	3.5±0.8
11c		0.9±0.4 (n = 6) 0.5±0.2[b]	27±11 (n = 6) 5.3±3.7[b]	>25	>25	11.2±3.2	12.9±6.1	9.2±0.4
11e		1.3 1.1[b]	10.1 9.0[b]	13.7±5.3	16.2±5.5	9.4±0.3	14.6±5.0	9.3±0.9

[a] Enzyme inhibition: IC_{50} (μM); [b] Enzyme inhibition: K_i (μM); [c] Cellular antiproliferative activity: IC_{50} (μM); [d] IC_{50} ± SD (μM), n = 3 unless otherwise specified. ALK, anaplastic lymphoma kinase; IRK, insulin receptor kinase; NA/BaF3, BaF3 murine lymphoid cell line engineered to express NPM-ALK; BaF3, parental interleukin-3-dependent BaF3 cell line; Karpas299, human NPM-ALK-positive anaplastic large cell lymphoma cell line; K562, human BCR-ABL-positive chronic myeloid leukemia cell line; Jurkat, human T-cell leukemia cell line. All assays were performed with two or more repetitions.

5.2.5 CONCLUSION

The discovery and optimization of novel pyridone chemotype as ALK inhibitors were demonstrated (*vide supra*) using high-throughput medicinal chemistry coupled with structure-based drug design tools (in this case homology modeling and virtual screening). After iterative optimization, pyridone **11a** was discovered, with an appropriate spacer and basic moiety as shown in Table 5.5. This compound has an enzymatic ALK IC_{50} of 380 nM and an ALK K_i of 200 ± 100 nM; it also possesses 10-fold selectivity over the IRK, 18-fold over the IGF1R, >10-fold over Flt3, and >50-fold over both Abl and Src (data not shown). This series of pyridone compounds has also been shown to possess non-selective inhibitory activity of tumor cell proliferation *in vitro,* with cellular IC_{50} values in the 4–18-µM range (Table 5.5).

5.3 CASE STUDY II: OPTIMIZATION OF MELANOCORTIN RECEPTOR AGONIST

5.3.1 INTRODUCTION

Melanocortin receptors are part of a family of seven-transmembrane G-protein-coupled receptors and are composed of five receptor subtypes (MC1R–MC5R). The first melanocortin receptor (MC1R) was identified in melanocytes.[23] Subsequently, the other subtypes were expressed in different tissue types: MC2R (adrenal gland),[24] MC3R (brain, gut, and heart),[25] MC4R (brain),[26] and MC5R (brain).[27] These receptors interact with their endogenous ligands, the melanocortins and corticotropins, to regulate a large number of physiologic functions in humans, including the control of feeling and sexual behavior as well as skin pigmentation and energy metabolism.[28] In recent years, research efforts in drug discovery have focused on identifying novel MC4R agonists for the potential treatment of obesity and sexual dysfunction.[29] The approach we have used to design templates for MC4R relies on a novel design of general peptidomimetics, which mimic beta turns,[30] and its extension to the critical insights reported by Garland and Deans.[31]

5.3.2 DESIGN AND OPTIMIZATION OF MC4R AGONIST

We designed our first exploratory library using the template branched out with building blocks that could mimic some of the common interactions from the sidechains of potent peptide MC4R agonists containing the small-molecule mimics in the His-(D)Phe-Arg-Trp sequence.[32] Using the chemistry outlined in Scheme 5.4, we prepared a set of compounds **14 series** (Table 5.6) for evaluation as potential

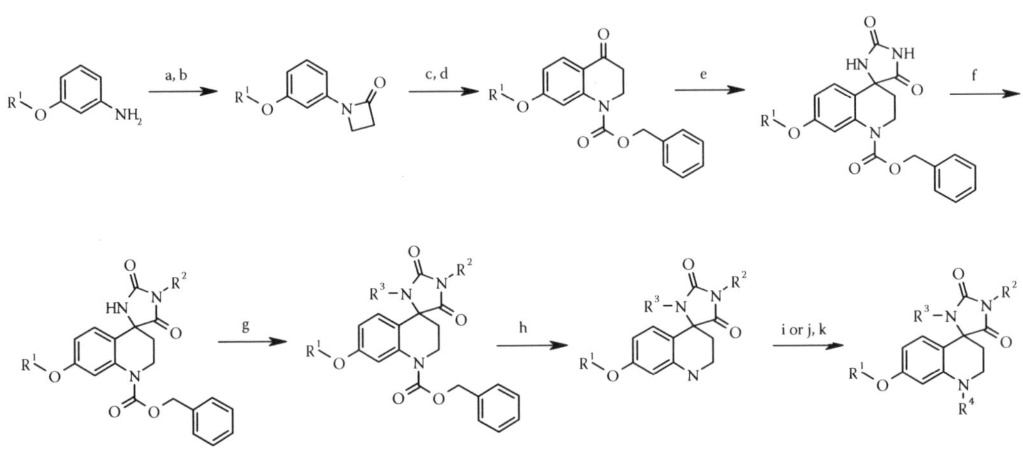

SCHEME 5.4 Reagents: (a) 3-bromopropionic acid/EDC/TEA/DMAP; DCM; (b) TBACl/KOH/DCM; (c) PPE/DCE; (d) CbzCl/DIEA/DCM; (e) KCN/NH$_4$)$_2$CO$_3$/EtOH/70°C; (f) R$_2$X/CsHCO$_3$ or Na$_2$CO$_3$/DMF; (g) R$_3$X/K$_2$CO$_3$/DMF or NaH/DMF; (h) H$_2$/Pd-C; (i) TFA/DCM; (j) R$_4$CHO/NaB(OAc)$_3$H/DCM; (k) R$_4$COOH/EDC/TEA/DMAP.

TABLE 5.6
The First-Generation MC4R Focused Library

Entry	R2	R3	R4	EC$_{50}$ MC4R (µM)
14a	INDE	H	Cbz	NA
14b	2-Picolyl	H	Cbz	NA
14c	4-Picolyl	H	Cbz	NA
14d	INDE	Methyl	Cbz	>50
14e	INDE	3-Picolyl	Cbz	43
14f	2-Picolyl	Methyl	Cbz	NA
14g	2-Picolyl	3-Picolyl	Cbz	28
14h	INDE	3-Picolyl	H	39
14i	INDE[H]	3-Picolyl	H	>50
14j	PIPM-2	Methyl	H	NA
14k	2-Picolyl	3-Picolyl	H	NA
14l	4-Picolyl	3-Picolyl	H	NA
14m	INDE	Methyl	H	NA
14n	INDE	Methyl	Acetyl	NA

Note: Abbreviations: NA = not active; Cbz = benzyloxycarbonyl; INDE = indole-3-(eth-2-yl); INDE[H] = indoline-3-(eth-2-yl); PIPM-2 = piperinemeth-2-yl.

MC4R agonists.[33] The most active compound in this first-generation focused library was compound **14g,** which showed weak agonist activity at MC4R and contained a 2-picolyl and a 3-picolyl group as substituents at R2 and R3, respectively. Presumably, the picolyl group with dual function could serve as both a basic group and an aryl group.

A series of focused libraries was designed to explore substitution patterns at all four positions of the template.[30] In one of the focused libraries, fivefold improvement in activity was achieved from compound **14g** after only six more compounds were synthesized. This second-generation library held the R1 through R3 constant and varied R4. As a result, compound **15** turned out to be the best compound, with an EC$_{50}$ of 5.6 µM for MC4R (Figure 5.3), and compound **16** as a partial agonist with an EC$_{50}$ of 9.8 µM (Figure 5.3). In addition, these compounds showed at least 20-fold selectivity over MC5R.

Further optimization of compounds **15** and **16** using HTMC resulted in submicromolar MC4R agonist after synthesis of 199 compounds.[30] This library primarily explored modifications at R3, while R4 was held constant as isobutyl, with some very limited modifications at R1 and R2. Eight compounds (**17a–17h** in Table 5.7) were confirmed as MC4R agonists with EC$_{50}$ values between 0.7 and 2.5 µM.

5.3.3 CONCLUSION

The novel chemistry developed that allows for the high-throughput synthesis (HTS) of both templates and libraries made fast optimization of MC4R agonist possible. Iteration of well-designed

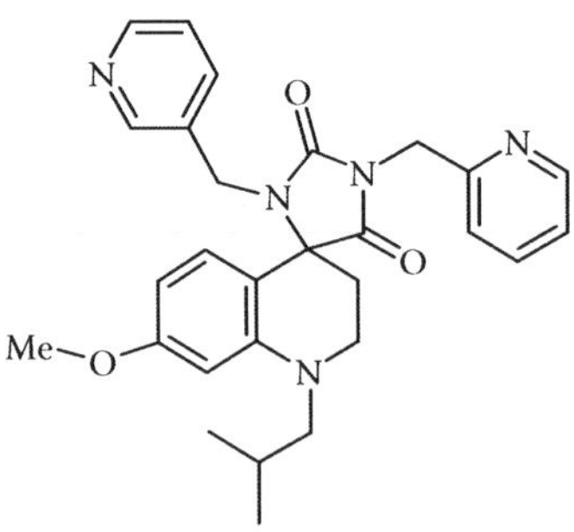

FIGURE 5.3 Early hits of MC4R agonists.

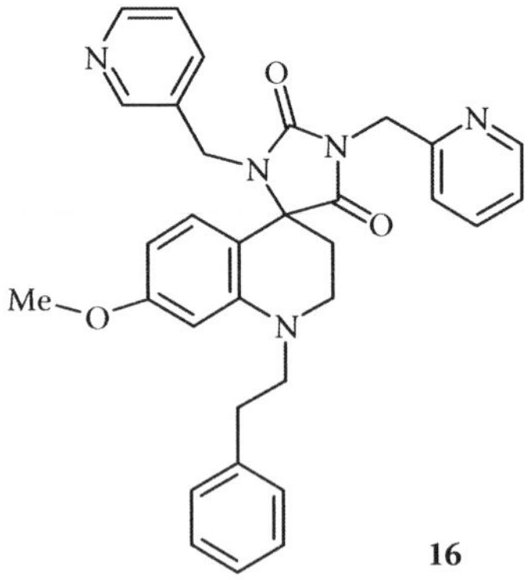

TABLE 5.7
Examples of the Third-Generation MC4R Focused Library

Entry	R1	R3	EC$_{50}$ MC4R (μM)
17a	Methyl	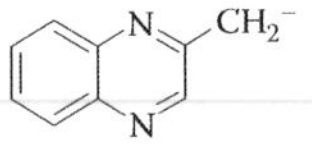	0.7
17b	Methyl	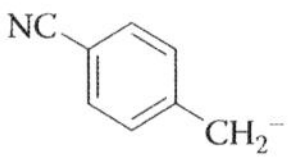	1.0
17c	Ethyl	2-Picolyl	1.2
17d	Methyl	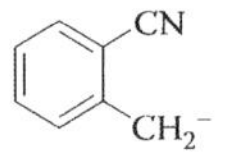	1.6
17e	i-Propyl	2-Picolyl	2.0
17f	Methyl	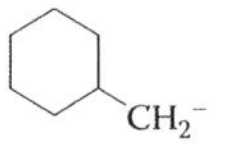	2.1
17g	Methyl	4-Picolyl	2.5
17h	Methyl	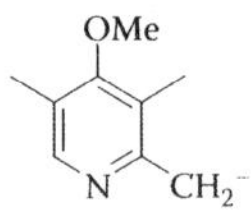	2.5

library expedites the understanding of structure-activity relationships for MC4R agonists. A 40-fold improvement in activity was achieved after a few iterations with only 240 compounds synthesized.

5.4 SUMMARY

In this chapter, combined techniques and enabling tools such as virtual screening and structure-based drug design using 3-D structure information of drug target (in this case homology modeling) are used to assist in the generation of novel ideas leading to "hit" or "lead" compounds, guide hit-follow-up and lead optimization, and SAR studies. As demonstrated, the turn-around time can be shortened even further when these tools are coupled with high-throughput medicinal chemistry and high-throughput screening. Since combinatorial chemistry started in the early 1990s for synthesis of compound library, in both the solid and solution phases, novel chemistry and chemistry "know-how" have accumulated and become assets for high-throughput medicinal chemistry. The library size changed drastically from a starting point of 10,000+ compounds to a few dozen compounds. The turn-around time has shrunk from months to weeks to days, while the quality of library compounds continues to improve. Purification of every compound by HPLC and complete characterization of library compounds became routine. These library compounds, whose quality is no less than par to the individually synthesized compounds, are assured to address and answer critical SAR questions. If the "know-how" built into the past 14 years of high-throughput medicinal chemistry can be used wisely in conjunction with other enabling tools, the cost of drug discovery can be reduced dramatically.

ACKNOWLEDGMENTS

The author thanks the ALK research team (ChemBridge Research Laboratories: Thomas R. Webb, Tong Zhu, Zheng Yan, Jian Wang, Danny McGee, Vidyasagar Reddy Gantla, Jason C. Pickens, Doug McGrath, and Alexander Chucholowski; St. Jude Children's Research Hospital: Stephan W. Morris, Liquan Xue, Qin Jiang, and Xiaoli Cui) and the MC4R research team (ChemBridge and ChemBridge Research Laboratories: Thomas R. Webb, Luyong Jiang, Sergey Sviridov, Jian Wang, Ruben E. Venegas, Anna V. Vlaskina, Doug McGrath, John Tucker, and Yang Yang; Chemical Computing Group: Alain Deschenes) for their contributions.

REFERENCES

1. Li, R., How to optimize reactions for solid-phase synthesis of combinatorial libraries using R_f tagged microreactors, in *Optimization of Solid Phase Combinatorial Synthesis,* Yan, B. and Czarnik, A.W., Eds., Marcel Dekker, New York, 2001, chap. 3.
2. Li, R., Xiao, X.-Y., and Nicolaou, K.C., High-output solid phase synthesis of discrete compounds using split-and-pool strategy, in *High-Throughput Organic Synthesis,* Sucholeiki, I., Ed., Marcel Dekker, New York, 2000, chap. 6.
3. Weinmann, H. and Metternich R., Drug discovery process for kinase inhibitors, *ChemBioChem.*, 2005, 6, 455.
4. Manning, G. et al., The protein kinase complement of the human genome, *Science,* 2002, 298, 1912.
5. Miles, A.F. et al., A small molecule-kinase interaction map for clinical kinase inhibitors, *Nat. Biotechnol.,* 2005, 23, 329.
6. Morris, S.W. et al., Fusion of a kinase gene, ALK, to a nucleolar protein gene, NPM, in non-Hodgkin's lymphoma, *Science,* 1994, 263, 1281.
7. Iwahara, T. et al., Molecular characterization of ALK, a receptor tyrosine kinase expressed specifically in the nervous system, *Oncogene,* 1997, 14, 439.
8. Morris, S.W. et al., *ALK,* the chromosome 2 gene locus altered by the t(2;5) in non-Hodgkin's lymphoma, encodes a neural receptor tyrosine kinase that is highly related to leukocyte tyrosine kinase (LTK), *Oncogene,* 1997, 14, 2175.
9. Pulford, K.P., Morris, S.W., and Turturro, F., Anaplastic lymphoma kinase proteins in growth control and cancer, *J. Cell. Physiol.*, 2004, 199, 330.
10. Bernards, A. and de la Monte, S.M., The ltk receptor tyrosine kinase is expressed in pre-B lymphocytes and cerebral neurons and uses a non-AUG translational initiator, *EMBO J.,* 1990, 9, 2279.
11. Krolewski, J.J. and Dalla-Favera, R., The ltk gene encodes a novel receptor-type protein tyrosine kinase, *EMBO J.,* 1991, 10, 2911.
12. Snijders, A.J., Haase, V.H., and Bernards, A., Four tissue-specific mouse ltk mRNAs predict tyrosine kinases that differ upstream of their transmembrane segment, *Oncogene,* 1993, 8, 27.
13. Li, R. and Morris, S.W., Development of anaplastic lymphoma kinase inhibitors for cancer therapy, *Med. Res. Rev.,* in press, 2007.
14. Webb, T.R. et al., Discovery of novel inhibitors of (ALK) small molecule tyrosine kinases, *221st ACS National Meeting*, MEDI-228, San Diego, CA, April 1–5, 2001.

15. Kuntz, I.D. Structure-based strategies for drug design and discovery, *Science,* 1992, 257, 1078.

16. Leach, A.R., Shoichet, B.K., and Peishoff, C.E., Prediction of protein-ligand interactions. Docking and scoring: Success and gaps, *J. Med. Chem.,* 2006, 49, 5851.

17. Mazur, A.K. and Abagyan, R.A., New methodology for computer-aided modelling of biomolecular structure and dynamics. 1. Non-cyclic structures, *J. Biomol. Struct. Dyn.,* 1989, 6, 815.

18. Abagyan, R. and Argos, P., Optimal protocol and trajectory visualization for conformational searches of peptides and proteins, *J. Mol. Biol.,* 1992, 225, 519.

19. Abagyan, R., Totrov, M., and Kuznetsov, D., ICM: a new method for protein modeling and design: applications to docking and structure prediction from the distorted native conformation, *J. Comp. Chem.,* 1994, 15, 488.

20. Li, R. et al., Design and synthesis of 5-aryl-pyridonecarboxamides as inhibitors of anaplastic lymphoma kinase. *J. Med. Chem.,* 2006, 49, 1006.

21. Katritzky, A.R. and Pozharski, A.F., Reactivity of Heterocycles. In *Handbook of Heterocyclic Chemistry.* Katritzky, A.R. and Pozharski, A.F., Eds.; Elsevier Science Ltd: Oxford, U.K., 2000; pp. 272–274.

22. Zhu, T. et al., Polymer-supported synthesis of pyridone-focused libraries as inhibitors of anaplastic lymphoma kinase, *J. Comb. Chem.,* 2006, 8, 401.

23. Chhajlani, V. and Wikberg, J.E.S., Molecular cloning and expression of the human melanocyte stimulating hormone receptor cDNA, *FEBS Lett.,* 1992, 309, 417.

24. Mountjoy, K.G. et al., The cloning of a family of genes that encode the melanocortin, *Science,* 1992, 257, 1248.

25. Chhajlani, V., Muceniece, R., and Wikberg, J.E.S., Molecular cloning of a novel human melanocortin receptor, *Biochem. Biophys. Res. Commun.,* 1993, 195, 866.

26. Gantz I. et al., Molecular cloning of a novel melanocortin receptor, *J. Biol. Chem.,* 1993, 268, 8246.

27. Gantz, I. et al., Molecular cloning, expression and gene localization of a fourth melanocortin receptor, *J. Biol. Chem.,* 1993, 268, 15174.

28. Sebhat, I. et al., Melanocortin-4 receptor agonists and antagonists: chemistry and potential therapeutic utilities. In *Annu. Rep. Med. Chem.,* Doherty, A.M., Ed., Elsevier Academic Press, San Diego, 2003, chap. 4.

29. Palucki, B.L. et al., Discovery of (2S)-N-[(1R)-2-[4-cyclohexyl-4-[[(1,1-dimethylethyl)amino]carbonyl]-1-piperidinyl]-1-[(4-fluorophenyl)methyl]-2-oxoethyl]-4-methyl-2-piperazinecarboxamine (MB243), a potent and selective melanocortin subtype-4 receptor agonist, *Bioorg. Med. Chem. Lett.,* 2005, 15, 171.

30. Webb, T.R. et al., Application of a novel design paradigm to generate general nonpeptide combinatorial templates mimicking beta. Synthesis of ligands for melanocortin receptors, *J. Comb. Chem.,* 2007, 9, 704–12.

31. Garland, S.L. and Dean, P.M., Design criteria for molecular mimics of fragments of the β-turn. 1. C atom analysis. *J. Comp. Aided Mole. Des.,* 1999, 13, 469.

32. Al-Obeidi, F. et al., Design of a new class of superpotent cyclic alpha-melanotropins based on quenched dynamic simulations, *J. Am. Chem. Soc.,* 1989, 111, 3413.

33. Webb, T.R., et al., Design and synthesis of MC4-focused target library around novel spiro[imidazolidinequinolin]-2,4-dione scaffold, *29th National Medicinal Chemistry Symposium,* Abstract #79, University of Wisconsin, Madison, June 27–July 1, 2004.

6 Rapid Lead Identification of Inhibitors of Adenine Nucleotide Translocase

A Case Study of Applying Combinatorial Chemistry Techniques in Drug Discovery

Yazhong Pei, Walter H. Moos, and Soumitra Ghosh

CONTENTS

6.1 INTRODUCTION

6.1.1 COMBINATORIAL CHEMISTRY

In the past two decades, the field of combinatorial chemistry has experienced several cycles of evolution. In the mid-1980s, Geysen[1] and Houghten[2] demonstrated the effectiveness of combinatorial synthesis by preparing large numbers of peptides using solid-phase synthetic techniques pioneered by Merrifield.[3] Peptide leads were quickly identified and optimized for various molecular targets, including ligands for transmembrane receptors and inhibitors for enzymes.[4] In the 1990s, the field of combinatorial chemistry grew exponentially. Numerous techniques and instrumentation platforms were developed for high-throughput synthesis and purification. This enabled chemists to carry out more complex organic syntheses in parallel fashion, both on solid supports and in solution, to prepare libraries of non-peptide small molecules.[5,6] The design of these libraries was typically based on chemistries that were amenable to solid-phase organic synthesis or reactions in solution to furnish the final products in a few steps with high overall yield and purity, rather than directed toward any specific protein target. These large libraries of random compounds brought to drug discovery the promise of rapid hit identification against seemingly any target. Computational techniques were often employed to select the building blocks to enhance the molecular diversity around any given template to increase the probability of producing hits for a variety of targets.[7] The increased chemistry throughput in combination with high-throughput screening generated leads and accelerated the overall drug discovery process, resulting in many successful examples for both lead discovery and lead optimization.[8] Nowadays, combinatorial chemistry has become an integral tool for medicinal chemists. Small and focused libraries are routinely designed and synthesized for lead identification and optimization against a given target.[9] In particular, the positional scanning strategy pioneered by Houghten has proven very effective in identifying ligand segments, from large mixtures of peptides and peptide mimetics, that contribute to the observed biological activity.[10] This approach simplifies the deconvolution process associated with mixture-based libraries, and, more importantly, provides structure-activity-relationship (SAR) information for the design of follow-up libraries. This method has been successfully applied to many projects and a wide range of targets.[11,12] This chapter describes in detail another successful application of solid-phase combinatorial synthesis, namely, in the discovery of small-molecule leads for the inhibition of adenine nucleotide translocase (ANT).[13] Using an analogous approach to positional scanning, we were able to follow SAR trends based on weak activity. Through library iterations, several novel ANT ligands, with comparable affinity to natural product leads, were identified.

6.1.2 ADENINE NUCLEOTIDE TRANSLOCASE (ANT)

Mitochondria serve as power plants for almost all living organisms and are responsible for the energy production needed for most cellular processes.[14,15] Adenine nucleotide translocase (ANT) is a mitochondrial inner membrane protein and is a key component in maintaining cellular energy homeostasis. Under normal conditions, ANT transports cytosolic ADP across the otherwise impermeable inner mitochondrial membrane into the matrix in exchange for ATP. It thus supplies ADP as substrate for ATP production through the mitochondrial oxidative phosphorylation process, and brings matrix ATP to the cytosol, providing high-energy phosphate source for other cellular processes. ANT is also a component of a protein complex called the mitochondrial permeability transition (MPT) pore.[16,17] During the onset of stroke or myocardial infarction, it has been observed that intracellular calcium levels are elevated, the mitochondria are subjected to oxidative stress, and adenine nucleotides are depleted. These events trigger the opening of the MPT pore, allowing cytoplasmic solutes with molecular weights less than 1500 daltons (Da) to equilibrate across the inner mitochondrial membrane. This leads to the subsequent collapse of mitochondrial membrane potential, shutdown of ATP production, and causes the swelling and eventual rupture of the outer mitochondrial membrane to release cytotoxic macromolecules, such as cytochrome c, that induce cell death.

6.1.3 NATURAL PRODUCT LEADS

There are only two natural products — bongkrekic acid (BKA) and atractyloside (ATR) — that have been reported to bind to ANT with high affinity. BKA is a natural toxin produced by the microorganism *Pseudomonas cocovenans*[18–20] and is a potent inhibitor of ANT-mediated ADP/ATP transport across the inner mitochondrial membrane, with a reported IC_{50} value of 20 nM.[21] A number of reports

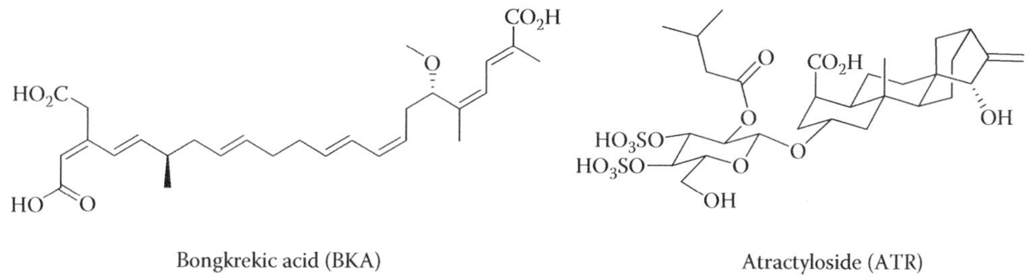

Bongkrekic acid (BKA)   Atractyloside (ATR)

FIGURE 6.1

indicate that BKA and ATR bind to ANT asymmetrically, inducing and stabilizing different conformations of ANT.[20] ATR binds to ANT from the cytosolic side of the inner mitochondrial membrane and induces the *c*-conformation, which leads to permeability transition. BKA interacts with ANT from the matrix side of the inner mitochondrial membrane, stabilizes the *m*-conformation, and prevents the permeability transition triggered by various stimuli. Although BKA and ATR do not bind to ANT at the same site, they can mutually displace each other by inducing ANT conformational changes. These observations indicate that small molecules that bind to ANT may have diverse therapeutic implications. Agents with ATR-like properties may be used as cytotoxic agents for hyperproliferative diseases. Compounds with BKA-like activities can maintain and stabilize mitochondrial function under pathological conditions, and provide potential therapies for the treatment of ischemia/reperfusion injury and certain neurodegenerative diseases.

6.2 DESIGN AND SYNTHESIS OF IMINODIACETIC ACIDS AS ANT INHIBITORS

6.2.1 DESIGN CONSIDERATIONS

It was our goal to identify small-molecule leads by exploring the structure-activity relationships around BKA. These compounds should interact with ANT only to prevent the permeability transition pore from opening without interfering with the ADP/ATP exchange required for normal cellular function. The stereocontrolled, convergent total synthesis of BKA was reported first by Corey et al. in 1984[23] and later by Shindo et al. in 2004.[24] Both processes are lengthy and involve complex differential protection/deprotection strategies. This makes it impractical to prepare BKA in large amounts, most importantly for SAR development, which requires reaction conditions to accommodate a wide range of structural and functionality variation. We did consider using computational techniques to aid our hit identification efforts. However, BKA is a linear and rather flexible molecule, and is capable of adopting many possible conformations when bound to ANT. Moreover, the absence of additional chemical leads made it difficult to generate reliable pharmacophore models by computational methods. From the protein target angle, the structure of ANT based on either x-ray data or sequence homology is not known. In addition, ANT is a dynamic protein and goes through conformational changes while interacting with different ligands. This lack of information made *in silico* screening using protein target-based pharmacophore models inapplicable. Therefore, we decided to utilize our medicinal chemistry intuition and combinatorial chemistry techniques to identify small-molecule ANT leads.

A critical design consideration was that the combinatorial synthesis of the target libraries, preferably on solid support. This would ensure our success by accessing a large number of compounds quickly and to identify SAR trends from weak biological activities. The dominant molecular features of BKA include three carboxylic acids at the ends connected by a flexible hydrophobic chain. We postulated that the high affinity of BKA for ANT was derived from the electrostatic interaction between the terminal carboxylic acids of BKA and positively charged amino acid residues of ANT and/or hydrophobic interaction between the middle portion of BKA and ANT. Based on these considerations,

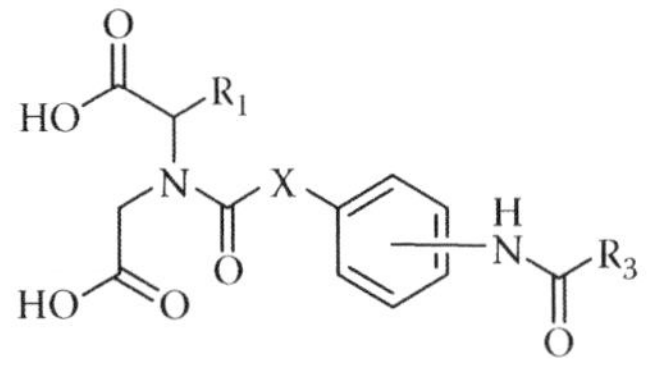

FIGURE 6.2 *N*-acyl iminodiaretic acids as potential BKA analogs.

N-acyl iminodiacetic acids were designed as potential BKA analogs (as shown in Figure 6.2). The *N*-acyl iminodiacetic acid moiety was expected to adopt a planar conformation due to conjugation between the lone pair electrons of the imino nitrogen and the amide carbonyl, thereby mimicking the α,β-unsaturated dicarboxylate end of BKA. The envisioned synthetic routes to these compounds allowed us to introduce substituents at the α-position of the iminodiacetic acid, to further enable us to define its conformation. To delineate the preferred conformation of BKA, aminophenyl carboxylic acids were incorporated into the middle section. The amino moiety served as an attachment point for mimics of the mono-acid end of BKA. These aminophenyl carboxylic acids allowed us to introduce rigidity and reduce the number of possible conformations, and at the same time enabled us to probe the bound geometry of di-acid and mono-acid ends by varying the relative locations of the acid and the amino groups on the phenyl ring. Furthermore, the phenyl group could serve as a mimic of C=C unsaturation in the middle section of BKA and to probe potential π–π and/or face-edge interactions between the ANT protein and small-molecule ligands. The phenyl group could also serve as a template for attaching additional functional groups for SAR exploration. These aminophenyl carboxylic acids can be transformed from their corresponding nitro derivatives, which are readily available from commercial sources. All of the proposed chemical transformations appeared to be straightforward and compatible with solid-phase organic synthesis.

6.2.2 COMBINATORIAL SYNTHESIS OF IMINODIACETIC ACIDS ON SOLID PHASE

The general synthetic pathways for these *N*-acyl iminodiacetic acids are illustrated in Scheme 6.1.[25] Bromoacetic acid was coupled to Wang resin using diisopropylcarbodiimide (DIC) and a catalytic amount of 4-(*N*,*N*-dimethylamino)pyridine (DMAP) in DMF to afford **1**. Resin-bound bromoacetate **1** was allowed to react with amino ester (R1 component) in DMSO in the presence of diisopropylethylamine (DIEA) to give **2**. Iminodiacetate **2** was acylated with nitrobenzoic acids/nitrobenzoyl chlorides (R2 component) to yield **3**. The nitro intermediate **3** was transformed to **4** by reduction using tin(II) chloride dihydrate in DMF. Acylation of **4** with R3 components furnished **5**. Compound **5** was cleaved from the resin using trifluoroacetic acid to give the final product **6**. All reactions were carried out in a combinatorial fashion in polypropylene syringes fitted with a polypropylene filter. The solid-phase bound intermediates were divided into equal portions (by weight) according to the number of building blocks used in the next step.

6.3 ITERATION OF LIBRARIES

6.3.1 LIBRARY 1

The building blocks used in Library 1 are listed in Table 6.1. At the R_1 position, we employed glycine *t*-butyl ester, aspartic di-*t*-butyl ester, and glutamic di-*t*-butyl ester. The *t*-butyl ester protecting groups were removed during the cleavage to reveal the carboxylic acid moiety. These building blocks allowed us to create the di-acid and tri-acid derivatives, respectively, to mimic the di-acid end of BKA. Three nitrobenzoyl chlorides with *ortho*-, *meta*-, and *para*-substitution patterns were used to explore the conformation of the middle section of the molecules. Seven carboxylic acids and derivatives representing aromaticity and hydrocarbon chains with varying length and endings were

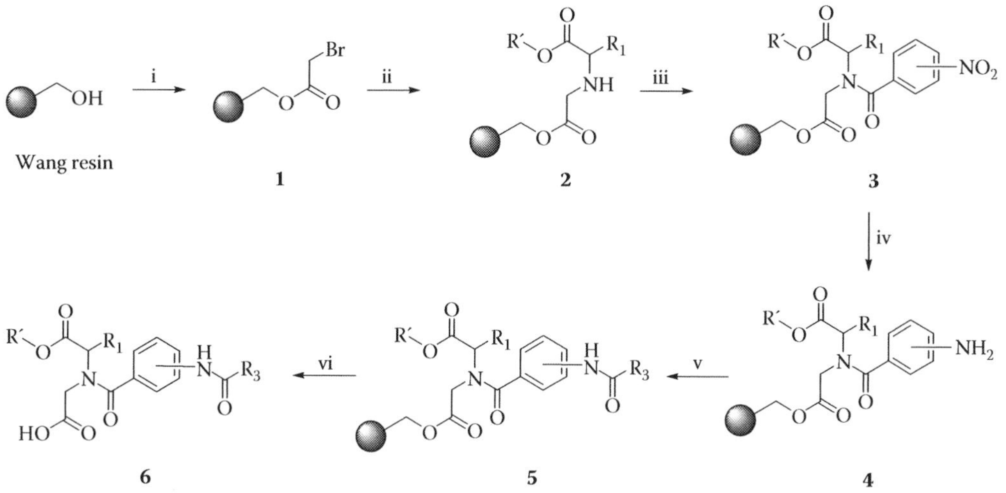

SCHEME 6.1 General synthesis of iminodiacetic acid derivatives.

TABLE 6.1
Building Blocks for Library 1

Code	R$_1$ component	R$_2$ component	R$_3$ component
A	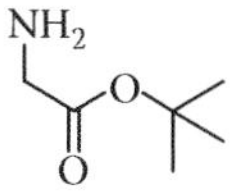 (structure: glycine t-butyl ester, NH$_2$)	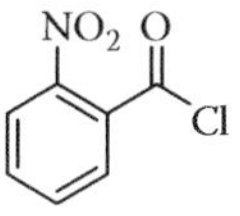 (structure: 2-nitrobenzoyl chloride, NO$_2$, C(=O)Cl)	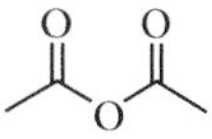 (acetic anhydride)
B	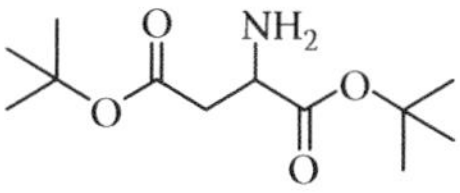 (structure: aspartic acid di-t-butyl ester, NH$_2$)	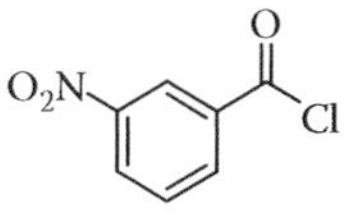 (structure: 3-nitrobenzoyl chloride, O$_2$N)	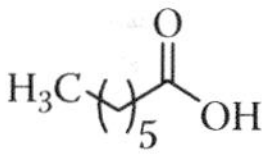 H$_3$C–(CH$_2$)$_5$–COOH
C	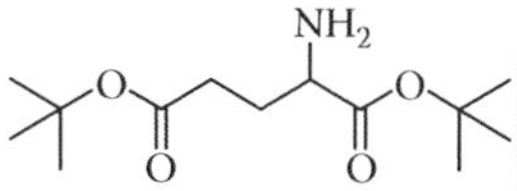 (structure: glutamic acid di-t-butyl ester, NH$_2$)	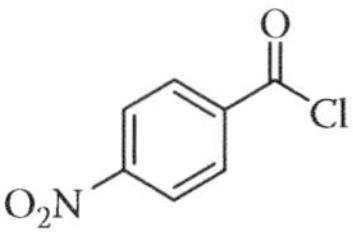 (structure: 4-nitrobenzoyl chloride, O$_2$N)	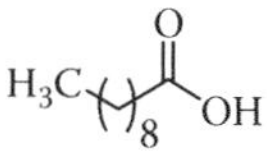 H$_3$C–(CH$_2$)$_8$–COOH
D			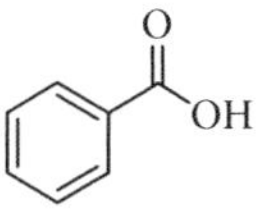 (benzoic acid)
E			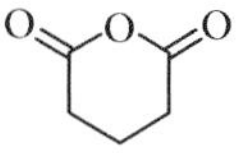 (glutaric anhydride)
F			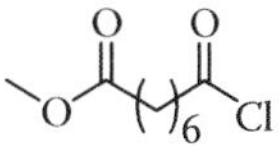 (structure: methyl ester acyl chloride, –(CH$_2$)$_6$–, Cl)
G			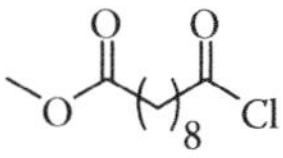 (structure: methyl ester acyl chloride, –(CH$_2$)$_8$–, Cl)

used as R_3 components. Most reactions worked well and gave the desired products in good yield and with sufficient purity. Final products with purity less than 85% (determined by LC-MS) were further purified using preparative HPLC. In all, 63 compounds were released for testing, representing all possible permutations of the building blocks.

These *N*-acyl iminodiacetic acids were evaluated for their affinity to ANT in a competitive binding assay using a radio-iodinated atractyloside derivative and mitochondria harvested from bovine heart a.[26] All compounds were screened in triplicate at a concentration of 10 µM for an initial evaluation. From the first library, none of the compounds showed inhibition of greater than 50%. However, interesting SAR trends were observed in the initial data. At the R_1 position, several compounds derived from glycine (**R_{1A}**) showed weak inhibition (20 to 30%, data not shown), while no derivatives from aspartate (**R_{1B}**) and glutamate (**R_{1C}**) showed appreciable activity. At the R_2 position, a few derivatives from *meta*- and *para*-nitrobenzoyl chlorides (**R_{2B}** and **R_{2C}**) showed weak activity, while *ortho*-nitrobenzoyl chloride (**R_{2A}**) did not produce any compounds with inhibitory activity. No derivatives with acetyl (**R_{3A}**) or benzoyl (**R_{3D}**) groups at the R_3 position showed any affinity to ANT, while a few analogs with longer chains at the R_3 position did show weak binding to ANT.

It is worthwhile to note that although Library 1 only yielded compounds with weak affinity to ANT, these closely related analogs did reveal valuable information on SAR trends and provided direction for further optimization in the design of the next round of libraries. These weak activities might have been lost in the noise of the assay (±15%) had only structurally unrelated compounds been screened initially.

6.3.2 LIBRARY 2

The building blocks used in Library 2 are listed in Table 6.2. Based on the data from Library 1, we decided to further explore the SAR around the R_1 position. Glycine methyl ester, glycine amide,

TABLE 6.2
Building Blocks for Library 2

Code	R_1 component	R_2 component	R_3 component
A	H₂N–CH₂–C(=O)–O– (glycine methyl ester)	O₂N–C₆H₄–C(=O)Cl (meta)	H₃C–(CH₂)₅–C(=O)–OH
B	H₂N–CH₂–C(=O)–NH₂	O₂N–C₆H₄–C(=O)Cl (para)	H₃C–(CH₂)₈–C(=O)–OH
C	H₂N–CH₂CH₂–OH		O=C–O–C=O (glutaric anhydride)
D			–O–C(=O)–(CH₂)₆–C(=O)–Cl
E			–O–C(=O)–(CH₂)₈–C(=O)–Cl

and aminoethanol were introduced at the R_1 position to vary the structure in terms of electrostatic, hydrogen bonding, and polarity. To reduce the size of Library 2, we only employed building blocks at the R_2 and R_3 positions that showed hints of activity in Library 1. The synthesis of Library 2 followed the same route as shown in Scheme 6.1. Reactions were carried out to enumerate all possible combinations of the building blocks listed in Table 6.2. The reactions worked well for glycine methyl ester ($\mathbf{R_{1A}}$) and amide ($\mathbf{R_{1B}}$) and furnished the 20 desired products in good yield and purity comparable to Library 1. No final product was obtained with aminoethanol at the R_1 position. The unprotected alcohol ($\mathbf{R_{1C}}$) appeared to be incompatible with the reaction sequence. In a similar fashion, these 20 compounds were screened at 10-µM concentration in the ANT binding assay, as mentioned above. Several of the compounds showed ANT inhibitory activity in the same range as that seen with Library 1 (20% to 30% inhibition).

6.3.3 LIBRARY 3

Based on these data, we decided to introduce various α-substituted amino acids to further explore the hydrophobicity and conformational constraints at the R_1 position. The building blocks used in Library 3 are listed in Table 6.3. Alanine benzyl ester, valine benzyl ester, leucine benzyl ester, and phenylalanine benzyl ester were used at the R_1 position. The R_2 and R_3 building blocks were kept the same as those of Library 2 to allow for easy comparison between the libraries. Using the chemistry as shown in Scheme 6.1, compounds representing all possible combinations of the building blocks were prepared. After cleavage from the solid support, each product was divided into two equal portions. One portion was kept as the mono-benzyl ester, and the other portion

TABLE 6.3
Building Blocks for Library 3

Code	R$_1$ component	R$_2$ component	R$_3$ component
A	NH$_2$ ... O–Ph (alanine benzyl ester)	O$_2$N– benzoyl chloride (meta)	H$_3$C–(CH$_2$)$_5$–COOH
B	NH$_2$ ... O–Ph (valine benzyl ester)	benzoyl chloride (para-O$_2$N)	H$_3$C–(CH$_2$)$_8$–COOH
C	NH$_2$ ... O–Ph (leucine benzyl ester)		glutaric anhydride
D	Ph ... NH$_2$ ... O–Ph (phenylalanine benzyl ester)		O–CO–(CH$_2$)$_6$–CO–Cl
E			O–CO–(CH$_2$)$_8$–CO–Cl

was hydrogenated in the presence of Pd/C to give the corresponding dicarboxylic acid. In all, 80 compounds were obtained in Library 3. A number of compounds, especially derivatives from leucine and phenylalanine, both mono-benzyl esters and dicarboxylic acids, at 10-μM concentrations showed ANT inhibitory activity (20% to 40%) in the binding assay. Although the potency of Library 3 did not improve significantly over Library 2, the increased number of compounds showing ANT affinity suggested that increasing the hydrophobicity/bulk at the R_1 position was beneficial.

6.3.4 LIBRARY 4

Library 4 was designed to examine the SAR around the middle portion of the molecules. The building blocks are listed in Table 6.4. Four sets of nitrophenyl carboxylic acids were employed at the R_2 position to explore the distance between the dicarboxylate end and the phenyl group, as well as additional constraints such as C=C bond and heteroaromatic rings. Various substitution patterns within each set allowed us to explore relative orientations between the R_1 and R_3 fragments. To minimize the size of Library 4, only preferred R_1, leucine benzyl ester, and phenylalanine benzyl ester substituents were used at the R_1 position. The building blocks for R_3 were kept constant. Following the general procedure outlined in Scheme 6.1, final products representing all possible combinations of the building blocks were synthesized. As in Library 3, each product was divided into two equal portions. One portion was kept as the mono-benzyl ester, and the other portion was converted to the corresponding dicarboxylic acid using catalytic hydrogenation in the presence of Pd/C. In all, 240 compounds in Library 4 were released for screening.

Within this library, several compounds at 10 μM showed significant inhibition (>50%) in the ANT binding assay, and were further evaluated in 10-point dose response studies at concentrations ranging from 5 nM to 30 μM. BKA was used as a reference standard in this assay and showed an IC_{50} of 0.2 μM. Several compounds were identified that displayed low micromolar binding to ANT in this assay. These compounds and their ANT inhibition activity are listed in Table 6.5.

The SAR trends observed in the iteration of libraries and the newly identified inhibitors indicated that di-acid moieties may not be required for ANT binding, because mono-methyl esters in Library 2 and benzyl esters in Library 3 and Library 4 yielded appreciable activity. Based on the assay conditions used, it is safe to assume that the esters were not hydrolyzed to the corresponding acids. The loss in possible ionic pairing may have been compensated by hydrophobic interactions between the protein and the esters or other parts of the molecules. Large hydrophobic substituents, such as isobutyl and benzyl groups, at the α-position of the di-acetate terminus were required for ANT inhibition. It was not clear whether these substituents improved binding through favorable hydrophobic interactions or by restricting the conformation of the di-acetate terminus. It was beneficial to increase the distance between the di-acetate terminus and the central phenyl ring, as noted in comparing leucine and phenylalanine derivatives in Library 3 and Library 4. Additional constraints, such as 2,5-furanyl and carbon-carbon double bonds, between the di-acetate terminus and the central phenyl ring further improved ANT inhibition (e.g., MITO-3152 vs. MITO-3207 and MITO-3331). Further SAR exploration is possible at the R_3 position beyond the original set of building blocks. Although the data reported herein hint that hydrocarbon chains may be preferred, analogs with improved affinity for ANT can be identified by incorporating building blocks with a wider range of structural features. As a cautionary note, the observations described above were only SAR trends based on weak activity and a limited number of hits that can be used in the design of new libraries for further optimization. Additional exploration is needed to better understand the contributions from each segment of the small-molecule ANT ligands and to establish more detailed structure-activity relationships.

TABLE 6.4

Building Blocks for Library 4

Code	R_1 component	R_2 component	R_3 component
A			
B			
C			
D			
E			
F			
G			
H			
I			
J			
K			
L			

TABLE 6.5
ANT Affinity of Selected BKA Analogs

Compound	Structure	IC$_{50}$ (µM)
BKA		0.2
MITO-3152		18.8
MITO-3207		4.7
MITO-3331		5.9
MITO-3332		12.1

6.4 SUMMARY

This chapter described the design and identification of iminodiacetate derivatives as ANT inhibitors based on the natural product (BKA) lead. We also developed a combinatorial synthesis strategy using straightforward chemistry that enabled quick access to libraries of these iminodiacetate derivatives. The SAR trends described in this report can serve as a starting point for additional exploration of BKA mimetics. It is worthwhile to reiterate the importance of screening libraries of structural closely related compounds to follow weak SAR trends. High-throughput screening programs focusing on screening large and diverse compound collections have proven successful in identifying new lead compounds for various biological targets. The cost of accessing a large, high-quality compound library is high and is a potential prohibitive factor for some. Therefore, screening well-designed, small, and focused libraries offers a viable alternative for rapid lead identification.

ACKNOWLEDGMENTS

The authors wish to thank Dr. Christen M. Anderson and Ms. Amy K. Carroll (MitoKor) for screening the compound libraries, and Dr. Neil Howell (Migenix) for his help in the preparation of this manuscript.

REFERENCES AND NOTES

1. Geysen, H.M., Meloen, R.H., and Barteling, S.J., Use of peptide synthesis to probe viral antigens for epitopes to a resolution of a single amino acid, *Proc. Natl. Acad. Sci. U.S.A.,* 1984, 81(13), 3998.
2. Houghten, R.A., General method for the rapid solid-phase synthesis of large numbers of peptides: specificity of antigen-antibody interaction at the level of individual amino acids, *Proc. Natl. Acad. Sci. U.S.A.*, 1985, 82(15), 5131.
3. Merrifield, R.B., Solid phase peptide synthesis. I. The synthesis of a tetrapeptide, *J. Am. Chem. Soc.,* 1963, 85(14), 2149.
4. Lebl, M., Parallel personal comments on "classical" papers in combinatorial chemistry, *J. Comb. Chem.,* 1999, 1(1), 3. A review and references cited therein.
5. Dolle, R.E., Comprehensive survey of combinatorial libraries with undisclosed biological activity: 1992–1997, *Mol. Div.*, Volume Date 1998, 4(4), 233, 2000.
6. Kirschning, A., Monenschein, H., and Wittenberg, R., Functionalized polymers — emerging versatile tools for solution-phase chemistry and automated parallel synthesis, *Angew. Chem., Int. Ed.*, 2001, 40(4), 650.
7. De Julian-Ortiz, J.V., Virtual Darwinian drug design: QSAR inverse problem, virtual combinatorial chemistry, and computational screening, *Comb. Chem. High Throughput Screen.,* 2001, 4(3), 295.
8. Adang, A.E.P. and Hermkens, P.H.H., The contribution of combinatorial chemistry to lead generation: an interim analysis, *Curr. Med. Chem.*, 2001, 8(9), 985.
9. Edwards, P.J., Allart, B., Andrews, M.J.I., Clase, J.A., and Menet, C., Expediting drug discovery: recent advances in fast medicinal chemistry — optimization of hits and leads, *Curr. Opin. Drug Disc. Dev.,* 2006, 9(4), 425.
10. Pinilla, C., Appel, J.R., Blanc, P., and Houghten, R.A., Rapid identification of high affinity peptide ligands using positional scanning synthetic peptide combinatorial libraries, *BioTechniques*, 1992, 13(6), 901.
11. Richardson, P.L., The determination and use of optimized protease substrates in drug discovery and development, *Curr. Pharm. Design*, 2002, 8(28), 2559.
12. Pinilla, C., Appel, J.R., Blondelle, S.E., Dooley, C.T., Eichler, J., Nefzi, A., Ostresh, J.M., Martin, R., Wilson, D.B., and Houghten, R.A., Synthesis and screening of positional scanning synthetic combinatorial libraries. In *Combinatorial Chemistry: A Practical Approach* (Fenniri, H., Ed.), Oxford University Press, Oxford, 2000, pp. 51–74.
13. Pei, Y., Carroll, A.K., Anderson, C.M., Moos, W.H., and Ghosh, S.S., Design and combinatorial synthesis of *n*-acyl iminodiacetic acids as bongkrekic acid analogues for the inhibition of adenine nucleotide translocase. *Synthesis*, 2003, 11, 1717.
14. Dykens, J.A., Davis, R.E., and Moos, W.H., Introduction to mitochondrial function and genomics, *Drug Dev. Res.*, 1999, 46(1), 2.
15. Moos, W.H., Dykens, J.A. and Davis, R.E., It was the best of mito times: a tale of two genomes, *Pharm. News*, 1999, 6(1), 15. A review and references cited therein.
16. Tsujimoto, Y., Nakagawa, T., and Shimizu, S., Mitochondrial membrane permeability transition and cell death, *Biochim. Biophys. Acta, Bioenergetics*, 2006, 1757(9-10), 1297.
17. Halestrap, A.P. and Brenner, C., The adenine nucleotide translocase : a central component of the mitochondrial permeability transition pore and key player in cell death, *Curr. Med. Chem.*, 2003, 10(16), 1507. A review and references cited therein.
18. Nugteren, D.H. and Berends, W., Bongkrekic acid, the toxin from *Pseudomonas cocovenenans*, *Rec. Trav. Chim. Pays-Bas Belg.*, 1957, 76, 13.
19. Lijmbach, G.W.M., Cox, H.C., and Berends, W., Elucidation of the chemical structure of bongkrekic acid. I. Isolation, purification, and properties of bongkrekic acid, *Tetrahedron*, 1970, 26(24), 5993.
20. Lijmbach, G.W.M., Cox, H.C., and Berends, W., Elucidation of the structure of bongkrekic acid. II. Chemical structure of bongkrekic acid and study of the UV, IR, NMR, and mass spectra, *Tetrahedron*, 1971, 27(10), 1839.
21. Stubbs, M., Inhibitors of the adenine nucleotide translocase, *Pharmacol. Ther.*, 1979, 7(2), 329.
22. Fiore, C. et al., The mitochondrial ADP/ATP carrier: structural, physiological and pathological aspects, *Biochimie*, 1998, 80(2), 137.
23. Corey, E.J. and Tramontano, A., Total synthesis of bongkrekic acid, *J. Am. Chem. Soc.*, 1984, 106(2), 462.
24. Shindo, M., Sugioka, T., Umaba, Y. and Shishido, K., Total synthesis of (+)-bongkrekic acid, *Tetrahedron Lett.*, 2004, 45(48), 8863.

25. General procedures for the solid phase synthesis of *N*-acyl iminodiacetic acids are described below.

Coupling of Bromoacetic Acid to Wang Resin: Polystyrene Wang resin (10.0 g, 1.25 mmol/g) was shaken at r.t. with bromoacetic acid (8.68 g, 62.5 mmol), DIC (9.79 mL, 62.5 mmol), and DMAP (100 mg) in DMF (60 mL) in a polypropylene bottle for 4 hr. The resin was collected via vacuum filtration using a 50-mL polypropylene syringe fitted with a polyethylene frit, and washed with DMF (3×40 mL), CH_3OH (3×40 mL), DMF (3×40 mL), methanol (3×40 mL), CH_2Cl_2 (3×40 mL), methanol (3×40 mL), and air dried. The resulting bromoacetate resin **1** (12.0 g) was used in the next step without further analysis.

Displacement of Bromide with Amino Esters: Bromoacetate resin **1** (4.0 g) was shaken with glycine *tert*-butyl ester HOAc salt (3.82 g, 20.0 mmol) and DIEA (7.2 ml, 75 mmol) in DMSO (13 mL) in a 20-mL polypropylene syringe fitted with a polyethylene frit at r.t. for 24 hr. The resin was washed with DMF (3×20 mL), CH_3OH (3×20 mL), DMF (3×20 mL), CH_3OH (3×20 mL), CH_2Cl_2 (3×20 mL), CH_3OH (3×20 ml), and air dried. The resulting resin **2** was used in the next step without further analysis.

Coupling of Nitrophenyl Acids: Resin **2** (1.5 g) was shaken with 2-nitrophenylacetic acid (1.25 g, 6.9 mmol), DIEA (2.0 mL, 11.5 mmol), and bromo-(tris-pyrrolidino) phosphonium hexafluorophosphate (PyBrop) (3.26 g, 7.0 mmol) in DMF (10 mL) at r.t. overnight. The resin was washed with DMF (3×10 mL). The reaction was repeated to ensure complete coupling. The resin was washed with DMF (3×10 mL), CH_3OH (3×10 mL), DMF (3×10 mL), and CH_3OH (3×10 mL), CH_2Cl_2 (3×10 mL), CH_3OH (3×10 mL), and air dried to yield Resin **3**. A small sample (ca. 50 mg) of the resulting resin **3** was treated with TFA-H_2O (95:5, 1.0 mL) for 1 hr at r.t. The solution was collected via filtration. The resin was washed with HOAc (3×1 mL). The combined solution was lyophilized. The residue was analyzed by mass spectrometry to confirm the identity and approximate purity of the intermediate.

Reduction of Nitro Groups to Amines: Resin **3** (1.7 g) was shaken with tin dichloride dihydrate (2.0 M, 20 ml) at r.t. overnight. The resin was washed with DMF (5×10 mL), CH_3OH (3×10 mL), DMF (3×10 mL), CH_3OH (3×10 mL), CH_2Cl_2 (3×10 mL), CH_3OH (3×10 mL), and air dried to yield resin **4**. A small sample (ca. 50 mg) of the resulting resin **4** was treated with TFA–H_2O (95:5, 1.0 mL) for 1 hr at r.t. The solution was collected via filtration. The resin was washed with HOAc (3×1 mL).

The combined solution was lyophilized. The residue was analyzed by mass spectrometry to confirm the identity and purity of the intermediate.

Coupling of $R_3C(=O)X$ to Resin: When a carboxylic acid was used, the coupling reaction was carried out in the presence of DIC. For example, resin **4** (0.275 g) was shaken with benzoic acid (0.31 g, 2.5 mmol), DIC (0.47 mL, 3.0 mmol), DIEA (0.87 mL, 5.0 mmol), and DMAP (10 mg) in DMF (5.0 mL) at r.t. overnight. The resin was washed with DMF (3×5 mL), CH_3OH (3×5 mL), DMF (3×5 mL), CH_3OH (3×5 mL), CH_2Cl_2 (3×5 mL), CH_3OH (3×5 mL), and air dried. The resulting resin **5** was treated with TFA-H_2O (95:5, 3.0 mL) for 1 hr at r.t. The solution was collected via filtration. The resin was washed with HOAc (3×5 mL). The combined solution was lyophilized to give the desired product **6**. Its purity and identity were assessed using LC-MS spectrometry. When an acyl anhydride or chloride was used, the coupling reaction was carried out without DIC. For example, resin **4** (0.275 g) was shaken with acetic anhydride (0.24 mL, 2.5 mmol), DIEA (0.87 mL, 5.0 mmol), and DMAP (10 mg) in DMF (5.0 mL) at r.t. overnight. The resin was washed with DMF (3×5 mL), CH_3OH (3×5 mL), DMF (3×5 mL), CH_3OH (3×5 mL), CH_2Cl_2 (3×5 mL), CH_3OH (3×5 mL), and air dried.

TFA Cleavage: Resin **5** (0.305 g) was treated with TFA-H_2O (95:5, 3.0 mL) for 1 hr at r.t. The solution was collected via filtration. The resin was washed with HOAc (3×5 mL). The combined solution was lyophilized to give the desired product **6**. Its purity and identity were assessed using LC-MS spectrometry.

26. Carroll, A. K., Clevenger, W.R., Szabo, T., Ackermann, L.E., Pei, Y., Ghosh, S.S., Glasco, S., Nazarbaghi, R., Davis, R.E., and Anderson, C.M., Ectopic expression of the human adenine nucleotide translocase, isoform 3 (ANT-3). Characterization of ligand binding properties, *Mitochondrion*, 2005, 5(1), 1.

7 Parallel Synthesis of Anticancer, Antiinflammatory, and Antiviral Agents Derived from L- and D-Amino Acids

Robert C. Reid and David P. Fairlie

CONTENTS

7.1 CHAPTER PREFACE

The synthesis of multiple organic compounds in parallel arrays is still perhaps the most powerful approach to drug discovery and optimization in the 21st century, despite the dramatic improvements that have been made in structure-based drug design, virtual screening, and diversity-oriented combinatorial chemistry. Described herein are parallel syntheses of small groups of inhibitors for the tumor growth-promoting human enzymes histone deacetylases, inhibitors of the pro-inflammatory mediator human secretory phospholipase A_2 group IIa, and inhibitors of the viral enzyme HIV-1 protease. The availability of close structural analogs of bioactive compounds facilitates the development of informative structure-activity relationships that can lead to identification of drug leads and clinical candidates with important and improved pharmacological activities. This chapter demonstrates how derivatization of some simple L- and D-amino acids through parallel solution- and solid-phase synthesis approaches has enabled us to produce nonpeptidic and peptidomimetic compounds as potent inhibitors of human and viral enzymes, with reasonable cell permeability and/or oral bioavailability and potentially valuable pharmacological properties *in vivo.*

7.2   ANTICANCER AGENTS DERIVED FROM L-AMINO ACIDS

7.2.1   INTRODUCTION

The incidence of cancer is escalating to the point of becoming the leading cause of death in industrialized countries. Chemotherapy in the 20th century relied heavily on the use of combinations of cytotoxic drugs that indiscriminately killed cancer cells but at the same time also destroyed normal human cell types, resulting in dose-limiting side effects. This approach eventually slowed the progression (up to 15 to 20 years) from discovery to market for new drugs due to the implementation of increasingly more stringent safety standards and larger hurdles for clinical approvals. Leading into the 21st century there consequently has been a greater focus on better understanding oncology at the molecular level and upon directing medicinal chemistry efforts toward more target-selective anticancer drugs that more effectively interfere with tumor development without damaging normal tissues.

In this context we have been derivatizing simple amino acids (e.g., L-cysteine) into new classes of orally active anticancer drugs that can kill a wide variety of cancer cell types (e.g., breast, cervical, melanoma, ovarian, prostate) without affecting the growth of normal cell types.[1,2] Our interest in these compounds stems from their apparent capacity to kill cancer cells at nanomolar concentrations while normal human cells (melanocytes, fibroblasts, etc.) survive in the presence of micromolar drug concentrations. The reason for this selectivity is unknown but the compounds do share[1,2] the capacity to inhibit histone deacetylases (HDACs), causing hyperacetylation of histones, inducing p21 expression, and modulating chromatin structure, altering gene expression and cell cycle regulation.[3–10] Interestingly, in addition to killing most cancer cells, the phenotypes of surviving tumor cells revert to a more normal morphology, a property that is likely to be most attractive in modern cancer chemotherapeutics. The new compounds are more cytoselective in their killing than HDAC inhibitors such as butyrate (a notable component of butter), retinoic acid, HMBA, ABHA, SBHA, SAHA, scriptaid, and sirtinol,[3–10] which are of low potency, non-selective *in vivo,* and cause cellular differentiation that can be readily reversed. They also appear to be more selective than more potent HDAC inhibitors such as trichostatin,[11] trapoxin,[12] apicidin,[13] and analogs, which usually have no more than a fivefold discrimination in their cytotoxicity for cancer cells over normal cells. Furthermore, most such inhibitors are not orally active and are compromised *in vivo* by low bioavailability and rapid metabolism. Nevertheless, many such HDAC inhibitors are in development,[1–10,14] and SAHA[10] was approved in late 2006 for use in humans.

We have used parallel synthesis approaches to optimize drug structures for potency, selectivity, and bioavailability to enhance prospects for broad-spectrum antitumor activity. Parallel synthesis in this application can potentially allow us to tease out the pharmacophores (if different) for selectivity versus cytotoxicity.

7.2.2   PARALLEL SYNTHESIS OF ANTICANCER AGENTS FROM L-AMINO ACIDS

The key starting point was the consideration of the reversible acetylation of DNA-binding lysine sidechains [$-(CH_2)_4-NH_2 \Leftrightarrow -(CH_2)_4-NH-COMe$] in nuclear proteins known as histones. We chose to mimic L-lysine with L-cysteine and create chiral compounds such as structure **1** with an analogous sidechain modified to match the acetylated lysine but with a better terminal zinc-binding hydroxamate ligand [$-CH_2-S-(CH_2)_3-CONHOH$] to enhance affinity for HDAC enzymes.[1] We used a parallel synthesis approach to develop chemical libraries such as **1**[1] and **2**[2] from the amino acids cysteine and 2-aminosuberic acid, fused at the C-terminus to amines, at the N-terminus to acids, and at the sidechain to alkyl hydroxamates and derivatives.

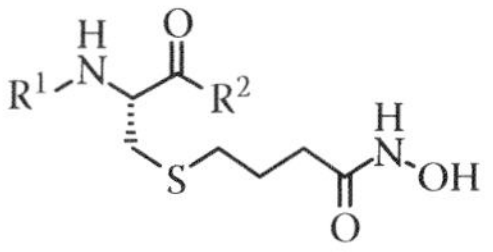

STRUCTURE 7.1 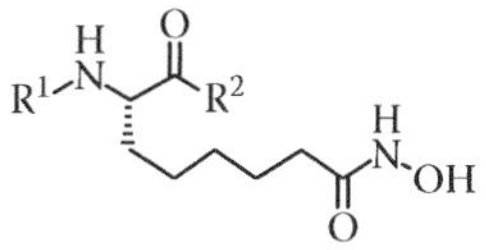 **STRUCTURE 7.2**

Scheme 7.1 illustrates the approach to a subset of a hydroxamate library **1**, with benzylamine condensed at the C-terminus of cysteine. The amino acid provided a convenient chiral scaffold, the sulfur affording some water solubility to library members, while the hydroxamate served as a zinc-binding ligand for the active site of histone deacetylase enzymes. Briefly, unprotected cysteine was alkylated at the sidechain sulfur nucleophile with 4-iodobutyric acid *tert*-butyl ester (Scheme 7.1) to produce an amino acid with an extended sidechain comparable in length to the active site hydrophobic zinc-binding cavity or channel in HDACs. Formation of disulfide or alkylated amine was avoided using basic deoxygenated conditions. Fmoc protection of the amine, formation of the allyl ester from the acid, removal of the *tert*-butyl ester with TFA, and coupling of the acid to 2-chlorotritylhydroxylamine resin using HATU proceeded in good yield (70%, 1 hr). Once attached to resin, the allyl ester was removed with Pd and allyl scavenger DMBA, and the acid was coupled to benzylamine or other amines. Deprotection of the N-terminal amine gave compounds of greater than 80% purity after cleavage from resin, and further purification was effected by reverse-phase HPLC.[1]

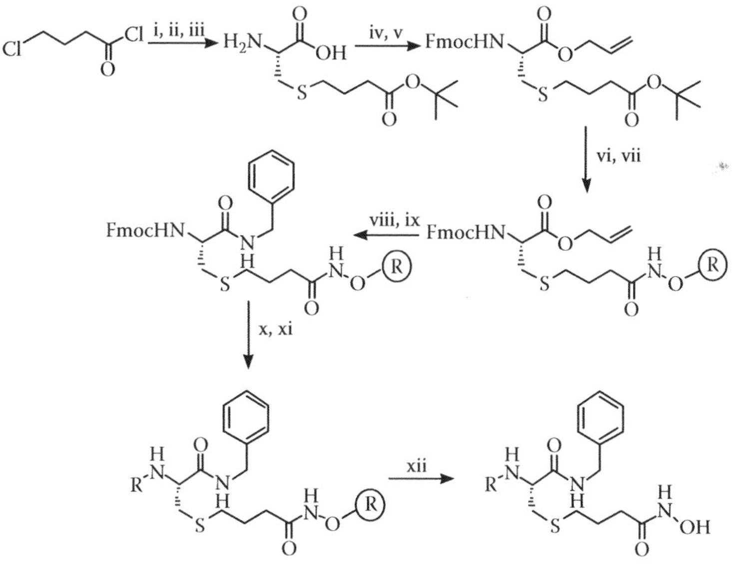

SCHEME 7.1 Parallel synthesis of anticancer agents based on cysteine: (i) *tert*-BuOH, pyridine; (ii) NaI, THF; (iii) Cys, NaOH, MeOH (iv); Fmoc-OSu, NaHCO₃, THF, water; (v) allyl bromide, DMF, K₂CO₃; (vi) TFA; (vii) HATU, DIPEA, 2-chlorotritylhydroxylamine resin, DMF; (viii) Pd(Ph3)4, DMBA; (ix) benzylamine, HBTU, DIPEA, DMF; (x) piperidine; (xi) R-CO₂H, HBTU, DIPEA, DMF; (xii) TFA.

Scheme 7.2 illustrates the parallel synthesis of a different library **2** in which 2-aminosuberic acid was the scaffold that was derivatized with varying substituents at the N- and C-terminus of the amino acid and hydroxamate was on the sidechain.[2] The key feature of the synthesis was the enzymatic resolution of the 2-aminosuberate.[2,15]

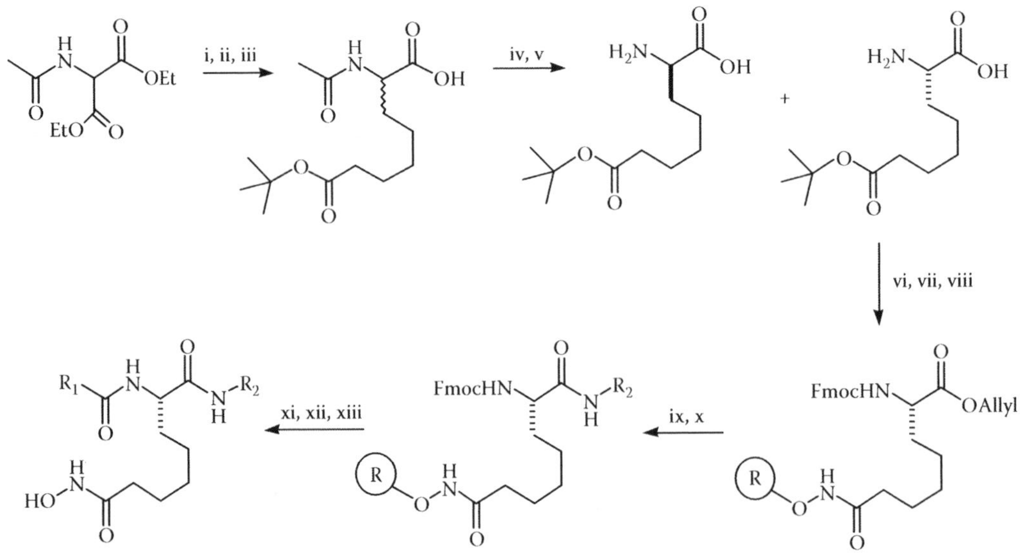

SCHEME 7.2   Parallel synthesis of anticancer agents based on 2-aminosuberic acid: (i) (a) NaH, DMF; (b) 6-iodo-hexanoic acid *tert*-butyl ester; (ii) LiCl-H_2O, DMSO, 160°C; (iii) LiOH, H_2O:EtOH; (iv) acylase I (aspergillus melleus), $CoCl_2$, phosphate buffer pH 7.2; (v) (a) Fmoc-OSu, $NaHCO_3$ THF:H_2O (b) rpHPLC separation; (vi) allyl bromide, NaHCO3, DMF; (vii) TFA:DCM 9:1; (viii) 2-(9H-fluoren-9-ylmethoxycarbonylamino)-octanedioic acid 1-allyl ester, 2-chlorotritylhydroxylamine resin, HATU, DIPEA, DMF; (ix) Pd(PPh$_3$)$_4$, DMBA, DCM; (x) R$_2$ = amine, HBTU, DIPEA, DMF; (xi) piperidine:DMF; (xii) R$_1$ = acid, HBTU, DIPEA, DMF; (xiii) TFA:DCM 95:5.

7.2.3   ANTICANCER ACTIVITY

A large number of cysteine-derived compounds have been prepared, featuring amines fused to the C-terminus, carboxylic acids fused to the N-terminus, and 4-butanoylhydroxamate fused to the S-terminus. Table 7.1 shows nine such compounds that were cytotoxic at nanomolar concentrations (IC_{50} ~200 nM) to human melanoma cells MM96L, three showing selectivities greater than 5:1 for human melanoma cells over normal human cells (e.g., neonatal fetal fibroblasts, NFF).[1] These compounds were also cytotoxic to other cancer cell types (melanomas SK-MEL-28, DO4, prostate DU145, breast MCF7, ovarian JAM, C180-13S). Compound **1b** was orally bioavailable in rats (5 mg/kg/p.o., C_{max} = 6 µg/mL, T_{max} = 15 min), with high levels of drug present in plasma for at least 4 hr.[1]

   Table 7.2 shows a selection of compounds of structure **2** similarly derived from 2-aminosuberic acid instead of cysteine. This compound series, in which a methylene group replaces the thioether, was based on recognizing that the S-substituent can position the hydroxamate carbonyl a little further away from the active site zinc in HDACs than the acetylated lysine sidechain of histones, two C–S bonds being longer than two C–C bonds. The resulting compounds **2** were up to tenfold more potent in killing melanoma MM96L cells (IC_{50} = 20 to 250 nM) than normal human cells (e.g., neonatal fetal fibroblasts, NFF), and up to tenfold more selective as well.[2] The significant selectivities (SI > 50:1; Selectivity Index (SI) = IC_{50} (NFF)/ IC_{50} (MM96L)) were very encouraging[2] and suggested that even higher selectivity might be obtained through further optimization using such simple parallel syntheses. Some of these compounds have been studied recently in animal trials as anticancer agents.

   Since this work was conducted, there have been a number of crystal structures of HDAC enzymes reported[16,17] as well as more information about the properties of some HDAC inhibitors that allow them to discriminate between cancer cells and normal cells.[18,19] Together with further chemical derivatization of compounds shown in Tables 7.1 and 7.2 using parallel syntheses approaches, ongoing mechanistic studies promise to deliver even more selective anticancer compounds for clinical development.

TABLE 7.1

Selected Cytotoxicity Data for Nonpeptidic Hydroxamates of Structure 1

Compound	R1	R2	*	LogD	IC$_{50}$(μM) NFF	IC$_{50}$(μM) MM96L	SI
1a	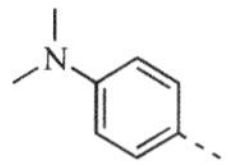	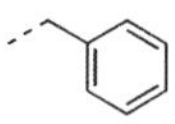	S	2.1	0.35(7)	0.14(9)	2.5
1b	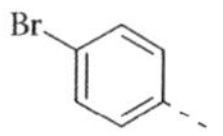	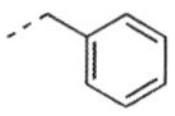	S	2.7	0.83(9)	0.2(1)	4.2
1c	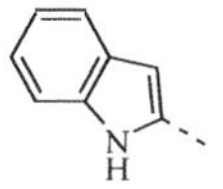	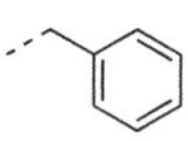	S	1.4	0.8(2)	0.13(9)	6.2
1d	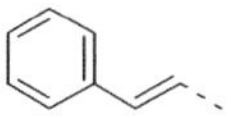	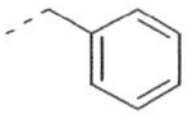	S	2.2	0.8(2)	0.2(1)	4.0
1e	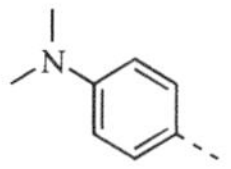	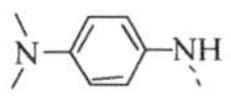	S	1.6	0.6(1)	0.1(1)	6.0
1f	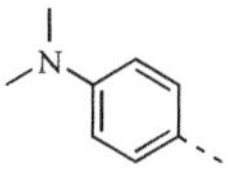	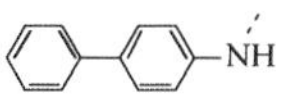	S	3.6	0.32(5)	0.17(5)	1.9
1g	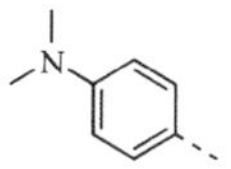	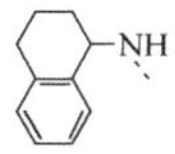	S	2.4	2.2(2)	0.2(1)	11.0
1h	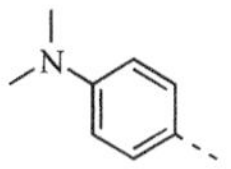	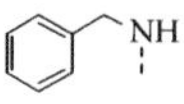	S	1.3	0.35(7)	0.14(9)	2.5
1i	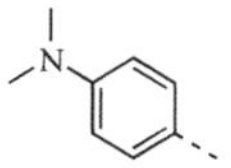	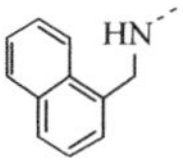	S	2.5	0.42(5)	0.20(2)	2.1

TABLE 7.2

Selected Cytotoxicity Data for Nonpeptidic Hydroxamates of Structure 2

Compound	R1	R2	*	LogD	IC$_{50}$(μM) NFF	IC$_{50}$(μM) MM96L	SI
2a	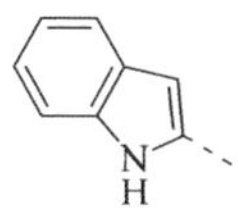	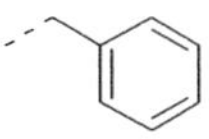	S	2.5	1.26(3)	0.25(2)	5
2b	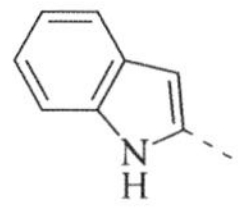	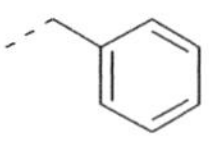	R	2.5	0.572(8)	0.080(6)	7
2c	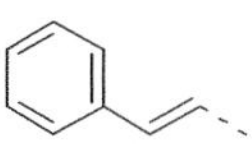	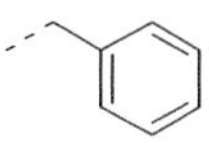	S	2.7	4.01(9)	0.46(4)	9
2d	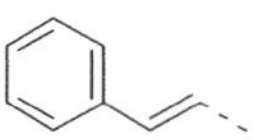	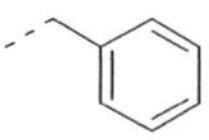	R	2.7	1.15(8)	0.25(1)	5

(continued)

TABLE 7.2

Selected Cytotoxicity Data for Nonpeptidic Hydroxamates of Structure 2 (continued)

Compound	R1	R2	*	LogD	IC$_{50}$(μM) NFF	IC$_{50}$(μM) MM96L	SI
2e			R	2.8	1.0(1)	0.16(8)	6
2f			S	2.8	0.57(7)	0.02(1)	28
2g			S	3.0	0.34(6)	0.021(4)	16
2h			rac	1.1	4.6(4)	0.2(1)	23
2i			S	3.2	1.24(7)	0.010(5)	59
2j			S	3.0	0.45(3)	0.043(6)	11
2k			S	2.6	1.22(2)	0.023(3)	53
2l			S	5.3	2.97(8)	0.074(5)	40

7.3 ANTIINFLAMMATORY LIPID ANALOGS DERIVED FROM D-AMINO ACIDS

7.3.1 INTRODUCTION

More than 100 human inflammatory conditions are now known to be mediated by the same networks of endogenous proteins, with numerous built-in redundancies and feedback regulatory mechanisms that continue to confound our understanding of disease progression.[20] Most antiinflammatory drugs available for use in humans are non-steroidal antiinflammatory drugs (NSAIDs) that typically block formation or action of inflammatory mediators (e.g., prostaglandins, leukotrienes, thromboxanes, platelet activating factor) produced near the bottom of the inflammatory cascade from arachidonic acid. The latter is produced via the degradation of membrane phospholipids, which is particularly accelerated after activation of macrophages in response to inflammatory stimuli (e.g., bacteria, viruses, parasites, endotoxins, autoantigens) and release of proinflammatory cytokines (e.g., TNFα, IL-1, IL-6, etc.), which in turn cause secretion of enzymes known as phospholipases A$_2$ (PLA$_2$).[21–25] This family of phospholipases is secreted from many cell types (e.g., synoviocytes, osteoblasts, chondrocytes, vascular smooth muscle, immune cells, etc.) and is distinguished from other lipases by the capacity to selectively hydrolyze ("sn – 2" cleavage) 1,2-diacylphospholipids at the 2-acyl position (Figure 7.1) to form fatty acid (R^{1}CO$_2$H) and lysophospholipid ([HOCH(CH$_2$OCOR^2)

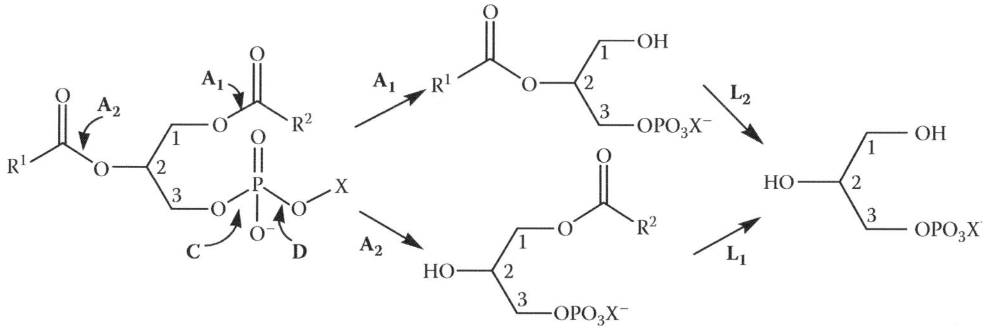

FIGURE 7.1 Hydrolysis by phospholipases (A_1, A_2, C) and lipases (L_1, L_2).

(CH_2O-PO_3X)]), usually from aggregated substrates such as monolayers, vesicles, micelles, or membranes.[26,27]

Clinical results for inhibitors of both cyclooxygenases and lipoxygenases suggested that inhibiting the formation of the precursor substrate, arachidonic acid, might prevent most eicosanoid formation, thereby leading to more effective antiinflammatory drugs. Although there are many isoforms of PLA_2, both calcium-dependent secretory ($sPLA_2$ groups I, IIa, IIc, V, X) and cytosolic ($cPLA2$ group IV) as well as calcium-independent ($iPLA2$ group VIa), no specific inhibitors of any kind of PLA_2 are yet available in humans.[21–28] Our interest focused on the ubiquitous group IIa human enzyme, long implicated in eicosanoid production and found in unusually high concentrations in people with inflammatory conditions such as rheumatoid and osteoarthritis, sepsis, asthma, pancreatitis, inflammatory bowel diseases, fibrotic disorders, psoriasis, atherosclerosis, burn injuries, etc.[24,25] Recent literature downplays the importance of $sPLA_2$ in the pathogenesis of inflammatory diseases,[24] but its importance in several animal models of inflammatory disease is incontrovertible, and there is a strong correlation between $sPLA_2$ (IIa) concentrations and severity of disease.[24,29] Our goal was to link the potency and specificity of inhibitor action on this enzyme *in vitro* with the potency of antiinflammatory action *in vivo*.

7.3.2 PARALLEL SYNTHESIS OF ANTIINFLAMMATORY AGENTS FROM D-AMINO ACIDS

Using a substrate-based approach from glycerophospholipid substrates, we had decided in the 1990s to generate inhibitors of the specific enzyme isoform $sPLA_2$ (IIa) through parallel synthesis methods, starting with simple D-amino acids.[29,30] Our preliminary modeling of various putative amino acid derivatives into the reported crystal structure[31–33] of $sPLA_2$ (IIa), from the synovial cavity of patients with rheumatoid arthritis, had suggested to us that such an approach might be successful in rapidly generating potent inhibitors. Representative of our attempts are the chiral substrate analogs of the simple D-amino acids (Figure 7.2) and a disclosed set of D-tyrosine derivatives of structure **3** (Table 7.3).

For example, benzyl-protected Boc-D-tyrosine was converted (Scheme 7.3) via the Weinreb amide to an aldehyde before chain extension with methyl/ethyl triphenylphosphoranylidene acetate (*S*-stereochemistry being confirmed for the amine derivative by Mosher analysis), before sequential deprotection, parallel coupling to phenyl/pyridl alkanoic acid derivatives, hydrogenation over Pd/C in ethyl acetate, and saponification with NaOH to produce inhibitors **3a–3f** and, by analogy, **3g**.[29] Two further steps involving removal of the benzyl group and alylation with alkyl halides in base led to larger analogs **3h–3p**. A crystal structure determination[29] showed that analog **3b** bound in the active site of recombinant human nonpancreatic secretory $sPLA_2$-IIa, the carboxylate and amide carbonyl oxygens chelating as a bidentate ligand to calcium, the amide NH hydrogen bonding to catalytic residue His48, and the phenylhepatanoyl and benzyloxy substituents occupying the large hydrophobic cleft of the enzyme active site. This information was obtained well after the

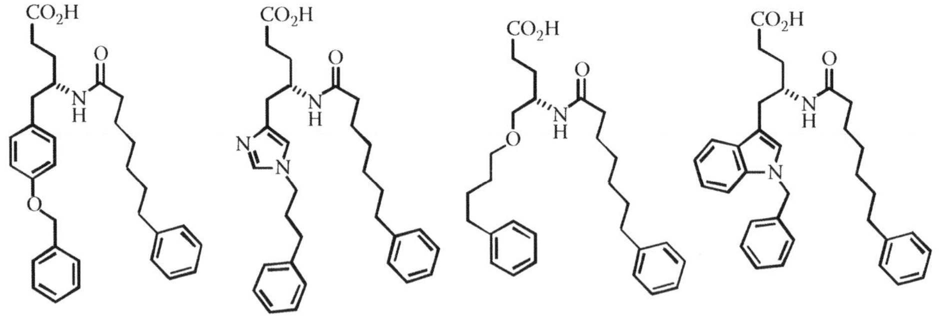

FIGURE 7.2 Representative inhibitors of sPLA$_2$ (IIa) derived from amino acids D-tyrosine, D-histidine, D-serine, and D-tryptophan.

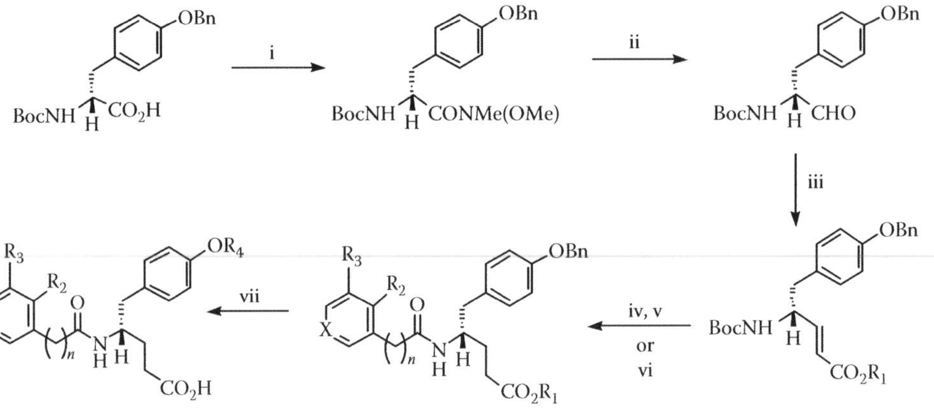

SCHEME 7.3 Parallel synthesis of antiinflammatory agents from D-tyrosine. Reagents: (i) BOP, DIPEA, DMF, NHMe(OMe).HCl; (ii) LiAlH$_4$, THF; (iii) Ph$_3$P=CHCO$_2$Me or Ph$_3$P=CHCO$_2$Et, THF; R1 = Me or Et; (iv) Pd/C, H$_2$, EtOAc; (v) (a) TFA, CH$_2$Cl$_2$, (b) BOP, DIPEA, DMF, 7-(3-nitrophenyl)-heptanoic acid, (c) aq. NaOH, THF, MeOH, (d) 1 M HCl; (vi) (a) TFA, CH$_2$Cl$_2$, (b) BOP, DIPEA, DMF, substituted or unsubstituted phenyl- or pyridyl-alkanoic acid, (c) H$_2$, Pd/C, EtOAc, (d) aq. NaOH, THF, MeOH, (e) 1 M HCl, R1 = H; (vii) R1 = Et or Me, (a) H$_2$, Pd/C, THF, HCl, (b) R4-Y (Y = Cl, Br, I), K$_2$CO$_3$, DMF, (c) NaOH, THF, MeOH.

structure-activity relationships had been determined for compounds **3a–3q** generated in parallel, but confirmed that the two hydrophobic tails in this series of compounds do fill the hydrophobic cleft of the enzyme.

7.3.3 ENZYME INHIBITION AND ANTIINFLAMMATORY ACTIVITY

Table 7.3 shows some structure-activity relationships for representative compounds **3a–q** as inhibitors of recombinant human nonpancreatic secretory sPLA$_2$-IIa.[29] The *in vitro* colorimetric assay[34] utilized a thioester substrate and Ellman's reagent to detect cleavage by enzyme, and the results from this assay have nicely correlated with our *in vivo* findings of antiinflammatory activity in rats (e.g., against arthritis, ischemia-reperfusion injury, fibrotic disorders, inflammatory bowel diseases, hypertension, uterine contractions, uveitis).[29,30] Thirteen compounds in Table 7.3 had substantial potency at submicromolar concentrations against the human enzyme, and ten compounds had IC$_{50}$ values below 250 nM (**3b, 3h–3p**).

Many compounds covered by our patent[30] have been found to be potent and selective inhibitors of sPLA$_2$-IIa over other sPLA$_2$ enzymes. Crystal structures solved for several compounds demonstrated similar binding modes to that for **3b**. A key feature was the location of an aromatic ring of the inhibitor within bonding distance of His6,[29] which is unique to the human enzyme (not present

TABLE 7.3

Inhibitory Potencies of Selected Nonpeptidic Analogs 3 Derived from D-tyrosine Against the Human Recombinant Enzyme sPLA$_2$ IIa[a]

3a-q

Structure	X	n	R$_2$	R$_3$	R$_4$	IC$_{50}$ (µM)	Mole fraction (Xi)
3a	CH	5	H	H	OBn	0.662	0.00044
3b	CH	6	H	H	OBn	0.029	0.000019
3c	CH	7	H	H	OBn	2.48	0.0017
3d	CH	6	OMe	H	OBn	1.82	0.0012
3e	CH	6	H	NHAc	OBn	4.05	0.0027
3f	N	6	H	H	OBn	0.761	0.00051
3g	CH	6	H	NO$_2$	OBn	0.536	0.00036
3h	CH	6	H	H	2-picolyl	0.214	0.00014
3i	CH	6	H	H	3-picolyl	0.247	0.00017
3j	CH	6	H	H	cyclohexylmethyl	0.067	0.000045
3k	CH	6	H	H	cyclopentylmethyl	0.057	0.000038
3l	CH	6	H	H	1-naphthylmethyl	0.019	0.000013
3m	CH	6	H	H	2-naphthylmethyl	0.039	0.000026
3n	CH	6	H	H	cinnamyl	0.116	0.000078
3o	CH	6	H	H	iso-butyl	0.170	0.00011
3p	CH	6	H	H	n-heptyl	0.086	0.000058
3q	CH	6	H	H	H	2.57	0.0017

[a] Chromogenic assay.[34]

in rat/mouse/dog sPLA$_2$-IIa). The most potent inhibitors also inhibited IL-1β– or TNFα-induced release of PGE$_2$ from WI38 human lung fibroblasts or human HeLa S3 cells, respectively, at submicromolar drug concentrations (unpublished data).

Using **3b** in particular as an *in vivo* probe, we have found antiinflammatory activity for these orally active enzyme inhibitors at 0.3 to 5 mg/kg in rat models of intestinal[35] and myocardial[36] ischemia-reperfusion injury, adjuvant[29] and monoarticular arthritis, TNBS-induced[37] inflammatory bowel disease, cardiac fibrosis,[38,39] hypertension, osteoporosis,[40] uterine contractions,[41] LPS-induced changes in blood pressure, and in a dog model of uveitis. Unlike aspirin, **3b** for example did not show any analgesic activity. No evidence of toxicity was observed for **3b** in rats following chronic oral dosing at 10 mg/kg for up to 4 weeks. When administered at 30–300 mg/kg/p.o. (10–100 times the estimated effective doses) in olive oil to mice and examined for acute toxic symptoms (i.e., convulsions, tremors, muscle relaxation, etc.) and death at up to 72 hr, there was again no evidence of toxicity. Unlike Eli Lilly indole compounds, our compounds did not impair renal function. Compound **3b** did not display any mutagenicity in the Ames test — with or without microsomal metabolic transformation. *In vitro* radio-ligand binding assays did not reveal any appreciable binding/inhibition (IC$_{50}$ >> 5 µM) to a wide range of receptors (adenosine, adrenergic, angiotensin, bradykinin, calcium channels, dopamine, estrogen, GABA, glucocorticoid, glutamate, glycine, histamine, insulin, muscarinic, neuropeptide Y, nicotinic acetylcholine, opiate, phorbol ester, K^+ channel, progesterone, tachykinin).

Over the years there have been thousands of scientific reports and hundreds of patents dealing with PLA$_2$ enzymes (most often of snake origin), or weak inhibitors with IC$_{50}$ > 1 μM, or topical inhibitors. However, it became clear to us by the mid-1990s that identification of sPLA$_2$ inhibitors was heavily compromised by (1) poor lipid-water enzyme assay methods (falsely suggesting inhibitors); (2) the unavailability of significant quantities of human sPLA$_2$ enzymes; (3) a poor understanding of differences between sPLA$_2$ enzymes, and mechanisms of action; (4) uncharacterized roles of PLA$_2$ surfaces in receptor-binding; and (5) poor bioavailabilities. Consequently, there are, in fact, still today only a handful of truly potent *bona fide* inhibitors (IC$_{50}$ ≤ 100 nM) of sPLA$_2$-IIa actually known.[24] We have made and tested many reported inhibitors of "sPLA$_2$," only to find that they do not significantly inhibit recombinant human nonpancreatic secretory PLA$_2$-IIa (IC$_{50}$ < 1 μM) mediated hydrolysis of lipid substrates. Parallel synthesis strategies have enabled us to find a wide range of genuine inhibitors of this enzyme, and in our hands such inhibitors also show significant activity *in vivo* in many animal models of inflammation. We find that our compounds bind with high selectivity to PLA$_2$-IIa without interacting either with other PLA$_2$ isoforms or with other human receptors, in contrast to indoles such as LY311727 and analogs that do seem to be less discriminatory for target receptors. Our data, in particular for compound **3b**, appears to support important roles for sPLA$_2$-IIa in the pathogenesis of a wide range of acute and chronic inflammatory diseases, and this appears to be promising in light of the clinical failure of indole-based inhibitors reviewed elsewhere.[24]

7.4 ANTIVIRAL MACROCYCLIC PEPTIDOMIMETICS DERIVED FROM L-AMINO ACIDS

7.4.1 INTRODUCTION

Proteolytic enzymes are pivotal components of the physiology of living organisms, and their controlled regulation represents a valuable approach to the development of new therapeutic agents,[42–48] with many in humans or in advanced clinical trials.[42, 48] Because only a 5 to 10 amino acid segment of a polypeptide substrate actually interacts with the active site of a proteolytic enzyme, protease inhibitors can be rationally developed either through derivatization of short peptide substrates or through computer-assisted design of small molecules that can compete with substrates and occupy the substrate-binding groove of a protease.[42–50] A particular feature that aids rational development of protease inhibitors is the observation that x-ray crystal structures of protease-inhibitor complexes consistently show the ligand in a beta-strand conformation,[50–53] and thus even when there is no known structure for a particular protease, we can fairly confidently design inhibitors based on adoption of a beta strand in the active site.

However, a major difficulty with small-molecule design is that there are important cooperative effects between the enzyme and ligand components, and small modifications to one end of a ligand can render other components non-optimal because of unpredictable changes in the ligand-enzyme induced fit.[54,55] This cooperativity means that ligand binding cannot be considered as the sum of independent contributions from the various molecular interactions, changes to part of an inhibitor producing reciprocal changes to the receptor, and as a result the shape of the active site at a more remote position may undergo unpredictable changes that can influence substrate/inhibitor binding. These problems are more pronounced with peptide inhibitors because of their conformational flexibility.

To address the issues of beta-strand mimicry and the induced fit, we looked for a different approach to develop nonpeptidic inhibitors from peptides, one that could transfer elements of molecular recognition gleaned from the interaction of native peptidic substrates with the surface features of the enzyme's active site, onto a suitably functionalized organic compound (template) bearing appropriate substituents that mimic these interactions.[56–60] We used parallel synthesis approaches, dividing peptidic substrates/inhibitors into two halves and separately constraining the C- and N-termini into rigid, beta-strand mimicking macrocycles[56,57] that preserved and stabilized the protease-binding peptide beta-strand motif. By removing conformational freedom from a tripeptide fragment, the macrocycle locks all of the interactions made with the receptor into one collective set that functions to anchor the inhibitor to the active site. The result is that the effects of induced fit are now limited to only one-half of the molecule and, by appending units to the macrocycle, it is possible to more rapidly discover complementary entities with drug-like properties.[61] From a statistical perspective the number of compounds required to explore the chemical space of a hexapeptide ligand, which binds with a high degree of cooperativity, would be x^6 compared with x^4 if the three terminal residues were constrained into a macrocycle, an improvement in the likelihood of finding a hit by two orders of magnitude. This approach is particularly well suited to the inhibition of aspartic proteases where the scissile amide bond represents a logical dividing position, separating the prime and non-prime regions, and allows for the introduction of a transition state isostere. Macrocycles have also been successfully applied to metallo and serine proteases[62] and are represented in HDACs by trapoxin[13] and apicidin.[14] Nature has also made use of constrained macrocycles in numerous biologically active natural products.[63] Here we describe a parallel synthesis of a series of macrocyclic inhibitors of the aspartic viral protease, HIV-1 protease, and show that the compounds have antiviral activity at nanomolar concentrations.

7.4.2 N-Terminal Macrocycles

The initial "proof-of-concept" studies confirmed that macrocycles such as **4a** can mimic the protease bound conformation of the peptidic inhibitor JG-365 **4b** as their protease-ligand crystal structures superimposed well. Potent inhibition of the HIV-1 protease enzyme ($K_i = 12$ nM) also demonstrated that structural mimicry was translated into functional activity.[56,59]

The synthesis of **4a** was achieved by assembling the C-terminal peptidic fragment Pro-Ile-Val on MBHA resin using Boc solid-phase peptide synthesis. Then the N-terminus was alkylated with the α-bromoketone **4c** obtained by extensive modification of the amino acid, L-tyrosine (Scheme 7.4). Reduction of the ketone on resin with NaBH$_4$ completed the hydroxyethylamine transition state isostere (**4d** as a 1:1 mixture of diastereomers). Then, following removal of the Boc group with TFA, the final Asn residue was coupled in place. After cleavage of the product from the resin with HF, the macrocyclic ring was closed by amide bond formation in solution, and the final product **4a** was purified (including separation of diastereomers) by reversed-phase HPLC.

Several analogous macrocyclic inhibitors of HIV-1 PR were prepared reliably by these methods.[56,64] However, in pursuit of enhanced bioavailability and antiviral activity, we sought to investigate non-peptidic appendages to the macrocycle that were not amenable to solid-phase peptide synthesis.[60,65] Having concluded that the macrocycle and hydroxyethylamine isostere were the key indispensable elements in this series of compounds, the synthesis of larger numbers of analogs could be expedited using an epoxide derivative **4e** in parallel solution-phase synthesis (Scheme 7.5) to generate focused combinatorial libraries.[61] Upon reaction of the diastereomerically pure epoxide **4e** with an amine, the hydroxyethylamine isostere would be created and the sidechain of the respective amine used would be expected to fill the enzyme S1′ pocket, and beyond if the appropriately substituted amine could be employed, as determined by molecular modeling. Alternatively, in the case where primary amines are used, the resulting secondary amine intermediate is amenable to further derivatization by acid derivatives, including sulfonyl chlorides and isocyanates, for example, leading to the introduction of considerable chemical diversity onto what is already proven to be a powerful template for protease inhibitor discovery.

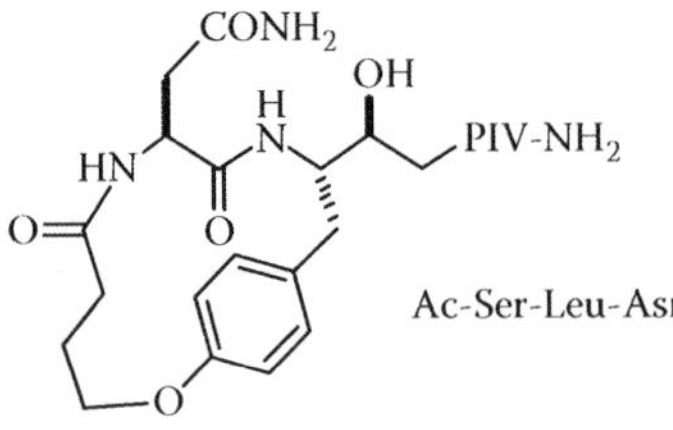

Ac-Ser-Leu-Asn-Phe-{*(S)*-CHOH-CH_2}-Pro-Ile-Val-OMe

4a **4b**

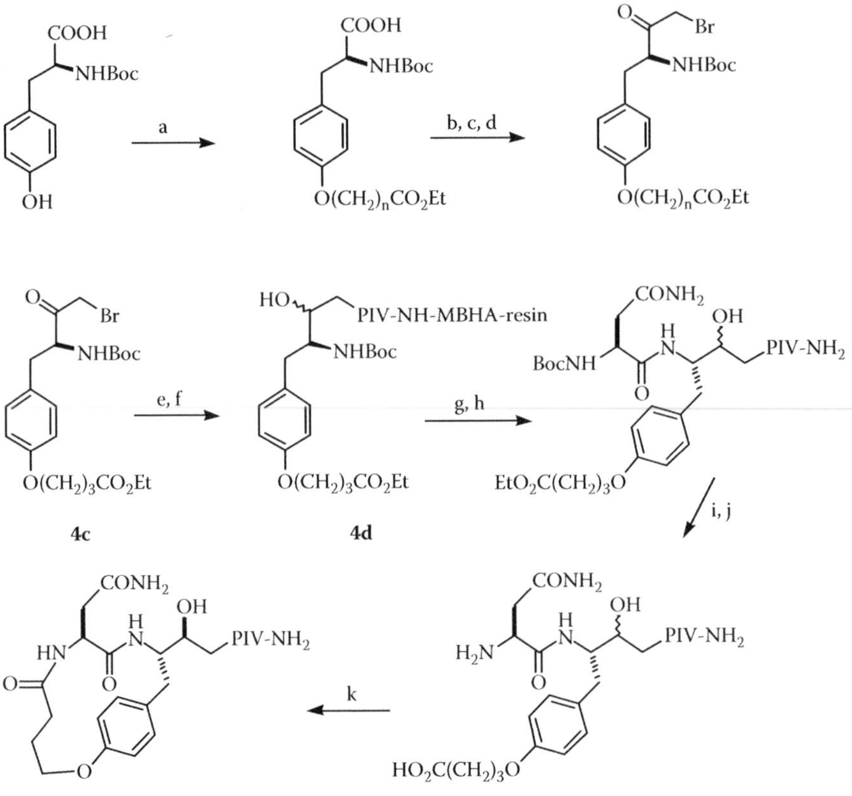

SCHEME 7.4 Synthesis of N-terminal macrocyclic peptidomimetics: (a) X-$(CH_2)_nCO_2Et$ (5 equiv), NaH (7 equiv); (b) NMO; iBuOCOCl; (c) CH_2N_2; (d) HBr (EtOAc solution); (e) H-Pro-Ile-Val-NH-MBHA resin; (f) $NaBH_4$; (g) TFA; (h) Boc-Asn-OH, HBTU, DIPEA; (i) HF; (j) NaOH; (k) BOP, DIPEA.

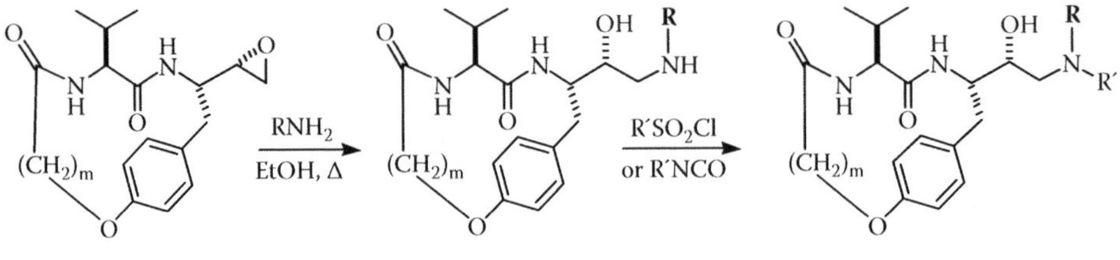

4e

SCHEME 7.5 Conversion of N-terminal macrocyclic epoxide to macrocyclic inhibitors.

Synthesis of the key macrocyclic epoxide **4e** has been achieved by two methods, which differ primarily in the way the macrocyclic ring is closed. Initially, as described in Scheme 7.5, the dipeptide **4f** was prepared in solution by coupling Boc-Val-OH to H-Tyr-OMe. Then, following removal of the Boc group with TFA, the final N-terminal acid unit was attached by reaction with 5-bromopentanoyl chloride. We have found it advantageous to then exchange the bromide **4g** with iodide using the Finkelstein reaction prior to cyclization both for reduced reaction time and improved yields. Using anhydrous K_2CO_3 in DMF cyclization proceeds cleanly at room temperature, giving the macrocycles **4h**. A side reaction is observed during attempted cyclization of the smallest macrocycles where the N-terminal amide NH competes with the phenolic oxygen, resulting in lactam formation. In the case of the 16-member macrocycles [$(CH_2)_4$], approximately a 1:1 ratio of products is obtained, requiring separation by chromatography, while the smaller 15-member macrocycle [$(CH_2)_3$] cannot be prepared by this method, instead giving the γ-lactam exclusively. For these two specific cases, the alternative mode of ring formation by amide bond closure could be applied successfully (Scheme 7.6).[66] The benzyl ester group was hydrogenated and the resulting acid **4i** was activated with isobutyl chloroformate and treated with diazomethane to give the diazoketone **4j** in good yield. Treatment with anhydrous HBr in EtOAc led to a smooth displacement of N_2 and the α-bromoketone **4k** so obtained was stereoselectively reduced with $NaBH_4$ to give the required (S)-bromohydrin **4l** as the major isomer in the ratio of 7:1 and was obtained pure by chromatography. Finally, cyclization of the bromohydrin **4l** to the key macrocyclic epoxide **4e** was accomplished with NaOMe/MeOH.[66]

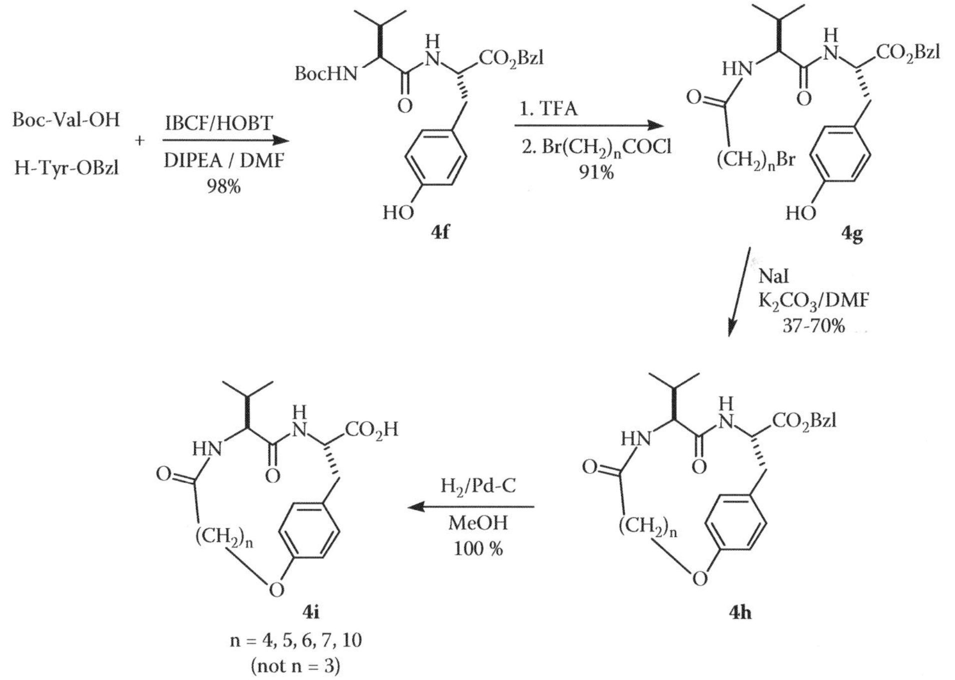

SCHEME 7.6   Conversion of valine to N-terminal macrocyclic acid.

We have made other macrocycles with differing ring sizes (e.g., $(CH_2)_n$ where n = 3, 4, 5, 6, 7, and 10)[66] to investigate the effects of constraint on structure and also to modify bioavailability by increasing the logP. The structure of the smallest macrocycle was determined using NMR spectroscopy and was a rigidly constrained beta strand.[66] Increasing the ring size does relax the level of constraint but the overall beta strand structure is retained even for the largest 22-member macrocycles studied. Compounds prepared from macrocycles with 15- to 17-member rings were similarly potent inhibitors of the HIV-PR enzyme; however, the greatest antiviral activity was found among the 17-member series[60] and we concentrated our parallel combinatorial synthesis program around these.

Facile and stereospecific ring opening of the epoxide **4e** with a large variety of amines was achieved by heating in EtOH solution and the resulting secondary amine products were then smoothly acylated by a variety of sulfonyl chlorides and isocyanates (Scheme 7.7). In published work[61] we investigated combinations of seven amines, ten sulfonyl chlorides, and four isocyanates, which together with some peptidic groups represented a library of more than 100 compounds. We have reported several inhibitors with low nanomolar activity,[61] and importantly such macrocycles are also resistant to nonspecific degradation by other proteases in the gut, blood, and cells.

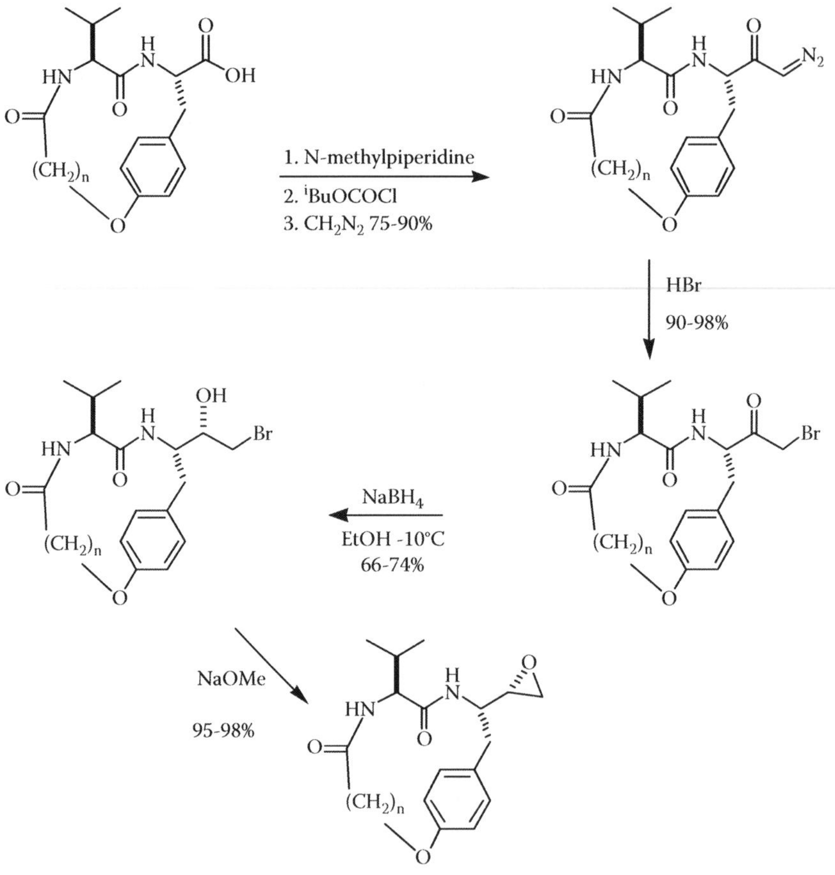

SCHEME 7.7   Conversion of N-terminal macrocyclic acid to epoxide.

7.4.3   C-Terminal Macrocycles

A logical extension to this methodology was to similarly constrain the C-terminal tripeptide sequence into a macrocyclic ring that again should lock in the observed enzyme bound beta strand conformation.[56,57] A C-terminal macrocyclic template would anchor the inhibitor to the active site and permit the exploration of novel non-peptidic N-terminal appendages with greatly reduced complications from the effects of cooperativity.

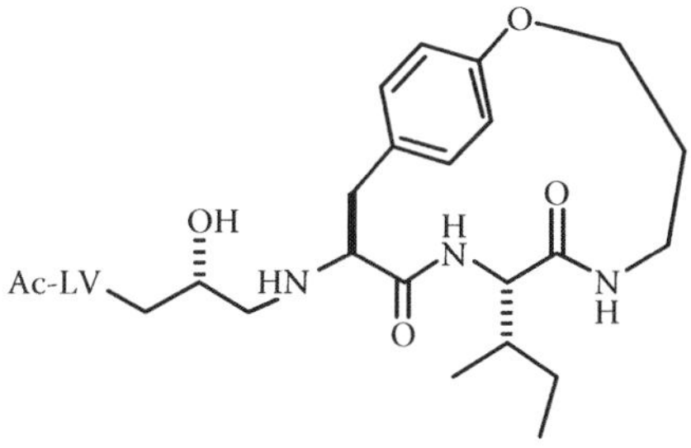

6a

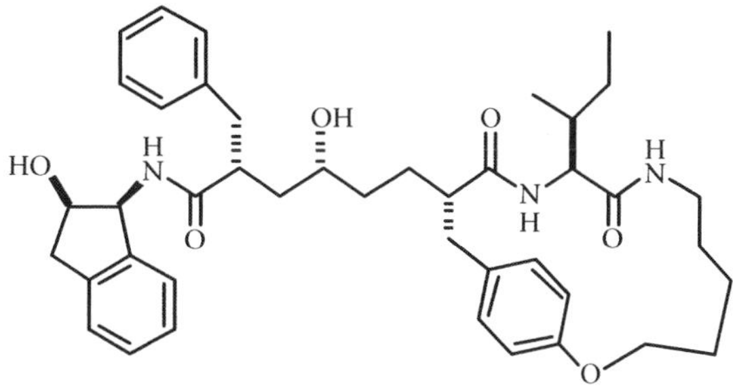

6b

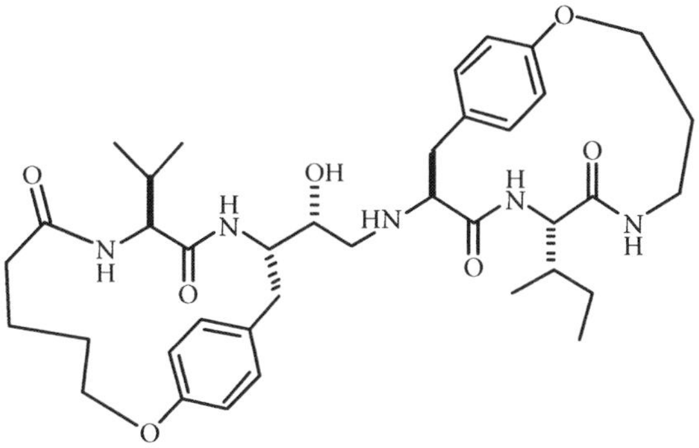

6c

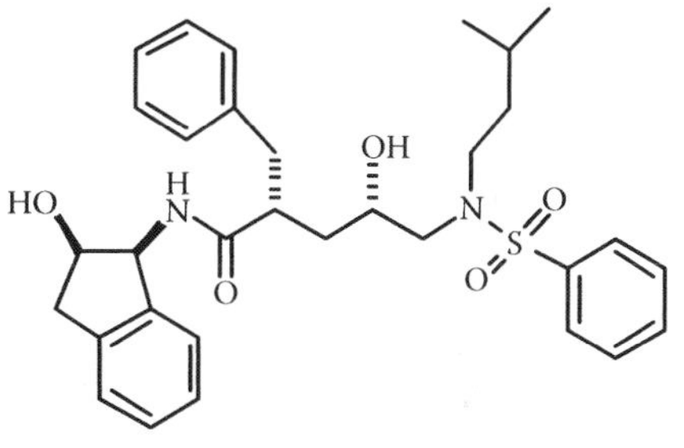

7

FIGURE 7.3   Structures **6a–6c** and **7**.

As detailed in Scheme 7.8, Boc-Ile-OH was coupled to 3-bromo-1-propylamine giving **5a**; then after removal of the Boc group with TFA, Boc-Tyr-OH was coupled to the new N-terminus giving **5b**. Cyclization of the Tyr sidechain phenol onto the bromopropyl chain under basic conditions then closed the macrocyclic ring of **5c** in an overall yield of 68% for the four steps. After removal of the Boc group, the macrocyclic amine **5d** was derivatized in parallel with a range of fragments either by epoxide ring-opening reactions that put in place the hydroxyethylamine isostere (Scheme 7.8), or reductive aminations resulting in the aminoethylene isostere linkage. Structural analysis by x-ray crystallography revealed that the macrocycle had indeed anchored the inhibitor in the active site and all of the sidechain locations and hydrogen bonds present in the complex with the linear peptidic inhibitor JG-365 **4b** were reproduced in the complex with **6** (Figure 7.3).[56,57,59] To confirm that the macrocycle was acting as an anchoring template that dictated the positioning of the entire molecule in the active site groove, we superimposed the x-ray crystal structures of a further three examples with widely varied N-terminal substituents. Gratifyingly, the location of the macrocycle in each case was conserved, thereby confirming our original hypothesis. Several of these compounds were potent inhibitors of the HIV-1 PR enzyme; and in particular, **6b** (Figure 7.3) possessed sub-nanomolar affinity and was also an active antiviral agent at nanomolar concentrations in cell culture.[60] There is also evidence (unpublished data) that compounds containing these macrocyclic strand mimetics are less prone to the development of resistance against HIV, perhaps because the macrocyclic component organizes the protease contours around it and thus dampens the effect of ligand-protease cooperativity.[59]

We next combined both N-and C-terminal macrocycles into the bicyclic structure **6c** (Figure 7.3) that formally mimics a hexapeptide that is totally constrained into a beta strand apart from some flexibility at the central hydroxyethylene transition state isostere.[58] Not surprisingly, **6c** was a potent inhibitor of HIV-1 PR (K_i = 3 nM) and its x-ray crystal structure superimposed well on those of both N- and C-terminal inhibitors described previously. A significant advantage of **6c** over **4a** was that the macrocycles conferred proteolytic stability to the molecule, therefore offering the

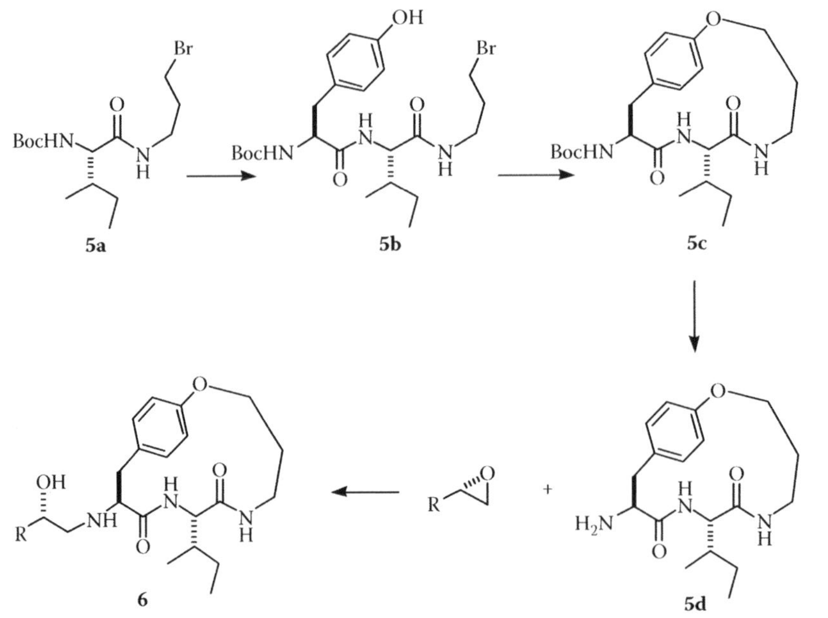

SCHEME 7.8 Synthesis of C-terminal macrocyclic inhibitors.

possibility for some meaningful *in vivo* pharmacological evaluation. At this point we were intrigued by the prospect of combining the non-peptidic N- and C-terminal appendages from the most potent inhibitors discovered thus far in this program. Compound **7**, comprised of the aminoindanol and isoamylbenzenesulfonyl fragments, was synthesized and confirmed as a potent HIV-PR inhibitor ($K_i = 7$ nM).

Thus, the use of constrained macrocycles to separately mimic the enzyme bound conformations of the N- and C-terminal sequences, and parallel synthesis, has allowed the progression from the peptidic lead with no bioavailability **4a**, to the potent non-peptide or peptidomimetic drug lead **7**.

This approach should be applicable to other proteases that share the common theme of recognition of their substrates (and inhibitors) in the beta-strand conformation.

7.5 FINAL COMMENTS

This chapter described the use of parallel synthesis approaches to elaborate simple L- and D-amino acids into non-peptidic and peptidomimetic inhibitors of enzymes implicated in cancer, inflammation, and viral infections. The amino acids represented convenient sources of chirality and useful "three-point" scaffolds for derivatization and elaboration into larger compounds with enzyme inhibitory activity. Although bearing polar substituents, the amino acids were readily converted into cell-permeable compounds with anticancer, antiinflammatory, and antiviral activities. We think that amino acids have been under-exploited as precursors to useful biological probes and drug leads and urge other researchers to give greater consideration to their merits in drug design and discovery.

ACKNOWLEDGMENTS

We thank the National Health and Medical Research Council for support of this research.

REFERENCES

1. Glen, M.P.; Kahnberg, P.; Boyle, G.M.; Hansford, K.A.; Hans, D.; Martyn, A.C.; Parsons, G.P.; and Fairlie, D.P. *J. Med. Chem.*, 2004, 47, 2984–2994.
2. Kahnberg, P.; Lucke, A.J.; Glenn, M.P.; Boyle, G.M.; Tyndall, J.D.A.; Parsons, P.G.; and Fairlie, D.P. *J. Med. Chem.*, 2006, 49, 7611–7622.
3. Rodriquez, M.; Aquino, M.; Bruno, I.; De Martino, G.; Taddei, M.; and Gomez-Paloma, L. *Curr. Med. Chem.*, 2006, 13, 1119–1139.
4. Rifkind, R.A.; Richon, V.M.; and Marks, P.A. *Pharmacol. Ther.*, 1996, 69, 97–102.
5. Leszczyniecka, M.; Roberts, T.; Dent, P.; Grant, S.; and Fisher P. B.; *Pharmacol. Ther.*, 2001, 90, 105–156.
6. Lin, H.Y.; Chen, C.S.; Lin, S.P.; Weng, J.R.; and Chen, C.S. *Med. Res. Rev.*, 2006, 26, 397–413.
7. Marks, P.A.; Rifkind, R.A.; Richon, V.M.; Breslow, R.; Miller, T.; and Kelly, W.K. *Nature*, 2001, 1, 194–202.
8. Parsons, P.G.; Hansen, C.; Fairlie, D.P.; West, M.L.; Danoy, P.A.C.; Sturm, R.A.; Dunn, I.S.; Pedley, J.; and Ablett, E.M. *Biochem. Pharmacol.*, 1997, 53, 1719–1724.
9. Qui, L.; Kelso, M.J.; Hansen, C.; West, M.L.; Fairlie, D.P.; and Parsons, P.G. *Br. J. Cancer*, 1999, 80, 1252–1258.
10. Marks, P.A.; Breslow, R., *Nature Biotechnology*, 2007, 25, (1) 84–90.
11. Yoshida, M.; Kijima, M.; Akita, M.; and Beppu, T. *J. Biol. Chem.*, 1990, 265, 17174–17179.
12. Kijima, M.; Yoshida, M.; Sugita, K.; Horinouchi, S.; and Beppu, T. *J. Biol. Chem.*, 1993, 268, 22429–22435.
13. Darkin-Rattray, S.J.; Gurnett, A.M.; Myers, R.W.; Dulski, P.M.; Crumley, T.M.; Allocco, J.J.; Cannova, C.; Meinke, P.T.; and Colletti, S.L. *Proc. Nat. Acad. Sci. U.S.A.*, 1996, 93, 13143–13147.
14. Atadja, P.; Gao, L.; Kwon, P.; Trogani, N.; Walker, H.; Hsu, M.; Yeleswarapu, L.; Chandramouli, N.; Perez, L.; Versace, R.; Wu, A.; Sambucetti, L.; Lassota, P.; Cohen, D.; Bair, K.; Wood, A.; and Remiszewski, S. *Cancer Res.*, 2004, 64, 689–695.
15. Rivard, M.; Malon, P.; and Cerovsky, V. *Amino Acids*, 1998, 15, 389–392.
16. Finnin, M.S.; Donigian, J.R.; Cohen, A.; Richon, V.M.; Rifkind, R.A.; Marks, P.A.; Breslow, R.; and Pavletich, N.P. *Nature*, 1999, 401, 188–193.
17. Vannini, A.; Volpari, C.; Filocamo, G.; Casavola, E.C.; Brunetti, M.; Renzoni, D.; Chakravarty, P.; Paolini, C.; De Francesco, R.; Gallinari, P.; Steinkühler, C.; and Di Marco. S. *Proc. Nat. Acad. Sci. U.S.A.*, 2004, 101, 15064–15069.
18. Qiu, L.; Burgess, A.; Fairlie, D.P.; Leonard, H.; Parsons, P.G.; and Gabrielli, B.G. *Mol. Biol. Cell.*, 2000, 11, 2069–2083.
19. Burgess, A.J.; Pavey, S.; Warrener, R.; Hunter, L.J.; Piva, T.J.; Musgrove, E.A.; Saunders, N.; Parsons, P.G.; and Gabrielli, B.G. *Mol. Pharmacol.*, 2001, 60, 828–837.
20. Hume, D. A.; and Fairlie, D.P. *Curr. Med. Chem.*, 2005, 12, 2925–2929.
21. Six, D.A.; and Dennis, E.A. *Biochim. Biophys. Acta*, 2000, 1488, 1–19.
22. Cho, W. *Biochim. Biophys. Acta*, 2000, 1488, 48–58.
23. Chen, Y.; and Dennis, E.A. *Biochim. Biophys. Acta*, 1998, 1394, 57–64.
24. Reid, R.C. *Curr. Med. Chem.*, 2005, 12, 3011–3026.
25. Glaser, K.B. Regulation of phospholipase A_2 enzymes. *Adv. Pharmacol.*, 1995, 32, 31–66.
26. Pruzanski, W.; and Vadas, P. *Immunol. Today*, 1991, 12, 143–146.
27. Scott, D.L.; White, S.P.; Otwinowski, Z.; Yuan, Y.; Gelb, M.H.; and Sigler, P.B. *Science*, 1990, 250, 1541–1546.
28. Winstead, M.V.; Balsinde, J.; and Dennis, E. A. *Biochim. Biophys. Acta*, 2000, 1488, 28–39.
29. Hansford, K.A.; Reid, R.C.; Clark, C.I.; Tyndall, J.D.A.; Whitehouse, M.W.; Guthrie, T.; McGeary, R.P.; Schafer, K.; Martin, J.L.; and Fairlie, D.P. *ChemBioChem.*, 2003, 4, 181–185.
30. Hansford, K.; Reid, R.C.; Clark, C.; McGeary, R.; and Fairlie, D.P. 2002, Patent WO200208189.
31. Thunnissen, M.M.G.M.; Eiso, A.B.; Kalk, K.H.; Drenth, J.; Dijkstra, B.W.; Kuipers, O.P.; Dijkman, R.; Dehaas, G.H.; and Verheij, H.M.E. *Nature*, 1990, 347, 689–691.
32. Scott, D.L.; White, S.P.; Browning, J.L.; Rosa, J.J.; Gelb, M.H.; and Sigler, P.B. *Science*, 1991, 254, 1007–1110.

33. Schevitz, R.W.; Bach, N.J.; Carlson, D.G.; Chirgadze, N.Y.; Clawson, D.K.; Dillard, R.D.; Draheim, S.E.; Hartley, L.W.; Jones, N.D.; Mihelich, E.D.; Olkowski, J.L.; Snyder, D.W.; Sommers, C.; and Wery, J.-P. *Nature Struct. Biol.,* 1995, 2, 458–465.

34. Reynolds, L.J.; Hughes, L.L.; and Dennis, E. A. *Anal. Biochem.,* 1992, 204, 190–197.

35. Arumgam, T.V.; Arnold, N.; Proctor, L.M.; Newman, M.; Reid, R.C.; Hansford, K.A.; Fairlie, D.P.; Shiels, I.A.; and Taylor, S.M. *Br. J. Pharmacol.,* 2003, 140, 71–80.

36. Kukuy, E.; John, R.; Szabolcs, M.; Ma, N.; Schuster, M.; Cannon, P.; Fairlie, D.P.; and Edwards, N.M. *J. Heart Lung Transplant.,* 2002, 21, 133.

37. Woodruff, T.M.; Arumugam, T.V.; Shiels, I.A.; Newman, M.L.; Ross, P.A.; Reid, R.C.; Fairlie, D.P.; and Taylor, S.M. *Int. Immunopharmacol.,* 2005, 5, 883–92.

38. Taylor, S.; Fairlie, D.; and Brown, L. *J. Mol. Cell. Cardiol.,* 2004, 37, 272–273.

39. Levick, S.; Loch, D.; Rolfe, B.; Reid, R.C.; Fairlie, D.P.; Taylor, S.M; and Brown, L. *J. Immunol.,* 2006, 176, 7000–7007.

40. Gregory, L.S.; Kelly, W.L.; Reid, R.C.; Fairlie, D.P.; and Forwood, M.R. *Bone,* 2006, 39, 134–142.

41. Shiels, I.A.; Taylor, S.M.; and Fairlie, D.P. Patent WO200205808.

42. Leung, D.; Abbenante, G.; and Fairlie, D.P. *J. Med. Chem.,* 2000, 43, 305–341.

43. Babine, R.E.; and Bender, S.L. *Chem. Rev.* 1997, 97, 1359–1472.

44. James, M.N.G.; and Sielecki, A.R., Eds. *Aspartic Proteinases and Their Catalytic Pathway.* John Wiley & Sons, New York, 1987, Vol. 3.

45. Powers, J. C.; and Harper, J. W., Eds. *Inhibitors of Serine Proteinases;* Elsevier, Amsterdam, 1986, Vol. 12.

46. Powers, J.C.; and Harper, J.W., Eds. *Inhibitors of Metalloproteases;* Elsevier, Amsterdam, 1986, Vol. 12.

47. Otto, H.-H.; and Schirmeister, T. *Chem. Rev.,* 1997, 97, 133–171.

48. Abbenante, G.; and Fairlie, D.P. *Med. Chem.,* 2005, 1, 71–104.

49. West, M.W.; and Fairlie, D.P. *Trends Pharmacol. Sci.,* 1995, 16, 67–75.

50. Loughlin, W.A.; Tyndall, J.D.A.; Glenn, M.P.; and Fairlie, D.P. *Chem. Rev.,* 2004, 104, 6085–6117.

51. Tyndall, J.; and Fairlie, D.P. *J. Molec. Recog.,* 1999, 12, 363–370.

52. Fairlie, D.P.; Tyndall, J.D.A.; Reid, R.C.; Wong, A.K.; Abbenante, G.; Scanlon, M.J.; March, D.R.; Bergman, D.A.; Chai, C.L.L.; and Burkett, B.A. *J. Med. Chem.,* 2000, 43, 1271–1281.

53. Tyndall, J.D.A.; Nall, T.; and Fairlie, D. P. *Chem. Rev.,* 2005, 105, 973–1000.

54. Koshland, D.E., Jr. *Proc. Nat. Acad. Sci. U.S.A.,* 1958, 44, 98–105.

55. Thoma, J.A.; and Koshland, D.E., Jr. *J. Am. Chem. Soc.,* 1960, 82, 3329–3333.

56. Abbenante, G.; March, D.R.; Bergman, D.A.; Hunt, P.A.; Garnham, B.; Dancer, R.J.; Martin, J.L.; and Fairlie, D.P. *J. Am. Chem. Soc.,* 1995, 117, 10220–10226.

57. March, D.R.; Abbenante, G.; Bergman, D.A.; Brinkworth, R.I.; Wickramasinghe, W.; Begun, J.; Martin, J.L.; and Fairlie, D.P. *J. Am. Chem. Soc.,* 1996, 118, 3375–9.

58. Reid, R.C.; March, D.R.; Dooley, M.J.; Bergman, D.A.; Abbenante, G.; and Fairlie, D.P. *J. Am. Chem. Soc.,* 1996, 118, 8511–8517.

59. Martin, J.L.; Begun, J.; Schindeler, A.; Wickramasinghe, W.A.; Alewood, D.; Alewood, P.F.; Bergman, D.A.; Brinkworth, R.I.; Abbenante, G.; March, D.R.; Reid, R.C.; and Fairlie, D.P. *Biochemistry,* 1999, 38, 7978–7988.

60. Tyndall, J.D.A.; Reid, R.C.; Tyssen, D.P.; Jardine, D.K.; Todd, B.; Passmore, M.; March, D.R.; Patten-den, L.K.; Bergman, D.A.; Alewood, D.; Hu, S.–H.; Alewood, P.F.; Birch, C.J.; Martin, J.L.; and Fairlie, D.P. *J. Med. Chem.,* 2000, 43, 3495–3504.

61. Reid, R.C.; Pattenden, L.K.; Tyndall, J.D.A.; Martin, J.L.; Walsh, T.; Fairlie, D.P. *J. Med. Chem.,* 2004, 47, 1641–1651.

62. Tyndall, J.D.A.; and Fairlie, D.P. *Curr. Med. Chem.,* 2001, 8, 893–907.

63. Fairlie, D.P.; Abbenante, G.; and March, D.R. *Curr. Med. Chem.,* 1995, 2, 654–686.

64. Abbenante, G.; Bergman, D.A.; Brinkworth, R.I.; March, D.R.; Reid, R.C.; Hunt, P.A.; James, I.W.; Dancer, R.J.; Garnham, B.; and Stoermer, M.L. *Bioorg. Med. Chem. Lett.,* 1996, 6, 2531–2536.

65. Glenn, M.P.; Pattenden, L.K.; Reid, R.C.; Tyssen, D.P.; Tyndall, J.D.A.; Birch, C.J.; and Fairlie, D.P. *J. Med. Chem.,* 2002, 45, 371–381.

66. Reid, R.C.; Kelso, M.J.; Scanlon, M.J.; and Fairlie, D.P. *J. Am. Chem. Soc.,* 2002, 124, 5673–5683.

8 Application of Solid-Phase Parallel Synthesis in Lead Optimization Studies

Bijoy Kundu

CONTENTS

8.1 INTRODUCTION

The rapid assembly of "drug-like" core templates is fundamental to the discovery phase of many medicinal chemistry programs. Provided that the syntheses of these core templates will allow for their ready functionalization, a meaningful SAR (structure-activity relationship) study can be efficiently conducted. In addition to ready assembly and core functionalization, the core template should have relatively low molecular weight and log P.[1] This will allow room for the molecule to "grow" as functionalization is incorporated, thus allowing the targeting of the many essential criteria of drug candidates including selectivity, potency, and efficacy. Combinatorial technology is at the forefront of medicinal chemistry and is receiving widespread attention as a powerful tool in drug discovery.[2] Many of the approaches devised to prepare such libraries rely on solid-phase synthesis techniques and exploit the efficient split/pool and parallel methods to assemble a statistical sampling of all possible combinations of a set of chemical building blocks.

Ironically, the philosophy behind combinatorial library design has changed radically since the early days of vast and diversity-driven libraries based on a single template. Maximizing diversity is rarely considered a sufficient design criterion for a library. In fact, a subtle change has taken place in recent years with the introduction of a medicinal chemistry component in the library-design stage in an attempt to decrease the high failure rate. Design and synthesis of focused libraries (or biased library) in parallel format with a significant medicinal chemistry design component is now a common strategy for lead optimization programs and has been successfully used in many areas of biomedical research for the purpose of lead optimization.[3] Bleicher et al.[4] have described the parallel synthesis of a set of spiroderivatives and analyzed their affinity to the neurokinin receptor, which has been implicated in anxiety and depression processes. As a result, novel receptor ligands with nanomolar affinities were identified. Haque et al.[5] developed potent inhibitors with high selectivity (>10,000-fold) of *Helicobacter pylori* DHODase using a solid-phase parallel synthetic approach. A further illustration of solid-phase parallel synthesis was the generation of an antimalarial sulfonamide library. The most potent compound displayed an activity 100-fold greater than chloroquine.[6] Recently, Guo et al.[7] using solid-phase parallel synthetic approach optimized the biological activity of 4-amino-2-biarylbutylurea. Their studies resulted in the identification of a derivative with potent single-digit nanomolar MCH 1R antagonists, a new class of therapeutics for the treatment of obesity. Therefore, a drug-discovery approach based on parallel synthesis of focused libraries designed for specific enzymes and receptors may offer an enhanced possibility of commercial success.

At the Central Drug Research Institute (CDRI) in Lucknow, India, we have been applying combinatorial techniques for lead optimization of compounds identified from the screening of our collections of compounds. Due to the space limitations of this chapter, only a few examples demonstrating this concept are discussed herein.

8.2 CASE STUDY 1

8.2.1 LEAD OPTIMIZATION STUDIES WITH A HEXAPEPTIDE AS LEAD MOLECULE

During the past two decades, the frequencies and types of life-threatening fungal infection have increased dramatically among cancer patients, transplant recipients, patients with AIDS, and patients receiving broad-spectrum antibiotic or parenteral nutrition. There has been a constant effort to develop more effective and safe antifungal drugs to combat these infections. Most of the current antifungal drugs simply reduce the level of growth of fungi. Furthermore, the development of resistant fungal strains in response to the widespread use of current antifungal drugs is likely to cause serious problems in the future. Therefore, there is an urgent need to develop safe, efficacious, and nontoxic antimycotic agents with a broad spectrum of antifungal activity.

In the past few years, membrane-active host defense molecules have been isolated from a variety of natural sources. Interestingly, they are small peptides or proteins, some of which have been shown to have antibacterial and antifungal activities.[8] Although native defense peptides themselves could not be used as therapeutic agents because of their low activity and poor bioavailability, these peptides have received attention because of their low levels of toxicity against mammalian cells and the ideal mechanism of perturbing the membrane of the pathogen.

Antifungal peptides are classified by their mode of action. The first group acts by lysis, which occurs via several mechanisms. Lytic peptides may be amphipathic, that is, molecules with two faces, with one being positively charged and the other being neutral and hydrophobic. Some amphipathic peptides bind only to the membrane surface and can disrupt the membrane structure without traversing the membrane. Finally, other amphipathic peptides aggregate in a selective manner, forming aqueous pores of variable sizes, and allowing passage of ions or other solutes. The second group of peptides interferes with cell wall synthesis or the biosynthesis of essential cellular components such as glucan or chitin.

The development of novel antifungal agents from peptides has been the center of attraction owing to their unique property of perturbing the cell membrane of the pathogen, thereby resulting in cell death.

8.2.2 Design of Peptide Libraries

Our laboratory initiated the screening of a variety of structurally diverse small synthetic peptides with the view to identify a lead peptide with a broad spectrum of antifungal activity. The studies led to the identification of a hexapeptide His-D-Trp-D-Phe-Phe-D-Phe-Lys-NH$_2$ (I) related to growth hormone releasing hexapeptide (GHRP-6) as a lead molecule with antifungal activity against five pathogenic fungi in the range of 6.2 to 100 µg/mL. The D-amino acids present at positions 2, 3, and 5 were found to be essential for antifungal activity, whereas amino acids at positions 1, 4, and 6 could be varied without any loss in activity. Thus, the lead peptide provided an interesting pharmacophore for the design of more potent antifungal agents. In the first instance, we decided to apply combinatorial approach to create a mixture-based library of the lead peptide I with the view to identify congeners with a higher antifungal activity. The libraries were designed in a manner wherein positions 1, 4, and 6 were subject to randomization whereas D-amino acids at positions 2, 3, and 5 were retained.[9] This led to the synthesis of three sets of sublibraries based on the lead peptide His-D-Trp-D-Phe-Phe-D-Phe-Lys-NH$_2$ (I) by subjecting positions 1, 4, and 6 to randomization. They were screened for their antifungal activity against *Candida albicans* and *Cryptococcus neoformans* to quantify inhibition at each step of the hexapeptide sublibrary iteration.

In the first instance, 19 pools of hexapeptide sublibraries based on hexapeptide motif I represented by a general formula O^{1}wfX^{4}fX^6-NH$_2$ have been synthesized. The position 1 of this library has been individually defined with each of the 19 genetically coded amino acids (excluding Cys) at a time, whereas X^4 and X^6 contained equimolar mixtures of 19 genetically coded amino acids. Each pool comprised 381 equimolar hexapeptides of different combinations. The libraries were tested for their antifungal activity by the method described earlier.[9]

Of the 19 pools, one of the sublibraries, RwfX^{4}fX^6-NH$_2$ with Arg at position-1, exhibited the best inhibition in comparison with other sublibraries with an IC$_{50}$ of 84 µM against *Candida albicans* (*Ca*) and an MIC of 28 µM against *Cryptococcus neoformansa* (*Cn*). The RwfX^{4}fX^6-NH$_2$ sublibrary was therefore selected for further iteration to define the remaining two mixture positions. In addition, iteration of HwfX^{4}fX^6-NH$_2$ (IC$_{50}$ = 114 µM against *Ca* and MIC = 57 µM against *Cn*) was also pursued for the purpose of comparison, as His is present at position-1 in the lead molecule. Thus, in the next step, two sets of sublibraries, each comprised of 19 pools (each pool consisting of 19 peptides) and represented by the general formula HwfO^{4}fX^6-NH$_2$ and RwfO^{4}fX^6-NH$_2$, were synthesized and evaluated for their antifungal activity.

From these 2 × 19 pools, the best inhibition was observed for Ile (IC$_{50}$ = 15.78 µM against *Ca* and MIC =14.09 µM against *Cn*) and Leu (IC$_{50}$ =18.04 µM against *Ca* and MIC =14.09 µM against *Cn*) with Arg at position-1. In contrast, HwfFfX^6-NH$_2$ (IC$_{50}$ = 30.72 µM against *Ca* and MIC = 27.72 µM against *Cn;* data not shown) corresponding to the lead molecule **I,** and also the best inhibitor among the 19 pools with His at position-1 exhibited higher IC_{50} and MIC values. Therefore, iteration for the His^1 sublibrary was dropped and deconvolution for $RwfIfX^6\text{-}NH_2$ was continued to define the third and final mixture position in an effort to complete the iterative process.

The final iteration of $RwfIfX^6\text{-}NH_2$ resulted in the synthesis of 19 single peptides with all the positions defined and assayed for their antifungal activity. It led to the identification of $RwfIfH\text{-}NH_2$ **II** as the most effective peptide inhibitor against *Ca* with IC_{50} value of 6.85 µM and was at least four times more potent than the lead molecule **I** with IC_{50} value of 28.56 µM. However, against *Cn* it was observed that His (MIC = 6.85 µM), Lys (MIC = 6.92 µM), Leu (MIC = 7.04 µM), and Arg (MIC = 6.71 µM) were the most acceptable ones at position-6, and their MIC values were more or less equal to the lead molecule **I** (MIC = 6.81 µM). Encouraged by the significant gains in the activity, further studies were initiated to evaluate the effect on antifungal activity when the chain length of the hexapeptide motif **II** is increased by introducing randomization either at the N-terminal region or at the C-terminal.

We next decided to examine the effect of increasing the chain length of **II** to nonapeptides on the biological activity, as they are known to adopt more ordered conformations than hexapeptides. This led us to choose **II** as a motif in a new library of nonapeptides with the view to identify congeners with higher antifungal activity. However, chain length of hexapeptide **II** to nonapeptide can be affected by increasing the chain length either at the N-terminal or C-terminal represented by general formulae $O^1\text{-}X^2\text{-}X^3\text{-}RwfIfH\text{-}NH_2$ **(III)** and $RwfIfHO^7X^8X^9\text{-}NH_2$ **(IV)**. Accordingly, 2 × 19 pools of sublibraries based on **III** and **IV** were synthesized and screened for their antifungal activity. As depicted in Table 8.1, the increase in chain length in the C-terminal region led to

TABLE 8.1
Antifungal Data for $RwfIfHO^7X^8X^9\text{-}NH_2$ and $AX^2X^3RwfIfH\text{-}NH_2$

$RwfIfHO^7X^8X^9\text{-}NH_2$			$AX^2X^3RwfIfH\text{-}NH_2$		
19 pools	*Ca* IC_{50} (µM)	*Cn* MIC (µM)	**19 pools**	*Ca* IC_{50} (µM)	*Cn* IC_{50} (µM)
$RwfIfHAXX\text{-}NH_2$	11.1	5.1	$AXXRwfIfH\text{-}NH_2$	82.37	53.45
$RwfIfHDXX\text{-}NH_2$	24.6	19.8	$DXXRwfIfH\text{-}NH_2$	79.49	79.49
$RwfIfHEXX\text{-}NH_2$	39.5	19.6	$EXXRwfIfH\text{-}NH_2$	78.61	78.61
$RwfIfHFXX\text{-}NH_2$	31.3	19.3	$FXXRwfIfH\text{-}NH_2$	77.51	77.51
$RwfIfHGXX\text{-}NH_2$	24.6	5.2	$GXXRwfIfH\text{-}NH_2$	83.33	52.91
$RwfIfHHXX\text{-}NH_2$	7.6	2.4	$HXXRwfIfH\text{-}NH_2$	78.12	43.28
$RwfIfHIXX\text{-}NH_2$	22.2	9.9	$IXXRwfIfH\text{-}NH_2$	79.61	21.17
$RwfIfHKXX\text{-}NH_2$	**3.7**	**2.4**	$KXXRwfIfH\text{-}NH_2$	27.14	22.50
$RwfIfHLXX\text{-}NH_2$	9.5	4.9	$LXXRwfIfH\text{-}NH_2$	79.61	33.91
$RwfIfHMXX\text{-}NH_2$	13.4	9.8	$MXXRwfIfH\text{-}NH_2$	78.49	24.01
$RwfIfHNXX\text{-}NH_2$	14.7	9.9	$NXXRwfIfH\text{-}NH_2$	79.55	48.05
$RwfIfHPXX\text{-}NH_2$	17.5	5.0	$PXXRwfIfH\text{-}NH_2$	80.64	15.16
$RwfIfHQXX\text{-}NH_2$	16.4	4.9	$QXXRwfIfH\text{-}NH_2$	78.67	49.80
$RwfIfHRXX\text{-}NH_2$	8.5	4.8	$RXXRwfIfH\text{-}NH_2$	**22.01**	**12.16**
$RwfIfHSXX\text{-}NH_2$	12.4	5.0	$SXXRwfIfH\text{-}NH_2$	81.30	28.94
$RwfIfHTXX\text{-}NH_2$	11.5	5.0	$TXXRwfIfH\text{-}NH_2$	80.38	35.85
$RwfIfHVXX\text{-}NH_2$	11.5	5.0	$VXXRwfIfH\text{-}NH_2$	80.51	30.27
$RwfIfHWXX\text{-}NH_2$	22.4	9.4	$WXXRwfIfH\text{-}NH_2$	75.24	23.70
$RwfIfHYXX\text{-}NH_2$	11.7	4.7	$YXXRwfIfH\text{-}NH_2$	55.36	47.09
$HwfFfK\text{-}NH_2$ (I)	28.65	6.81	$HwfFfK\text{-}NH_2$ (I)		
$RwfIfH\text{-}NH_2$ (II)	6.85	6.86	$RwfIfH\text{-}NH_2$ (II)		

a significant gain in the activity in comparison to increasing the chain length at the N-terminal. Of the 19 pools based on **III**, RwfIfHK^{7}X^{8}X^9-NH$_2$ exhibited the best inhibitions in comparison to other sublibraries, with an IC$_{50}$ value of 3.7 μM against *Ca* and an MIC value of 2.4 μM against *Cn*. Two other peptide mixtures with His (IC$_{50}$ = 7.6 μM against *Ca* and MIC = 2.4 μM against *Cn*) and Arg (IC$_{50}$ = 8.5 μM against *Ca* and MIC = 4.8 μM against *Cn*) at position-7 had activity within twofold of the most active peptide mixture. The RwfIfHKX^{8}X^9-NH$_2$ sublibrary was therefore selected for further deconvolution to define the remaining two mixture positions. Thus, in the next step, a second set of sublibraries comprised of 19 pools (each pool consisting of 19 equimolar peptides) and represented by the general formula RwfIfHKO^{8}X^9-NH$_2$ were synthesized and evaluated for their antifungal activity (Table 8.2). From these pools, the best inhibition was observed for the sublibrary RwfIfHKK^{8}X^9-NH$_2$ with an IC$_{50}$ value of 3.5 μM against *Ca* and an MIC value of 1.22 μM against *Cn*. Surprisingly, from the second sublibrary we could identify a mixture with only marginal gain in the antifungal activity against *Ca* in comparison to the parent sublibrary RwfIfHKX^{8}X^9-NH$_2$. However, against *Cn* a gain of twofold in the antifungal activity was observed. The only other mixture close to RwfIfHKK^{8}X^9-NH$_2$ in terms of activity was sublibrary RwfIfHKR^{8}X^9-NH$_2$ with Arg at position-8 (IC$_{50}$ = 4.15 μM and MIC = 2.39 μM). Lys was therefore selected for position-8; however, iteration of RwfIfHKR^{8}X^9-NH$_2$ was also pursued for the purpose of comparison. Thus, in the next step, two sets of sublibraries — RwfIfHKK^{8}O^9-NH$_2$ and RwfIfHKR^{8}O^9-NH$_2$ — were synthesized and evaluated for their antifungal activity. The final iteration resulted in the synthesis of two sets of 19 single peptides with all the positions defined. After assaying the antifungal activity, a number of peptides exhibited a high order of antifungal activity at this iterative step (Tables 8.3 and 8.4). From these sublibraries, the best inhibition against *Ca* was observed for Arg (IC$_{50}$ = 1.64 μM) and Ile (IC$_{50}$ = 1.65 μM) nonapeptides with Lys at position-8. Interestingly, against *Cn* although several nonapeptides were found to exhibit equipotent activities with MIC values in the range of 2.32 to 2.51 μM, the most active peptides were RwfIfHKKW-NH$_2$ and RwfIfHKKY-NH$_2$ with MIC values of 1.14 and 1.16 μM, respectively. In contrast, the best inhibition in the second set of 19 peptides with Arg at position-8 was observed for RwfIfHKRK-NH$_2$, with an IC$_{50}$ value of 3.34 μM against *Ca* and an MIC of 2.36 μM against *Cn*.

Our studies thus led to the identification of several nonapeptides with potent antifungal activity against *Ca* and *Cn*. Although the most active peptides — RwfIfHKKR-NH$_2$, RwfIfHKKI-NH$_2$, RwfIfHKKW-NH$_2$, and RwfIfHKKY-NH$_2$ — exhibited approximately four- to five-fold increase in the antifungal activity over that of the hexapeptide **II**, RwfIfHKKR-NH$_2$ (IC$_{50}$ = 1.64 μM) and

TABLE 8.2
Antifungal Data for RwfIfHKO^{8}X-NH$_2$

Iterative 2		Ca IC$_{50}$ (μM)	Cn MIC (μM)	Iterative 2		Ca IC$_{50}$ (μM)	Cn MIC (μM)
RwfIfHKAX-NH$_2$	(A)	10.54	2.55	RwfIfHKNX-NH$_2$	(N)	7.34	2.47
RwfIfHKDX-NH$_2$	(D)	19.7	4.94	RwfIfHKPX-NH$_2$	(P)	8.43	2.51
RwfIfHKEX-NH$_2$	(E)	24.07	4.88	RwfIfHKQX-NH$_2$	(Q)	6.92	2.44
RwfIfHKFX-NH$_2$	(F)	7.4	2.40	RwfIfHKRX-NH$_2$	(R)	4.15	2.39
RwfIfHKGX-NH$_2$	(G)	7.41	1.29	RwfIfHKSX-NH$_2$	(S)	7.15	2.52
RwfIfHKHX-NH$_2$	(H)	>20.0	>5.0	RwfIfHKTX-NH$_2$	(T)	6.12	2.49
RwfIfHKIX-NH$_2$	(I)	6.61	1.23	RwfIfHKVX-NH$_2$	(V)	15.86	2.45
RwfIfHKKX-NH$_2$	**(K)**	**3.5**	**1.22**	RwfIfHKWX-NH$_2$	(W)	6.94	2.33
RwfIfHKLX-NH$_2$	(L)	5.54	2.47	RwfIfHKYX-NH$_2$	(Y)	11.09	2.37
RwfIfHKMX-NH$_2$	(M)	7.07	2.43	HwfFfK-NH2 (I)	(I)	7.34	6.86
RwfIfH-NH$_2$	(II)	6.85	6.86				

TABLE 8.3
Antifungal Data for RwflfHKKO⁹-NH₂

Iterative 3		Ca IC₅₀ (µM)	Cn MIC (µM)	Iterative 3		Ca IC₅₀ (µM)	Cn MIC (µM)
RwflfHKKA-NH₂	(A)	3.17	2.48	RwflfHKKN-NH₂	(N)	3.35	2.40
RwflfHKKD-NH₂	(D)	3.52	4.80	RwflfHKKP-NH₂	(P)	3.89	2.43
RwflfHKKE-NH₂	(E)	3.57	2.37	RwflfHKKQ-NH₂	(Q)	2.58	2.37
RwflfHKKF-NH₂	(F)	>7.00	>7.00	RwflfHKKR-NH₂	(R)	**1.64**	2.32
RwflfHKKG-NH₂	(G)	2.77	2.51	RwflfHKKS-NH₂	(S)	1.77	2.45
RwflfHKKH-NH₂	(H)	1.90	2.36	RwflfHKKT-NH₂	(T)	2.38	2.44
RwflfHKKI-NH₂	(I)	**1.65**	2.40	RwflfHKKV-NH₂	(V)	3.08	2.43
RwflfHKKK-NH₂	(K)	1.83	2.37	RwflfHKKW-NH₂	(W)	2.61	**1.14**
RwflfHKKL-NH₂	(L)	2.18	2.40	RwflfHKKY-NH₂	(Y)	2.74	**1.16**
RwflfHKKM-NH₂	(M)	2.62	2.37	HwfFfK-NH₂	(I)	28.65	6.81
RwflfH-NH₂	(II)	6.85	6.86				

TABLE 8.4
Antifungal Data for RwflfHKRO⁹-NH₂

Iterative 3		Ca IC₅₀ (µM)	Cn MIC (µM)	Iterative 3		Ca IC₅₀ (µM)	Cn MIC (µM)
RwflfHKRA-NH₂	(A)	5.99	4.96	RwflfHKRN-NH₂	(N)	4.64	2.39
RwflfHKRD-NH₂	(D)	7.71	4.79	RwflfHKRP-NH₂	(P)	4.70	2.42
RwflfHKRE-NH₂	(E)	4.23	4.74	RwflfHKRQ-NH₂	(Q)	3.92	4.74
RwflfHKRF-NH₂	(F)	5.50	4.67	RwflfHKRR-NH₂	(R)	4.15	4.64
RwflfHKRG-NH₂	(G)	4.48	2.50	RwflfHKRS-NH₂	(S)	4.38	4.89
RwflfHKRH-NH₂	(H)	4.11	2.35	RwflfHKRT-NH₂	(T)	5.97	4.84
RwflfHKRI-NH₂	(I)	4.64	2.39	RwflfHKRV-NH₂	(V)	4.88	4.85
RwflfHKRK-NH₂	(K)	**3.34**	**2.36**	RwflfHKRW-NH₂	(W)	6.35	4.54
RwflfHKRL-NH₂	(L)	4.37	2.39	RwflfHKRY-NH₂	(Y)	7.08	4.62
RwflfHKRM-NH₂	(M)	4.62	2.36	HwfFfK-NH₂	(I)	28.65	6.81
RwflfH-NH₂	(II)	6.85	6.86				

RwflfHKKI-NH₂ (IC₅₀ = 1.65 µM) were found to be approximately 17 times more active than the lead peptide **I** (IC₅₀ = 28.65 µM) against Ca,[10] thus suggesting that extending the chain length in motif **II** from hexapeptides to nonapeptides with randomization at the C-terminal region had a favorable effect on the biological activity.

8.2.2 ANTIFUNGAL ASSAY

The libraries were tested for their antifungal activity using the twofold microbroth dilution technique.[11] Serial dilutions of the libraries from the millimolar (mM) to micromolar (µM) range were used to determine IC₅₀ and MIC values. The optical density was recorded on a microplate reader at 492 nm. The MIC was defined as the lowest concentration of compound that substantially inhibited the growth of *C. albicans* (*Ca*, SKF) after 24 hr and *C. neoformans* (*Cn*-17) after 48 hr. The inhibition of *Ca* by libraries has been expressed as IC₅₀, whereas for *Cn* it has been expressed as MIC. In the case of *Cn*, it was not possible to determine IC₅₀ values using twofold dilution technique protocol due to a sudden drop in inhibition from one dilution to another.

8.3 CASE STUDY 2

8.3.1 LEAD OPTIMIZATION STUDIES WITH A NATURAL PRODUCT: LUPEOL

Natural products play an important role in drug discovery, and many approved therapeutics as well as drug candidates have been derived from natural sources. They also represent promising scaffolds for diversification by using combinatorial techniques, as they have been selected by nature for their ability to undergo transformations in a three-dimensional space. Library construction around such scaffolds thus has the potential for both lead discovery and lead optimization. Nevertheless, any such diversification is usually functional group based and the original structure of the template remains unchanged. Indeed, libraries based on natural products have received wide attention because they give easy access to analogs having therapeutic potential that is superior to that of the parent scaffold.[12]

Of the variety of structurally diverse compounds obtained from tropical plants, terpenoids are often found in significant quantities, and a wide array have been isolated and characterized. The lupane type of triterpenoids and their derivatives represent a unique and one of the more important classes of biologically active natural products. Among this class of compounds, the pentacyclic triterpene lupeol (lup-20(29)-en-3β-ol), obtained in excellent yield from the stem bark of *Crataeva nurvula*, is of particular interest because of its wide spectrum of exhibited biological activity. The activities include urolithic, anticalciuric,[13] and antimalarial activity[14] against the chloroquine-resistant *P. falciparum*. Thus, lupeol provides an interesting template for diversification, which may result in the identification of more potent analogs. We targeted the template-directed synthesis of libraries based on based on 3- and 30-substituted lup-20(29)-ene derivatives of lupeol with the view to enhance its antimalarial activity.[15]

8.3.2 LINKING STRATEGY FOR LUPEOL

In the first step of our approach, we developed a linking strategy to produce polymer-linked lupeol to allow maximum introduction of diversity while maintaining a reliable and efficient cleavage procedure. We proposed to achieve this by incorporating bifunctional linkers, which on one side will remain linked to the resin, whereas on the other side can be linked with the hydroxyl group of lupeol at the C-3 position. Among the variety of commercially available bifunctional moieties, we selected aliphatic dicarboxylic acid anhydrides, as they can be conveniently used to derivatize the amide-based resin. The aliphatic chain present in these linkers may also enhance the lipophilicity of lupeol, thereby enhancing its biological activity. For our solid-phase synthesis, we preferred to use an amide-based resin, as hydroxy-based resins such as Wang resin, when acylated with aliphatic dicarboxylic acid anhydrides, resulted in poor loading. The synthesis was initiated by loading various anhydrides onto the Rink/Sieber amide resin (0.50 mmol/g) using pyridine in DMF at room temperature (r.t.) for 12 hr to obtain resin 1. The completion of the reaction was monitored by a negative Kaiser test. This was followed by coupling of lupeol to the carboxyl function on the resin using the standard DIC/DMAP procedure to afford polymer linked 3β-O-(resin-alkanoyl)-lup-20(29)-ene 3 as outlined in Scheme 8.1. The completion of the reaction was monitored by the carboxyl group test.[16]

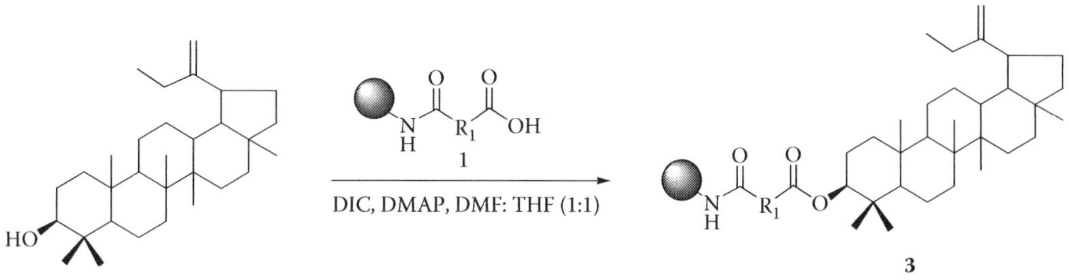

SCHEME 8.1

8.3.3 Synthesis of Scaffold I (3β-O-(Resin-Alkanoyl)-30-Bromo-Lup-20(29)-ene)

The linking strategy to produce the polymer-bound triterpenoid scaffold led to the provision for an additional site C-30 for introducing diversity. For this, the resin-bound **3** was treated with *N*-bromo-succinimide to give 3β-O-(resin-alkanoyl)-30-bromo-lup-20(29)-ene (Scheme 8.2).

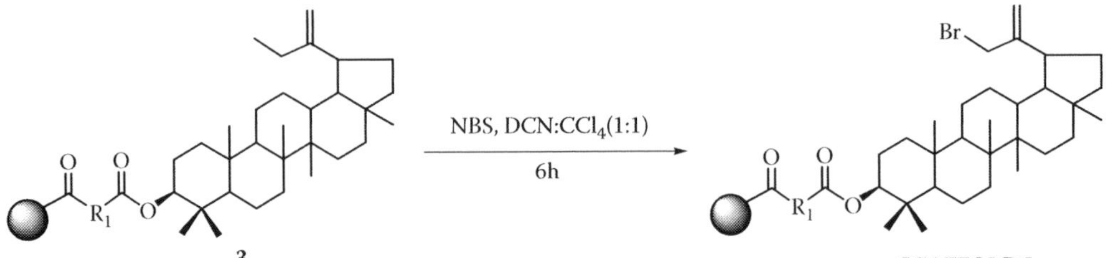

SCHEME 8.2

8.3.4 Generation of Library from Scaffold I

Scaffold **I** has been derivatized to produce a library using a variety of robust solid-phase substitution reactions — for example, reaction with a variety of primary and secondary amines and reaction with aromatic alcohols to give compounds **4** and **5** (Scheme 8.3), respectively. In the first instance, a library of 48 compounds has been generated in parallel format using automation with 36 compounds represented by **4** and 12 compounds represented by **5,** respectively.

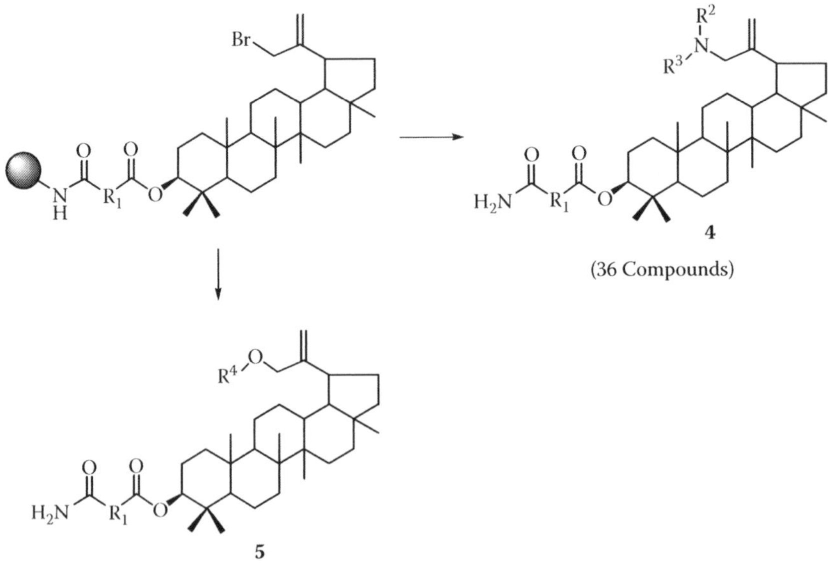

SCHEME 8.3 Reagents and conditions: (a) primary or secondary amine, DBU, 16 hr; (b) benzyl alcohol derivatives, DBU, 16 hr; (c) 1% TFA-DCM, 10 min. n = 2 (**X1**), 3 (**X2**), 4 (**X3**), 6 (**X4**), R_2 = cyclohexyl-, R_3 = H (**Y1**), R_2 = benzyl-, R_3 = H (**Y2**), R_2 = furfuryl-, R_3 = H (**Y3**), R_2 = 2-pyridyl ethyl-, R_3 = H (**Y4**), R_2 R_3 = piperidenyl- (**Y5**), R_2R_3 = morpholinyl- (**Y6**), R_2R_3 = N-methylpiperazinyl (**Y7**), R_2 = *N,N*-diethylamin-oethyl- R_3 = H (**Y8**), R_2 = *N*-cyclohexyl-, R_3 = *N'*-cyclohexyl-, (**Y9**) R_4 = 4-bromobenzyl- (**Y10**), benzyl- (**Y11**), 4-methoxybenzyl- (**Y12**).

8.3.5 Synthesis of Scaffold II (3β-O-(Resin-Alkanoyl)-30-Amino-Lup-20(29)-ene)

Scaffold I can be transformed to yet another intermediate **II**, which can be used for the generation of a second set of structurally diverse compounds based on lupeol. The method involves the treatment of **scaffold I** with sodium azide in DMSO for 24 hr at 80°C to obtain an intermediate 3β-O-(resin-alkanoyl)-30-azido-lup-20(29)-ene that is then reduced to give **scaffold II** by treatment with stannous chloride and thiophenol in the presence of triethylamine for 2 hr at room temperature (Scheme 8.4).

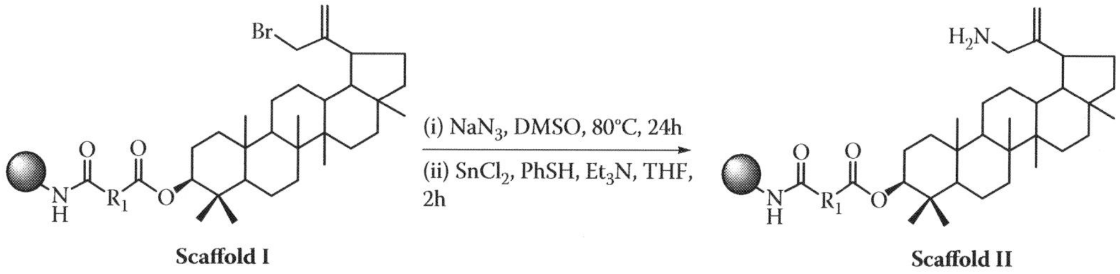

SCHEME 8.4

8.3.6 Generation of Library from Scaffold II

The second library based on lupeol involved the generation of compounds by derivatizing **II** with a variety of isocyanates, amino acids/aromatic acids, and aromatic aldehydes and led to the synthesis of structurally diverse derivatives **6**, **7**, and **8** (Scheme 8.5). A library of 48 compounds was generated in parallel format using automation with 8 compounds represented by **6**, 32 compounds represented by **7**, and 8 compounds represented by **8**.

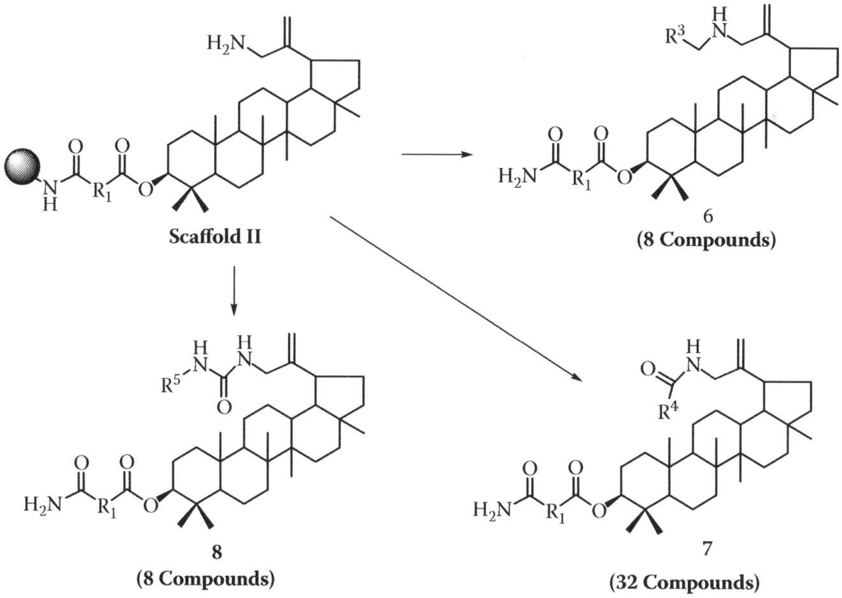

SCHEME 8.5 Reagents and conditions: (a) R_3CHO, TMOF: DMF (2:1), 2 hr, then NaCNBH$_3$, 1% AcOH in TMOF, 1 hr; (b) R_4COOH, HOBt, DIC, DMF, 6 hr; (c) R_5NCO, DCM, 24 hr; (d) 1% TFA-DCM, 10 min. n = 2 (**X1**), 3 (**X2**), 4 (**X3**), 6 (**X4**) R_3 = 4-hydroxyphenyl- (**Z1**), 3-pyridyl- (**Z2**), R_4CO = 4-chlorobenzoyl- (**Z3**), pyridyl-3-acetyl- (**Z4**), 4-hydroxycinnamoyl- (**Z5**), 2-prolyl- (**Z6**), acetyl- (**Z7**), benzoyl- (**Z8**), 2-amino-3-phenyl propionyl- (**Z9**), 7-aminoheptanoyl- (**Z10**), R_5 = benzyl (**Z11**), p-tolyl- (**Z12**).

8.3.7 GENERAL EXPERIMENTAL PROCEDURE FOR 4 OR 5

To a suspension of Sieber amide resin (100 mg, 0.05 mmol) linked to succinic acid in DMF:THF (2:1) mixture was added 4-DMAP (0.1 equiv, 0.05 mmol), lupeol (3 equiv, 0.15 mmol), and DIC (3 equiv, 0.15 mmol). After shaking the mixture for 16 hr at r.t., the solvent was drained and the resin was washed sequentially with DMF, MeOH, DCM (5 × 3 mL each), and dried *in vacuo* to afford **3**. In the next step, the 3β-O-(resin-alkanoyl) lupeol **2** (100 mg, 0.05 mmol) was suspended in DCM:CCl$_4$ (1:1) and to it was added NBS (4 equiv, 0.20 mmol) and the reaction was shaken at 600 rpm for 6 hr. Thereafter, the resin was sequentially washed with DMF, DCM, and ether (5 × 3 mL each), and dried to yield 3β-O-(resin-alkanoyl)-30-bromo-lup-20(29)-ene **scaffold I**. This was suspended in DMSO and to it was added DBU (2 equiv, 0.1 mmol) and primary or secondary amine or substituted benzyl alcohol (10 equiv, 0.5 mmol). The reaction was shaken for 16 hr at r.t. and then the resin was washed with DMF, MeOH, DCM, ether (5 × 3 mL each) and then dried *in vacuo*. The desired compound was cleaved from the resin using 1% TFA in DCM to obtain prototypes **4** or **5**. 3β-O-(4-amido succinyl)-30-*N* [benzyl amino]-lup-20(29)-ene [X1Y1; 4]: FAB MS 631 (M+H); ^{1}H NMR (300 MHz, CDCl$_3$): δ 5.45 (brs, 1H, H-29-α), 5.66 (brs, 1H, H-29-β), 4.50 (m, 1H, H-3), 3.45 (s, 2H, H-30), 1.25 (s, 3H, H-26), 1.15 (s, 3H, H-27), 1.13 (s, 3H, H-25), 0.92 (s, 3H, H-23), 0.88 (s, 3H, H-24), 0.78 (s, 3H, H-28); benzyl amino moiety, 4.03 (s, 2H, CH$_2$), 7.31(m, 5H, C$_6$H$_5$), succinyl moiety, 2.55 (t, 2H, J = 12 Hz, CH$_2$), 2.67 (t, 2H, J = 12 , CH$_2$).

8.3.7.1 General Procedure for Scaffold II and 8

To the 3β-O-(resin-alkanoyl)-30-bromo-lup-20(29)-ene **scaffold I** (100 mg, 0.05 mmol) on Sieber amide resin suspended in DMSO was added to sodium azide (10 equiv, 0.5 mmol) and the reaction mixture was shaken at 80°C for 16 hr. The solvent was drained and the resin was washed with DMF (5 × 3 mL), H$_2$O (5 × 3 mL), MeOH (5 × 3 mL), DCM (5 × 3 mL), and dried *in vacuo*. Thereupon, 3β-O-(resin-alkanoyl)-30-azido-lup-20(29)-ene so obtained was suspended in THF and treated sequentially with triethylamine (25 equiv, 1.25 mmol), thiophenol (20 equiv, 1.0 mmol), and stannous chloride (5 equiv, 0.25 mmol). After shaking at r.t. for 2 hr, the solvent was drained and the resin was washed with MeOH (5 × 3 mL), DCM (5 × 3 mL), MeOH (5 × 3 mL), DCM (5 × 3 mL), and then dried *in vacuo* to yield 3β-O-(resin-alkanoyl)-30-amino-lup-20(29)-ene, **scaffold II.** This was suspended in DMF:DCM (1:1) and then treated with 4-methyl phenyl isocyanate (10 equiv, 0.5 mmol) at r.t. for 16 hr. The solvent was drained and the resin was washed sequentially with DMF (5 × 3 mL), MeOH (5 × 3 mL), DCM (5 × 3 mL), and then dried *in vacuo*. The desired compound was cleaved from the resin using 1% TFA in DCM to obtain prototype **8**. 3β-O-(4-amido succinyl)-30-*N* [*N*′-(4-methyl phenyl) urea]-Lup-20(29)-ene [X1Z12]: FAB MS 674 (M+H); ^{1}H NMR (300 MHz, CDCl$_3$): δ 5.48 (brs, 1H, H29-α), 5.70 (brs, 1H, H29-β), 4.45 (m, 1H, H-3), 3.78 (s, 2H, H-30), 1.34 (s, 3H, H-26), 1.32 (s, 3H, H-27), 1.27 (s, 3H, H-25), 1.25 (s, 3H, H-23), 1.22 (s, 3H, H-24), 1.20 (s, 3H, H-28), *N*′-[4-methylphenyl]urea moiety, 2.32 (s, 3H, CH$_3$), 6.95 (d, 2H, J = 6 Hz), 7.11 (d, 2H, J = 6 Hz), succinyl moiety, 2.55 (t, 2H, CH$_2$, J = 12 Hz), 2.71 (t, 2H, CH$_2$, J = 12 Hz).

8.3.8 ANTIMALARIAL ACTIVITY AGAINST *PLASMODIUM FALCIPARUM IN VITRO*

The antiparasitic activity of the compounds was assessed by evaluating the minimum inhibitory concentration (MIC) against *P. falciparum in vitro*.[17] Asynchronous parasites obtained from the cultures of *P. falciparum* (strain NF-54) were synchronized after 5% sorbital treatment so as to obtain only ring stage parasites.[18] Parasite suspension medium RPM 1-1640 at 1% to 2% parasitemia and 3% hematocrit was dispersed into sterile 96-well plates, and test compounds were serially diluted in duplicate wells to obtain final concentrations of 50 μg/mL, 10 μg/mL, and 2 μg/mL. The culture plates were incubated in a candle jar at 37°C for 30 to 40 hr. Thin blood smears from each well were microscopically examined and the concentration, which fully inhibited the maturation of the ring parasites into schizont stage, was recorded as MIC. The *in vitro* results of some of the most active compounds are summarized in

Table 8.5. Of the 96 compounds screened for their biological activity, 15 compounds were found to be more active than lupeol; others were either inactive or equipotent to lupeol (data not shown). Out of the 15 compounds a majority of the compounds were derived from 3β-O-(resin-alkanoyl)-30-bromo-lup-20(29)-ene **4** intermediate and the most active compound was found to be X4Y10 (Table 8.5) with a MIC value of 13.07 μM in comparison to the MIC value of 117 μM for lupeol. The standard drug chloroquine exhibited a MIC value of 0.24 μM. The compound X4Y10 with suberic acid and 4-bromobenzyl alcohol as monomers was found to be at least 9 times more potent than lupeol.

8.3.9 Antimalarial Activity against *Plasmodium berghei in vivo*

To study the *in vivo* efficacy of the congeners identified using a combinatorial approach, one of the compounds (X1Z11) was evaluated for *in vivo* activity against *P. berghei*. Swiss mice (25 ± 1 g) of either sex were inoculated with 1×10^6 *P. berghei* parasitized cells on day 0. A group of six mice was administered an aqueous suspension of the compound X1Z11 at 100 mg/kg dose from day 0 to day 3 via an intraperitoneal route, while another six mice were administered the vehicle alone. Thin blood smears from the treated mice were observed daily to record the degree of parasitemia until the animals died.

The results summarized in Table 8.6 show that the mean percent parasitemia in treated group on day 4 was 1.7 ± 0.6%, as against 4.4 ± 0.5% in the control group. Thus, compound X1Z11 exhibited nearly 60% suppression after 4-day treatment; however, the subsequent progression of parasitemia was not significantly altered.

In summary, we have demonstrated that lupeol-based libraries can be generated employing robust synthetic methodologies on solid supports. Naturally available lupeol can be successfully transformed into various key scaffolds, followed by generation of libraries using an automated synthesizer. Screening of the library for antimalarial activity against *Plasmodium falciparum in vitro* led to the identification of compounds with seven- to nine-fold increases in the biological

TABLE 8.5
Antimalarial Activity of Lupeol Derivatives against *P. falciparum*

Compound	MIC (μM)	Compound	MIC (μM)
X1Y1	16.00	X4Y10	13.07
X1Y3	16.10	X4Y11	13.92
X1Y4	15.40	X1Z11	14.83
X1Y6	16.36	X1Z12	14.83
X2Y8	15.31	X2Z3	14.43
X2Y11	15.82	X2Z10	14.66
X2Y12	15.50	Lupeol	117.00
X3Y1	15.36	Chloroquine	0.24
X3Y9	13.66		

TABLE 8.6
***In vivo* Antimalarial Activity of Compound X1Z11 against *P. berghei* in Swiss Mice**

Group	Dose (mg/kg/day)	% Parasitemia on Day (Mean ± S.E.)		Survival Time in Days (Mean ± S.E.)
		Day 4	Day 6	
Compound X1Z11	100	1.7 ± 0.6	6.8 ± 1.6	16.5 ± 3.3
Vehicle (control)	—	4.4 ± 0.5	9.4 ± 1.5	13.3 ± 1.3

activity in comparison to lupeol. Thus, for the first time using a combinatorial approach, we have demonstrated an appreciable increase in the antimalarial activity of lupeol, thereby confirming the validity of combinatorial approach for lead optimization. Although in the present investigation we have used a limited set of monomers, these 15 compounds lay the foundation for the introduction of more structural diversity using a variety of carefully selected monomers. Further modifications are nevertheless necessary to optimize our lead structure to enhance its antimalarial activity.

8.4 CASE STUDY 3

8.4.1 LEAD OPTIMIZATION STUDIES WITH GLYCOCONJUGATES

The effective treatment regimens to decrease microfilaramia have been primarily responsible for the recent designation of lymphatic filariasis as a disease that can be eliminated and for the resolution by the World Health Assembly to eliminate lymphatic filariasis as a public health problem globally.[19] Human lymphatic filariasis, which is caused by helminths *Wuchereria bancrofti* (in 90% of cases) and *Brugia malayi* (in 10% of cases), affects approximately 120 million people, with one billion people considered at risk of becoming infected.[20] It has been found in 76 countries throughout the regions of South and Central America, Central Africa, the Eastern Mediterranean, Southeast Asia, and the Western Pacific. The challenge of drug discovery lies in the identification of novel therapeutic targets from the myriad of parasite enzymes, receptors, genome data, and metabolic pathways.[21] The potential antifilarials known today either act on the membrane receptors or on the metabolism of the organism. DNA topoisomerases are cellular enzymes that are intricately involved in maintaining the topographic structure of DNA, transcription, and mitosis.[22] They have been identified as important biochemical targets in cancer chemotherapy and microbial infections.

Intracellular bacteria have been detected[23] in most filarial worms, and it is contemplated that by eradicating the endobacteria, filarial parasites may also die due to disturbance in the endosymbiosis. In fact, tetracyclin therapy not only reduces the number of *Walbachia* in filarial oocytes and embryos, but also inhibits filarial embryogenesis.[24] Furthermore, we have recently identified DNA topoisomerase II (topo II) of the filarial parasites as a target for the development of antifilarial compounds and hence studied the effect of various inhibitors of topoisomerases and antifilarials. Only nalidixic acid and novobiocin were found to be strong inhibitors of topo II activity of *Setaria cervi*.[25]

A number of other compounds belonging to different classes have been also identified as inhibitors of DNA topoisomerases.[26] The established classes are quinolones, coumarins, and *m*-AMSA (4′-(9-acridinyl amino) methansulfon-*m*-anisidine, and a glycine-rich peptide Microcin B17. Further, a number of sugar derivatives with aglycon pharmacophores have been reported to inhibit DNA topoisomerases; however, carbohydrates as such have not been studied for their capacity to inhibit the above enzymes. In our laboratory, we initiated the screening of a variety of structurally diverse, small synthetic molecules with the view to identify potent inhibitors of topo II activity. These studies led to the identification of a glycosylated β-amino acid derivative (**I**; Figure 8.1) as a lead molecule with weak topo II inhibitory activity.

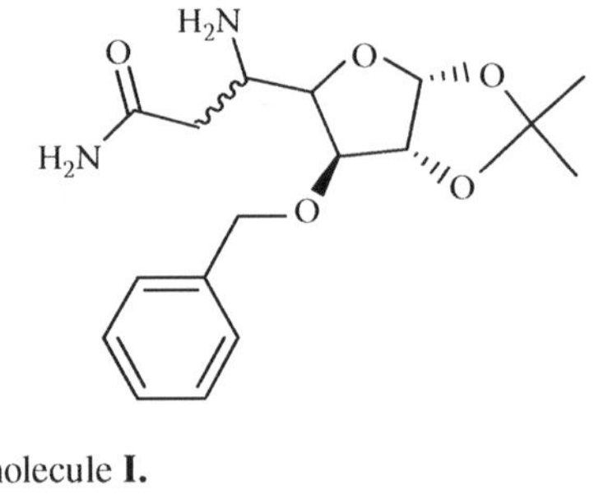

FIGURE 8.1   Structure of lead molecule **I**.

Sugar derivatives are known to offer better stability, better pharmacokinetic parameters, and help in the transport of drugs and at the same time are less toxic.[27] Thus, the lead molecule (**I**) provides an interesting pharmacophore for the design of more potent inhibitors of topo II. This prompted us to introduce diversity at the amino function of the lead molecule **I** in the first instance and we synthesized a library of glycoconjugates with the view to identify congeners with higher inhibitory activity.[28] For our studies the library represented by general formula X_1-X_2- carbohydrate (Table 8.7) is designed in a

TABLE 8.7

Structure of Variants X_1-X_2 Introduced at the Amino Function of the Carbohydrate Moiety

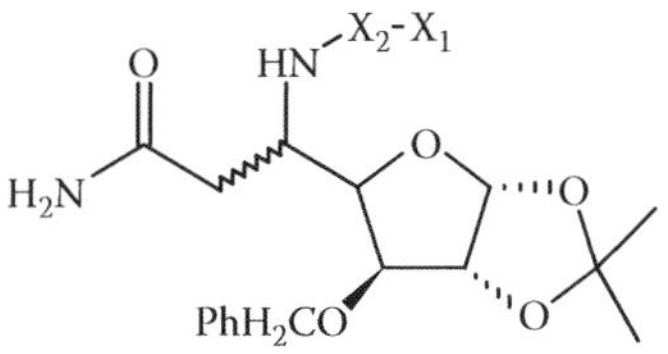

I

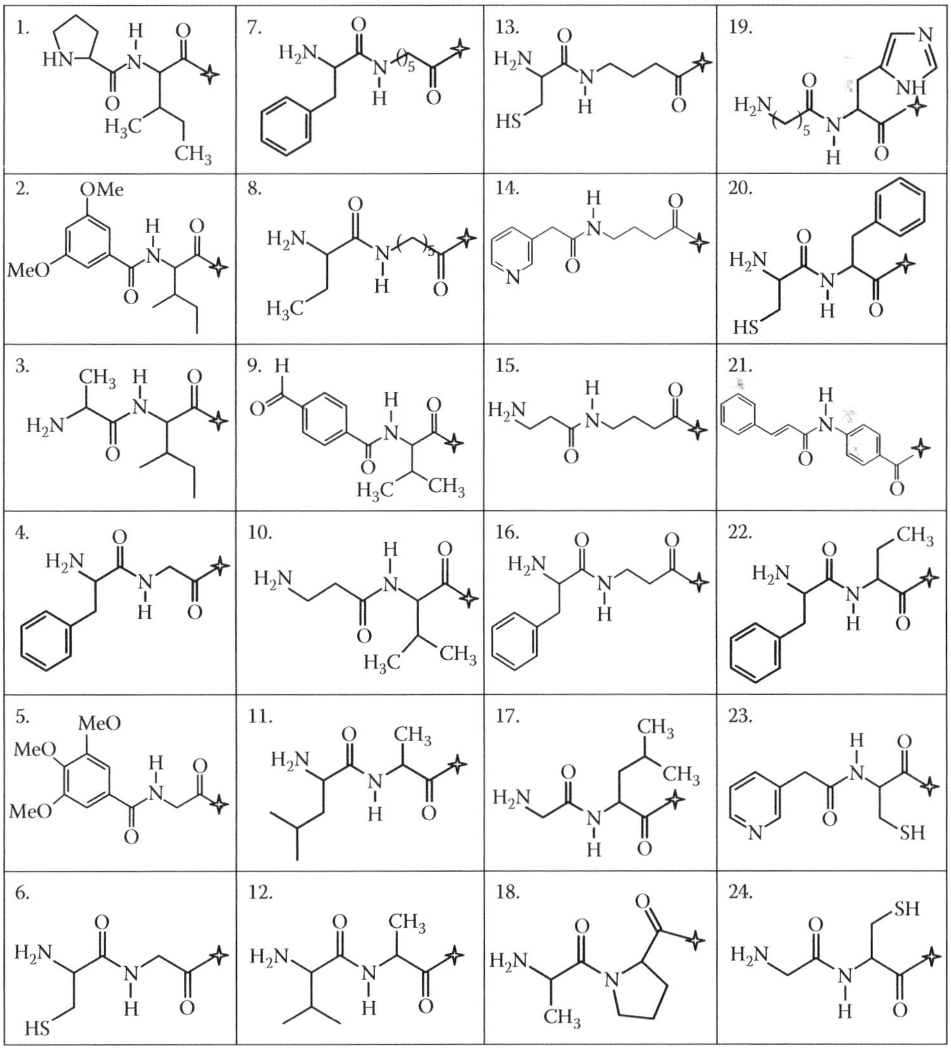

✦ Indicates point of attachment to the carbohydrate moiety **I**

manner wherein various bifunctional amino acids (with no functionality in their sidechain) have been introduced at position X_2; whereas at position X_1, both substituted aromatic acids and bifunctional amino acids have been introduced. This section reports on the topo II inhibitory activity (*Setaria cervi*) of novel glycoconjugates based on pharmacophore **I**.

8.4.2 SYNTHESIS OF LEAD MOLECULE I

The lead molecule **I** was synthesized by 1,4-conjugate addition of ammonia on the olefinic ester derived from sugar, followed by hydrolysis of the amino ester with aqueous ethanolic triethylamine at room temperature, and it was used as diastereoisomeric mixture. The method is described below.

8.4.2.1 Fmoc-3-Amino-(3'-O-Benzyl-1', 2'-Di-O-Isopropylidene-α-D-1', 4'-Pentofuranos-4'-yl)-Propanoic Acid

3-Amino-(3'-O-benzyl-1', 2'-di-O-isopropylidene-α-D-1', 4'-pentofuranos-4'-yl)-propanoic acid (2.1 g, 6.23 mmol) and Na_2CO_3 (0.72 g, 6.85 mmol) were dissolved in water (10 mL). Fmoc-Osu (2.3 g, 6.23 mmol) was added to it in dimethoxyethane (35 mL) and the mixture was stirred for 16 hr at room temperature. Solvent was evaporated to dryness and the residue was taken up in water (100 mL). The aqueous layer was extracted with ether (2 × 25 mL) and then acidified with $KHSO_4$. A white precipitate was obtained, which was then extracted in ethylacetate (3 × 50 mL), followed by washing with water (2 × 25 mL). The organic layer was dried with sodium sulfate and evaporated to dryness to yield Fmoc-3-amino-(3'-O-benzyl-1', 2'-di-O-isopropylidene-α-D-1', 4'-pentofuranos-4'-yl)-propanoic acid as a diastereomeric mixture. Yield 3.44 g (98%), FABMS: 660 (M+H)⁺. ¹H NMR, 300 MHz (CDCl₃):δ 1.32, 1.48 ppm (two s, each 3H, C(CH₃)₂); 2.66 (m, 2H, CH₂CO₂H); 3.94, 4.05 (two d, 1H, J = 6 Hz, J = 3 Hz, diastereomeric CH(4')); 4.19 (m, 3H, NHCH(3), OCH₂ Fmoc group); 4.43 (d, 1H, J = 3 Hz, CH(3'); 4.47 (d, J = 12 Hz, 1H, COCHᵃPh); 4.64 (d, J = 3.0 Hz, 1H, CH(2); 4.69 (d, 2H, J = 12 Hz, COCHᵇPh); 5.29 (bs, 1H, NH); 5.97 (d, 1H, J = 3.0 Hz, CH(1'); 7.34 (m, 9H, Ar-H); 7.59 (m, 3H, Ar-H); 7.76 (m, 2H, Ar-H).

8.4.3 PARALLEL SYNTHESIS OF GLYCOCONJUGATES LIBRARIES ON SOLID-PHASE FMOC-3-AMINO-(3'-O-BENZYL-1', 2'-DI-O-ISOPROPYLIDENE-α-D-1', 4'-PENTOFURANOS-4'-YL)-PROPANOYL-SIEBER AMIDE RESIN

The resin-bound lead molecule **I** was synthesized as follows. Sieber amide resin (1.1 g, 0.68 mmol) was placed in a solid-phase reaction vessel and treated twice with 20% piperidine-DMF (10 mL each) for 5 min and 25 min at room temperature under N_2 agitation. After this, the resin was successively washed with DMF (3 × 2 min), iPrOH (1 × 2 min), and DMF (3 × 2 min). Fmoc-3-amino-(3'-O-benzyl-1', 2'-di-O-isopropylidene-α-D-1', 4'-pentofuranos-4'-yl)-propanoic acid (1.14 g, 2.05 mmol) was then added to the resin, followed by the addition of DIPEA (0.67 mL, 4.09 mmol), TBTu (0.66 g, 2.05 mmol), and HOBt (0.31 g, 2.05 mmol) in DMF (10 mL) with N_2 agitation for 3 hr at room temperature. Resin was washed with DMF (3 × 2 min) and again treated with Fmoc-3-amino-(3'-O-benzyl-1', 2'-di-O-isopropylidene-α-D-1', 4'-pentofuranos-4'-yl)-propanoic acid (1.14 g, 2.05 mmol) in the presence of DIC (0.32 mL, 2.04 mmol) and HOBT (0.31 g, 2.05 mmol) for 3 hr. The resin was washed with DMF (3 × 2 min), MeOH (3 × 2 min), and DCMF (3 × 2 min). The completion of reaction was monitored by the Kaiser test.

8.4.4 PARALLEL SYNTHESIS

Both Fmoc amino acids and substituted aromatic acids were used as monomers to build the library in parallel format. The former has been introduced at both the X (1) and X (2) positions of the dipeptide moiety being linked to the carbohydrate unit, whereas substituted aromatic acids have been introduced only at position X (1). Following Fmoc, amino acids and substituted aromatic acids have been utilized to construct the library:

- *Fmoc-Amino acids:* Fmoc-Ile-OH, Fmoc-Leu-OH, Fmoc-Val-OH, Fmoc-Gly-OH, Fmoc-Pro-OH, Fmoc-Cys(Trit)-OH, Fmoc-N^ε-caproic acid, Fmoc-βAla-OH, Fmoc-γ-Abu-OH, Fmoc-αAbu-OH, Fmoc-Phe-OH, Fmoc-His(Trit)-OH, Fmoc-*p*-aminobenzoic acid.
- *Substituted aromatic acids:* 3-pyridylacetic acid, 3,5-dimethoxybenzoic acid, formylbenzoic acid, cinnamic acid.

8.4.4.1 Procedure for the Synthesis of X_1-X_2-Carbohydrate Moieties

The glycoconjugates (**1** to **24**; Table 8.7) were synthesized in parallel format in a 96-well reaction block using the 496 MOS Ω automated synthesizer. Fmoc-3-amino-(3′-O-benzyl-1′, 2′-di-O-isopropylidene-α-D-1′, 4′-pentofuranos-4′-yl)-propanoic acyl-Sieber amide resin (0.05 g, 0.031 mmol) was placed in 24 wells in the reaction block and treated with 20% piperidine-DMF (1 mL) twice for 5 min and 25 min. After this, the resin in each well was washed with DMF (3 × 2 min), iPrOH (1 x× 2 min), and DMF (3 × 2 min). This was followed by transfer of one of the Fmoc amino acids (0.093 mmol) from the list shown above to the desired well and subsequently treated with DIC (0.10 mL, 0.093 mmol) and HOBt (0.09 g, 0.093 mmol) in DMF for 3 hr. After completion, the resins were again washed using the same wash chemfile as described previously. The resin in each well was then treated with the desired Fmoc amino acid using the DIC/HOBt method for 3 hr. The completion of reaction was monitored by the negative ninhydrin test. This was followed by removal of the Fmoc group in each well with 20% piperidine-DMF and treatment with the desired substituted aromatic acids or Fmoc amino acid to complete the synthesis. The reaction was again monitored by negative ninhydrin test. Next, the resins in wells **1**, **3**, **4**, **6**, **7**, **8**, **10**, **11**, **12**, **13**, **15**, **16**, **17**, **18**, **19**, **20**, **22**, and **24** were treated with 20% piperidine-DMF to remove the Fmoc group, followed by washing as described above. Finally, the glycoconjugates were cleaved from the resin using a mixture of 2% TFA, 94% dichloromethane, and 4% triisopropylsilane. The final compounds obtained after deprotection and cleavage were lyophilized after dissolving them in *t*-butanol/water. The compounds were characterized using HPLC, FAB, and PMR spectral data. The glycoconjugates (**1** to **24**) were obtained in 85% to 90% purity, which was sufficient for *in vitro* evaluation.

8.4.5 DNA Topoisomerase II Estimation

The reaction catalyzed by DNA topoisomerases II was measured as reported previously.[28] The standard topoisomerase II reaction mixture in a final volume of 20 µL contained 50 mmol Tris-HCl (pH 7.5), 50 mmol KCl, 10 mmol $MgCl_2$, 1 mmol ATP, 0.1 mmol EDTA, 0.5 mmol DDT, 30 µg /mL BSA, 0.25 µg pBR322, and enzyme protein. Reaction was stopped by adding 5 µL of stock buffer followed by electrophoresis of samples on 1% agarose gel in Tris-acetate buffer for 18 hr at 20 volts. Gels were stained with ethidium bromide (0.5 µg/mL), visualized and photographed on GDS 7500 UVP (Ultra Violet Products, U.K.) transilluminator. The effect of inhibitors on the enzyme activity was measured by incubating the enzyme protein with inhibitor (40 µg/reaction mixture) for 10 min at 37°C and starting the reaction by adding pBR322. The percent inhibition was measured by microdensitometry of the gel using the Gel Base/Gel Blot Pro Gel analysis software program. The results obtained for various glycoconjugates are summarized in Table 8.8 and compared with the lead compound **I**, novobiocin, and naldixic acid.

8.4.6. *In Vitro* Topo II Inhibitory Activity

Out of the 24 glycoconjugates screened for their ability to inhibit DNA topoisomerase II, 9 congeners exhibited low to marked order of inhibitory activity. At the N-terminal region of glycoconjugates, Cys appeared to be the most preferred moiety responsible for the inhibitory activity. It is

TABLE 8.8

Effect of Glycoconjugates on DNA Topoisomerase II of Filarial Parasite *Setaria cervi*

Compound No.	Percent Inhibition
Enzyme alone	nil
Enzyme + **1**	nil
Enzyme + **2**	35
Enzyme + **3**	nil
Enzyme + **4**	nil
Enzyme + **5**	nil
Enzyme + **6**	95
Enzyme + **7**	2
Enzyme + **8**	nil
Enzyme + **9**	nil
Enzyme + **10**	nil
Enzyme + **11**	nil
Enzyme + **12**	nil
Enzyme + **13**	50
Enzyme + **14**	75
Enzyme + **15**	n.d.
Enzyme + **16**	11
Enzyme + **17**	15
Enzyme + **18**	15
Enzyme + **19**	22
Enzyme + **20**	23*
Enzyme + **21**	n.d.
Enzyme + **22**	20
Enzyme + **23**	26*
Enzyme + **24**	23*
Enzyme + Lead compound **I**	30
Enzyme + Novobiocin (at 100 nmols)	100
Enzyme + Naldixic acid (at 100 nmols)	90
Enzyme + DEC (at 100 nmols)	80
Enzyme + Ivermectin (at 100 nmols)	40

* The compounds formed complex with DNA; n.d. = not determined. The compounds **1** to **24** were used at 40-μg/reaction mixture concentration.

interesting to note that none of the congeners with Cys next to the carbohydrate moiety exhibited any activity.

Compounds **13** and **14** exhibited moderate inhibitory activity of 75% and 50%, respectively; however, they were more active than the lead compound **I,** which exhibited only 30% inhibition of DNA topo II (Table 8.8). It is interesting to note that replacement of Cys in compound **13** with the pyridylacetyl moiety resulted in compound **14,** which exhibited a drop in inhibition from 75% to 50%, thereby suggesting that Cys at the N-terminal region plays a crucial role in the inactivation of the enzyme. The most active compound of the series, compound **6** with Cys-Gly adornment, exhibited 95% inhibition of DNA topo II. Thus, the introduction of the Cys-Gly moiety at the amino function in the lead molecule **I** led to at least a threefold increase in the topo II inhibitory activity compared to the lead molecule **I**.

Compounds **20, 23,** and **24** formed complexes with the DNA, resulting in enzyme inhibition. This can be attributed to the nonavailability of the binding sites for the enzyme on the DNA. Compound **6** was further tested at different concentrations and it exhibited 90% to 100% inhibition at 21.8 nmol, whereas 50% inhibition was observed at 10.9 nmol. In contrast, naldixic acid and novobiocin exhibited 90% and 100% inhibition at 100 nmol. Similarly, the established antifilarials DEC and Ivermectin showed 80% and 40% inhibition, respectively, at 100 nmol. Thus, compound **6** appears to be at least 5 times more potent than the known inhibitors naldixic acid and novobiocin. In summary, our studies using the combinatorial approach thus provide a novel low-molecular-weight pharmacophore with potent DNA topo II inhibitory activity. Because no satisfactory inhibitor for topo II of *Setaria cervi* is currently available, glycoconjugate **6** lays the foundation for the design of more potent inhibitors.

8.4.7 CONCLUSION

In summary, the efficient use of combinatorial technology for drug discovery is demonstrated with the help of four examples. For the past two decades, combinatorial chemistry has taken a giant leap from an academic interest to a full-fledged tool in drug discovery in the pharmaceutical industry. Both diversity-directed and target-directed libraries have proven to be rich sources of lead compounds and complementary approaches for the lead discovery and optimization. As library construction moves inextricably to a spatially addressable format, more success stories from this technology will be commonplace in the near future.

REFERENCES

1. For reviews, see: (a) Lipinski, C.A. Drug-like properties and the causes of poor solubility and poor permeability, *J. Pharmacol. Toxicol. Meth.*, 2000, 44, 235. (b) Teague, S.J. et al. The design of leadlike combinatorial libraries, *Angew. Chem. Int. Ed.*, 1999, 38, 3743. (c) Lipinski, C.A. et al. Experimental and computational approaches to estimate solubility and permeability in drug discovery and development settings, *Adv. Drug Deliv. Rev.*, 1997, 23, 3. (d) Hann, M. M.; and Oprea, T.I. Pursuing the lead-likeness concept in pharmaceutical research., *Curr. Opin. Chem. Biol.*, 2004, 8, 255. (e) Lipinski, C.A. Lead- and drug-like compounds: the rule-of-five revolution, *Drug Disc. Today: Technol.,* 2004, 1, 337. (f) Leeson, P.D.; Davis, A.M.; and Steele, J. Drug-like properties: guiding principles for design — or chemical prejudice?, *Drug Disc. Today: Technologies*, 2004, 1, 189. (g) Fecik, R.A. et al. The search for orally active medications through combinatorial chemistry, *Med. Res. Rev.*, 1998, 18, 149. (h) Swanson, R.P. The entrance of informatics into combinatorial chemistry, *Am. Soc. Inf. Sci. Tech.,* 2004, 203–211. (i) Feuston, B.P. et al. Web enabling technology for the design, enumeration, optimization and tracking of compound libraries, *Curr. Top. Med. Chem.*, 2005, 5, 773. (j) Jónsdóttir, S.O. et al. Prediction methods and databases within chemoinformatics: emphasis on drugs and drug candidates, *Bioinformatics*, 2005, 21, 2145. (k) Selick, H.E.; Beresford, A.P.; and Tarbit, M.H. The emerging importance of predictive ADME simulation in drug discovery, *Drug Discovery Today*, 2002, 7, 109.
2. Leading reviews of applications in drug discovery: (a) Dolle, R.E. Comprehensive survey of chemical libraries yielding enzyme inhibitors, receptor agonists and antagonists, and other biologically active agents: 1992 through 1997, *Mol. Div.*, 1998, 3, 199. (b) Coffen, D.L. et al. Molecular diversity, biological activity and common ground shared by both, *Med. Chem. Res.*, 1998, 8, 206. (c) Kerwin, J.F., *Combinatorial Chemistry Molecular Diversity Drug Discovery*; Wiley-Liss, New York, 1998, p. 475. (d) Weber, L. Evolutionary combinatorial chemistry: application of genetic algorithms, *Drug Disc. Today*, 1998, 3, 379. (e) Kubinyi, H. The design of combinatorial libraries, *Drug Disc. Today*, 2002, 7, 503. (f) Plunkett, M.J.; and Ellman, J.A. Combinatorial chemistry and new drugs, *Sci. Am.*, 1997, 276, 68. (g) Terret, N.K. et al. Combinatorial synthesis — the design of compound libraries and their application to drug discovery, *Tetrahedron*, 1995, 51, 8135. (h) Musonda, C.C.; and Chibale, K. Application of combinatorial and parallel synthesis chemistry methodologies to antiparasitic drug discovery, *Curr. Med. Chem.*, 2004, 11, 2518. (i) Kundu, B. Solid phase strategies for the design and synthesis of heterocyclic molecules of

medicinal interest, *Curr. Opin. Drug Disc. Dev.*, 2003, 6, 815. (j) Furka, A. Combinatorial chemistry: 20 years on…, *Drug Disc. Today*, 2002, 7, 1. (k) Rose, S. Statistical design and application to combinatorial chemistry, *Drug Disc Today*, 2002, 7, 133.

3. (a) For review, see: Golebiowski, A.; Klopfenstein, S.R.; and Portlock, D.E. Lead compounds discovered from libraries. Part 2, *Curr. Opin. Chem. Biol.*, 2003, 7, 308. (b) Golebiowski, A.; Klopfenstein, S.R.; and Portlock, D.E. Lead compounds discovered from libraries, Part 1, *Curr. Opin. Chem. Biol.*, 2001, 5, 273. (c) Steven D. et al. Acyl dipeptides as reversible caspase inhibitors. Part 1. Initial lead optimization, *Bioorg. Med. Chem. Lett.*, 2002, 12, 2969. (d) Gopalsamy, A. et al. Pyrazolo[1,5-a]pyrimidin-7-yl phenyl amides as novel anti-proliferative agents: parallel synthesis for lead optimization of amide region, *Bioorg. Med. Chem. Lett.*, 2005, 15, 1591. (e) Zajdel, P. et al. Parallel solid-phase synthesis and characterization of new sulfonamide and carboxamide proline derivatives as potential CNS agents, *Bioorg. Med. Chem.*, 2005, 3029.

4. Bleicher, K.H. et al., Parallel solution- and solid-phase synthesis of spiropyrrolo-pyrroles as novel neurokinin receptor ligands, *Bioorg. Med Chem. Lett.*, 2003, 12, 3073.

5. Haque, T.S. et al. Parallel synthesis of potent, pyrazole-based inhibitors of *Helicobacter pylori* dihydroorotate dehydrogenase, *J. Med. Chem.*, 2002, 45, 4669.

6. Ryckebusch, A. Parallel synthesis and antimalarial activity of a sulfonamide library, *Bioorg. Med. Chem. Lett.*, 2002, 11, 2595.

7. Guo, T. et al., Discovery and SAR of 4-amino-2-biarylbutylurea MCH 1 receptor antagonists through solid-phase parallel synthesis, *Bioorg. Med. Chem. Lett.*, 2005, 15, 3691.

8. De Lucca, A.J.; and Walsh, T.J. Antifungal peptides: novel therapeutic compounds against emerging pathogens, *Antimicro. Agents Chemother.*, 1999, 1.

9. Kundu, B. et al. Combinatorial approach to lead optimization of a novel hexapeptide with antifungal activity, *Bioorg. Med. Chem. Lett.*, 2000, 10, 1779.

10. (a) Kundu, B. et al. Identification of novel antifungal nonapeptides through the screening of combinatorial peptide libraries based on a hexapeptide motif, *Bioorg. Med. Chem. Lett.*, 2002, 12, 1473. (b) Kumar, M. et al. Identification of a novel antifungal nonapeptide generated by combinatorial approach, *Int. J. Antimicrob. Agents,* 2005, 25, 313.

11. Iwata, K.; and Bossche, H.V. *In vitro and in vivo Evaluation of Antifungal Agents*, Elsevier Science, Amsterdam, 1986, p. 31.

12. (a) Hall, D.G.; Manku, S.; and Wang, F. Solution- and solid-phase strategies for the design, synthesis, and screening of libraries based on natural product templates: a comprehensive survey, *J. Comb. Chem.*, 2001, 3, 125. (b) Lee, M.L.; and Schneider, G. Scaffold architecture and pharmacophoric properties of natural products and trade drugs: application in the design of natural product-based combinatorial libraries, *J. Comb. Chem.*, 2001, 3, 284. (c) Nicolaou, K.C. et al. Natural product-like combinatorial libraries based on privileged structures. 1. General principles and solid-phase synthesis of benzopyrans, *J. Am. Chem. Soc.*, 2000, 122, 9939. (d) Watson, C. Polymer-supported synthesis of non-oligomeric natural products, *Angew. Chem. Int. Ed.*, 1999, 38, 1903. (e) Abreu, P.M.; and Branco, P.S. Natural product-like combinatorial libraries, *J. Brazilian Chem. Soc.*, 2003, 14, 675. (f) Cordell G.A.; Quinn-Beattie, M.L.; and Farnsworth, N.R., The potential of alkaloids in drug discovery, *Phytother. Res.*, 2001, 15, 183. (g) Wessjohann, L.A.; and Ruijter, E. Strategies for Total and Diversity-Oriented Synthesis of Natural Product(-Like) Macrocycles, *Topics in Current Chemistry: Natural Product Synthesis I: Targets, Methods, Concepts* (J. Mulzer, Ed.), Springer-Verlag GmbH, 243, 137, 2005. (h) Nielsen, J. Combinatorial synthesis of natural products, *Curr. Opin. Chem. Biol.*, 2002, 6, 297.

13. Anand, R. et al. Antioxaluric and anticalciuric activity of lupeol derivatives, *Ind. J. Pharmacol.*, 1995, 27, 265.B.

14. De Almeida Alves, T.M. et al. *In vitro* and *in vivo* evaluation of betulinic acid as an antimalarial, *Planta Med.*, 1997, 63, 554.

15. Srinivasan, T. et al. Solid-phase synthesis and bioevaluation of lupeol-based libraries as antimalarial agents, *Bioorg. Med. Chem. Lett.*, 2002, 12, 2803.

16. Attarde, M.E.; Porcu, G.; and Tadderi, M. Malachite green, a valuable reagent to monitor the presence of free COOH on the solid-phase, *Tetrahedron Lett.*, 2000, 41, 7391.

17. Bruce-Chwatt, L.-T. et al. *Chemotherapy of Malaria*, 2nd edition, World Health Organization, Geneva, 1986.

18. Lambros, C.; and Vanderberg, J.P. Order and specificity of the *Plasmodium falciparum* hemoglobin degradation pathway, *J. Parasitol.,* 1979, 65, 418.

19. Ottesen, E.A. *Infectious Diseases and Public Health*, Angelico, M. and Rocchi, G., Eds., Balaborn Publishers, 1998, p. 58.

20. Michael, E.; and Bundy, D.A.P. Global mapping of lymphatic filariasis, *Parasitol. Today,* 1997, 13, 472.

21. Martin, R.J.; Robertson, A.P.; and Bjorn, H. Target sites of anthelmintics, *Parasitology,* 1997, 114, 111.

22. Wang, J.C. DNA topoisomerases: why so many? *J. Biol. Chem.,* 1991, 266, 6659.

23. Sironi, M. Molecular evidence for a close relative of the arthropod endosymbiont *Wolbachia* in a filarial worm, *Mol. Biochem. Parasitol.,* 1995, 74, 223.

24. Hoerauf, A. et al. Tetracycline therapy targets intracellular bacteria in the filarial nematode *Litomosoides sigmodontis* and results in filarial infertility, *J. Clin. Invest.,* 1999, 103, 11.

25. Pandaya, U. et al. DNA topoisomerases of filarial parasites: effect of antifilarial compounds, *Med. Sci. Res.,* 1999, 27, 103.

26. (a) Wang, H.K.; Morris-Natscke, S.L.; and Lee, H.K., Recent advances in the discovery and development of topoisomerase inhibitors as antitumor agents, *Med. Res. Rev.,* 1997, 17, 367. (b) Olikubo, M. et al. Synthesis and biological activities of NB-506 analogues modified at the glucose group, *Bioorg. Med. Chem. Lett.,* 2000, 10, 419.

27. (a) Namane, A. et al. Improved brain delivery of AZT using a glycosyl phosphotriester prodrug, *J. Med. Chem.,* 1992, 35, 3039. (b) Fischer, J.F. et al. Peptide to glycopeptide: glycosylated oligopeptide renin inhibitors with attenuated *in vivo* clearance properties, *J. Med. Chem.,* 1991, 34, 3140. (c) Negre, E. et al. Antileishmanial drug targeting through glycosylated polymers specifically internalized by macrophage membrane lectins, *Antimicrobial Agents Chemother.,* 1992, 36, 2228.

28. Tripathi, R.P. et al. Identification of inhibitors of DNA topoisomerase II from a synthetic library of glycoconjugates. *Comb.Chem. Highthroughput Screen.,* 2001, 4, 237.

9 High-Throughput Solid-Phase Synthesis of Nucleoside-Based Libraries in the Search for New Antibiotics

Dianqing Sun and Richard E. Lee

CONTENTS

9.1 INTRODUCTION

There is an urgent need for the development of novel, more effective antibiotic drugs, with new mechanisms of action to combat emerging drug-resistant bacterial pathogens such as penicillin-resistant *Streptococcus pneumoniae,* vancomycin-resistant *enterococci*, methicillin-resistant *Staphylococcus aureus*, multi-drug-resistant *Mycobacterium tuberculosis, Pseudonomas aeruginosa,* and *Acinetobacter baumanii*.[1] The recent advent of genomic and proteomic sciences has led to the discovery of many new potential targets; however, few of these targets when screened have produced suitable lead candidates.[2] An equally attractive approach to lead identification in the early stages of the drug discovery process is to screen new chemical classes of compounds, often based around natural product templates, in phenotypic screens, followed by the application of genomics strategies to determine the exact target. This approach is particularly applicable to the discovery of new anti-microbial agents.

Natural products are the dominant source for successful bioactive compounds in the antimicrobial therapeutic area.[3] In nature, nucleoside scaffolds are commonly present in many naturally occurring antibiotics, such as mureidomycins, pacidamycins, and napsamycins, which target the early stages of bacterial cell wall biosynthesis (Figure 9.1).[4] Nucleoside analogs have shown moderate to good antibacterial activity against specific strains and are thus an important class of compounds worthy of further investigation.[5] A number of mureidomycin and pacidamycin analogs have been synthesized and evaluated in an effort to discover more effective antibacterial therapeutics and gain further understanding of the structure-activity relationship (SAR) from this class of natural products.[6]

Over the past 15 years, combinatorial chemistry and high-throughput synthesis (HTS) techniques have had a great impact on drug discovery.[7] By applying these technologies, a large number of compounds can be produced in a very efficient manner. After Merrifield's pioneering work in solid-phase peptide synthesis in 1963,[8] solid-phase organic synthesis (SPOS) has been employed extensively to accelerate the synthesis of small molecule libraries in a short time period for biological screening.[9] It offers the major advantage of simple purification of products from excess reactants and side products through filtration, allowing for the rapid preparation of vast arrays of compounds for drug testing.[10] It has thus become an invaluable tool for the identification of biologically active compounds and pharmacophores in the modern drug discovery paradigm.

As part of a continuing endeavor to discover new antibiotics, in particular new anti-tuberculosis agents that target the mycobacterial cell wall[11] and deoxythymidinediphosphate (dTDP)-L-rhamnose biosynthesis,[12,13] two nucleoside-based libraries were designed and synthesized in our laboratory: (1) a thymidinyl and 2′-deoxyuridinyl peptidomimetic Ugi library **1**; and (2) a thymidinyl dipeptide urea library **2** (Figure 9.1). This chapter reviews the high-throughput solid-phase synthesis strategies applied to make these nucleoside-based libraries using 96-well filter plate and IRORI directed sorting technologies.

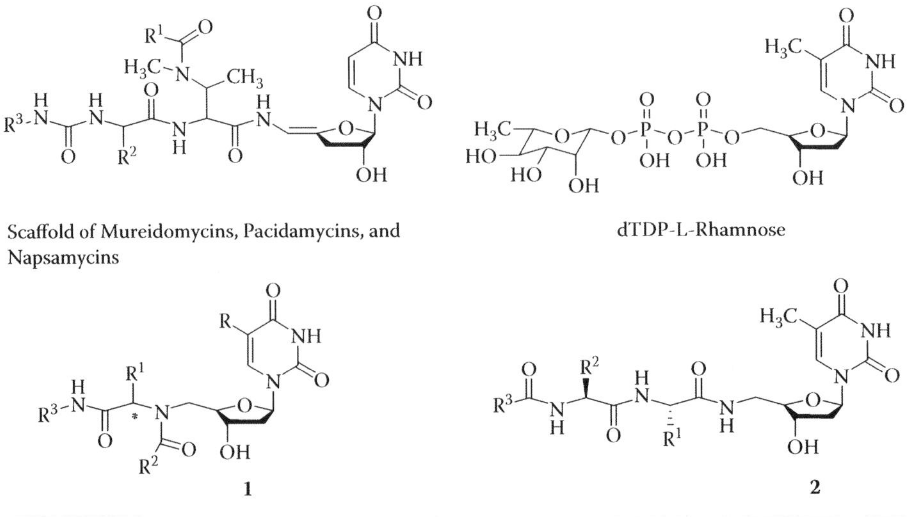

Scaffold of Mureidomycins, Pacidamycins, and Napsamycins

dTDP-L-Rhamnose

1

2

FIGURE 9.1   Naturally occurring antibiotics and our targeted libraries **1** and **2**.

9.2   CASE STUDY I: SOLID-PHASE SYNTHESIS OF A 1344-MEMBER NUCLEOSIDE UGI LIBRARY 1[14]

The Ugi reaction is a powerful multi-component reaction (MCR) that uses a four-component condensation (4CC) reaction involving amine, aldehyde, carboxylic acid, and isocyanide reactants.[15] Each reactant introduces a point of diversity (R, R[1], R[2], and R[3]), which allows for the generation of a vast library from relatively few reagents via a one-pot reaction (see **1** in Figure 9.1). Since the first examples were reported in 1959,[16] the Ugi reaction has been widely used for compound library generation due to the highly attractive peptidomimetics produced.[17] Various solid-phase synthesis strategies have been adopted to prepare Ugi libraries.[17a–d,h,18] The primary advantage of using solid-phase synthesis for Ugi chemistry is that it allows for rapid and facile removal of the multiple reactants and any by-products from the desired product, including the unreacted isocyanides that commonly have a strong and unpleasant odor. The design and synthesis of a thymidinyl and 2-deoxyuridinyl peptidomimetic Ugi library **1** are discussed in this section.

9.2.1   SOLID-PHASE SYNTHESIS STRATEGY SUMMARY

The solid-phase synthesis strategy developed for Ugi library **1** is illustrated in Scheme 9.1. First, the polystyrene butyl diethylsilane (PS-DES) resin was activated by reaction with 1,3-dichloro-5,5-dimethylhydantoin in anhydrous CH_2Cl_2 at room temperature for 2 hr to give the activated silyl chloride resin **3**. Second, in the presence of imidazole the 3'-hydroxyl of a nucleoside azide **4**[19] was attached to the solid support, followed by reduction of the azide **5** using $SnCl_2$:$HSPh$:$N(Et)_3$ (1:4:5) in THF,[20] yielding the polymer-bound 5'-deoxy-5'-aminonucleoside **6**. Third, this immobilized amine **6** was further treated with various aldehydes, carboxylic acids, and isocyanides to afford solid-supported

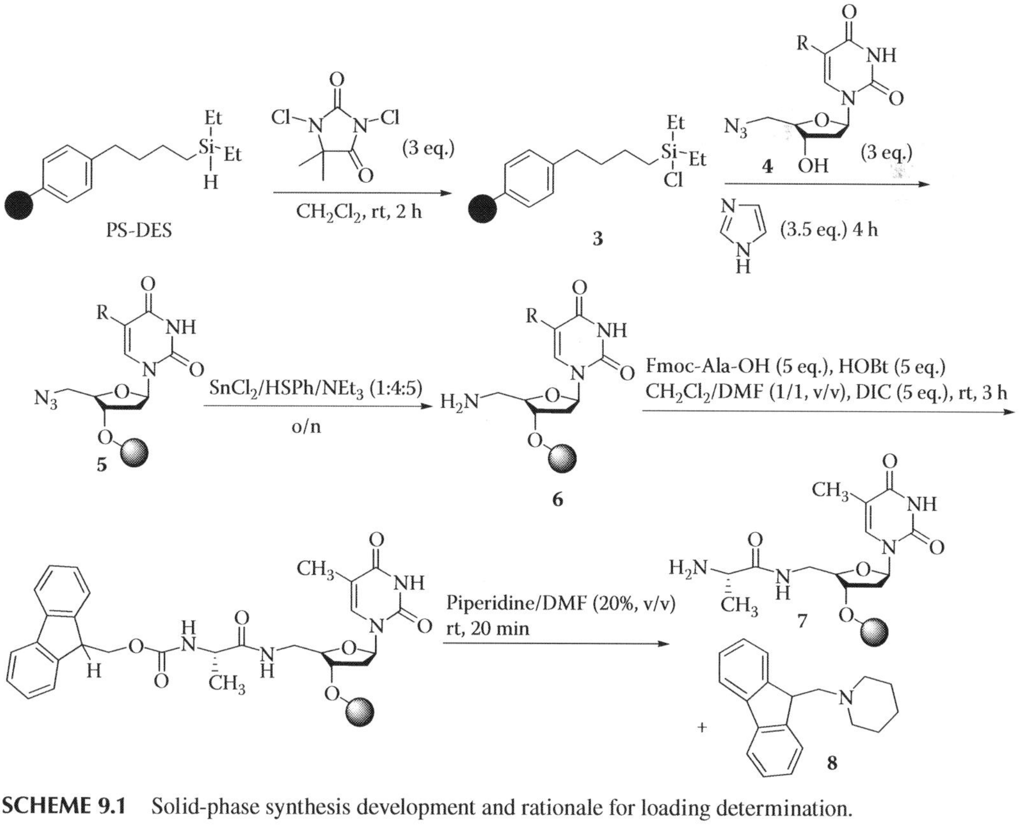

SCHEME 9.1   Solid-phase synthesis development and rationale for loading determination.

Ugi product **9**, followed by acidic cleavage with HF:pyridine in THF to afford the 1344-member Ugi library **1**, with no evidence of degradation of the products (Scheme 9.2). Development of the final library synthesis involved linker selection, optimization of loading capacity and reaction conditions, selection of the final building blocks, and quality assessment of each reaction and the final products. Each of these developmental steps is described in the following subsections.

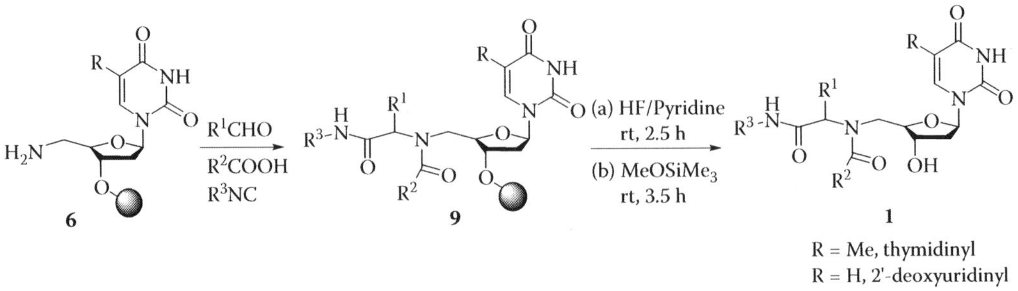

R = Me, thymidinyl
R = H, 2′-deoxyuridinyl

SCHEME 9.2   Synthesis of Ugi library 1.

9.2.2 SOLID SUPPORT AND LINKER CHEMISTRY

Solid support and linker selection plays a vital role in developing a practical solid-phase synthesis strategy.[21] This linker can be considered analogous to a protecting group in solution-phase synthesis, which functions as an anchor connecting the solid support and the first building block. It should be stable to all reaction conditions used during the construction of the library, allow the use of orthogonal chemistries to build the target library, and require cleavage at the end of the synthesis without degradation of final products. With these factors in mind, the 3′-hydroxy functional group of the nucleoside's ribosyl moiety was an obvious choice for the linker attachment site. After reviewing a number of potential linkers used previously to attach secondary alcohols, a silane-based linker strategy was selected and the PS-DES resin developed by Porco and co-workers[22] was chosen as a solid support. This resin can tolerate both neutral and basic reaction conditions, and can be cleaved selectively under mild acidic conditions such as AcOH or using a fluoride source (e.g., tetrabutylammonium fluoride or HF:pyridine). These conditions were compatible with the likely chemistries to be used in the library construction.

9.2.3 MONITORING OF REACTION PROGRESS

A primary challenge in developing any solid-phase synthesis is monitoring reaction progress such that each reaction can be allowed to reach completion in as short a time period as possible. In this case, we initially utilized cleavage of a small sample of resin at each step, followed by LC-MS analysis of the product to determine reaction success. However, this procedure was lengthy and there was concern that intermediates may not be completely stable to the cleavage conditions. Fortunately, during this study we obtained access to a Perkin-Elmer Spectrum 100 FT-IR spectrometer equipped with a universal attenuated total reflection (ATR) sampling accessory and a diamond/ZnSe contact crystal that allowed us to perform fast and convenient on-bead analysis using a small number of flattened beads. Using diagnostic peaks in the IR spectra, it was possible to easily follow the reaction progress. PS-DES resin at 2096 cm^{-1} has a characteristic stretch for the Si–H bond (Figure 9.2A). After 5′-azidonucleoside **4** was loaded, the formation of the bound azidonucleoside had diagnostic carbonyl and azide bands at 1691 and 2101 cm^{-1}, respectively, indicating that the reaction went to completion (Figure 9.2B and C). The subsequent disappearance of the azido band at 2101 cm^{-1} indicated the completion of the azide reduction (Figures 9.2D and E). The ca. 2100 cm^{-1} band in Figure 9.2B and C was shown to originate from azide rather than unreacted silane resin as the peak

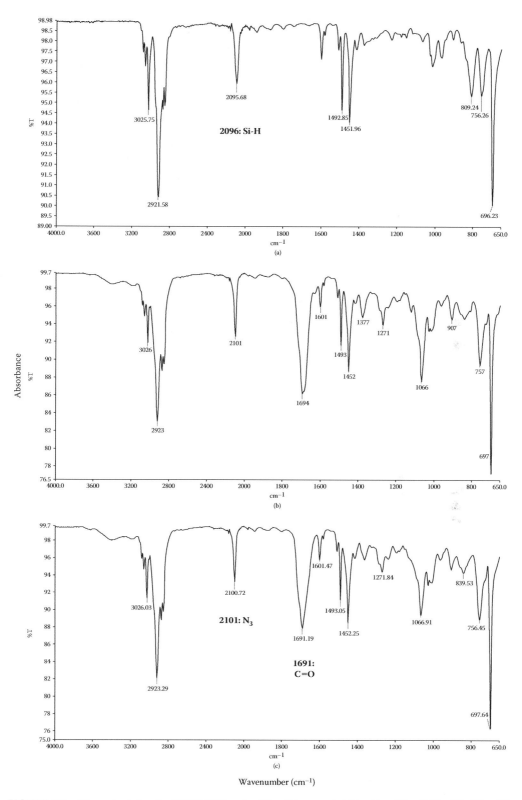

FIGURE 9.2 IR spectra taken for (A) PS-DES resin, (B) solid supported 5′-azidouridine **5** (R = H), (C) solid supported 5′-azidothymidine **5** (R = CH₃).

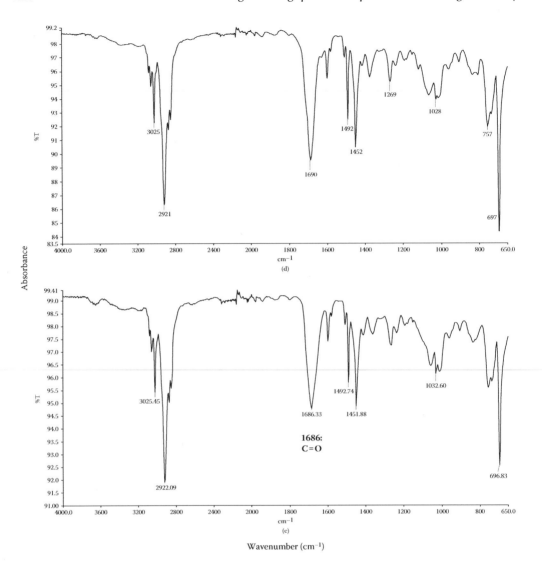

FIGURE 9.2 (D) solid supported 5′-aminouridine **6** (R = H), and (E) solid supported 5′-aminothymidine **6** (R = CH$_3$).

completely disappeared upon azide reduction. It was also possible to monitor the azide reduction step and subsequent acylations using the colorimetric Kaiser test as a confirmatory method.[23]

9.2.4 Optimization of Loading Capacity

To maximize the yield of the synthesis, it was necessary to ensure maximal loading of the 5′-azidonucleoside **4** to the PS-DES resin via the hindered 3′-hydroxyl. To do so, the maximum loading capacity was determined by quantitative 9-fluorenylmethoxycarbonyl (Fmoc) removal analysis.[24] A parallel loading experiment was designed and performed under a variety of potential loading conditions (entries 1 to 10 in Table 9.1), followed by azide reduction, Fmoc-alanine acylation, base-catalyzed Fmoc deprotection, and spectrophotometric quantitation of the released piperidine-dibenzofulvene adduct **8** (Scheme 9.1). Solid-supported 5′-aminothymidine **6** (R = CH$_3$) was acylated with Fmoc-alanine using the *N*,*N*′-diisopropylcarbodiimide (DIC) and 1-hydroxybenzotriazole (HOBt) coupling protocol.[25] Subsequent Fmoc deprotection was performed in 20% piperidine in DMF. The

TABLE 9.1

Survey of Loading Conditions Based on Temperature and Solvent Effects

Entry	Solvent	Temp (°C)	Calculated Loading (mmol/g)	Overall Yield (%)
1	CH$_2$Cl$_2$:DMF	25	0.40	54
2	DMF	25	0.48	65
3	THF	40	0.28	38
4	CH$_2$Cl$_2$:DMF	40	0.47	63
5	THF:DMF	40	0.31	42
6	DMF	40	0.43	57
7	NMP	40	0.49	67
8	CH$_2$Cl$_2$:NMP	40	0.52	70
9	DMF	80	0.45	61
10	NMP	80	0.31	42

UV absorbance of **8** at 301 nm was measured, and the effects of temperature and solvent type on the loading yield of 5′-azidothymidine were evaluated. The results from this study are shown in Table 9.1. The optimal loading condition was found to be in CH$_2$Cl$_2$:NMP at 40°C, which afforded 70% overall yield (after resin activation, loading, reduction, acylation, and Fmoc deprotection) based on the theoretical loading of PS-DES resin. Nevertheless, from a practical standpoint, DMF at room temperature loading conditions still provided comparable results with 65% overall yield, especially when bulk synthesis is carried out for the library synthesis in a solid-phase peptide synthesizer. Further elevating the temperature to 80°C in DMF or NMP did not yield a better result, due to some decomposition of activated resin **3** during the loading process. THF was not a good solvent for this loading (38% and 42% yield), probably due to the poor solubility of **4** in this solvent.

9.2.5 OPTIMIZATION OF UGI REACTION CONDITIONS ON SOLID SUPPORT

Optimization of reaction conditions is a key factor for successful solid-phase synthesis. In this study we evaluated the effects of temperature, solvent type, and the molar ratios of the reactants to increase the purity and yield of the products. A preliminary, exploratory experiment was carried out to evaluate the temperature effect on Ugi chemistry by utilizing the aminonucleoside-bound resin **6** as a support (Scheme 9.2). Two Ugi reactions were performed at room temperature and 40°C for 24 hr (Table 9.2). A mixture of CH$_2$Cl$_2$ and MeOH was used to obtain necessary resin swelling and favored Ugi reaction conditions, respectively. After the reaction was complete, the resin was filtered, washed, cleaved, and the products were analyzed by HPLC-MS (Table 9.2). A temperature of 40°C was found to give better yields than room temperature. The only impurity found was the unreacted starting material (5′-aminothymidine), which is inactive in our bioassays.

To reduce the amount of unreacted 5′-aminothymidine in the final product and thus increase the purity and yield of the desired products, the effects of solvent type and molar ratios of the reactants on product yields were further evaluated using a parallel array in a Robbins Flexchem 96-well filter plate. Twelve carboxylic acids {**1** to **12**}, three aldehydes {**1** to **3**}, and one isocyanide {**1**} (Figure 9.3) were used to give 12 different Ugi products (columns 1 through 12) under eight different reaction conditions of solvent and molar ratios (rows A through H). Methyl isocyanoacetate was chosen to optimize our method because this isocyanide was reported to lead to low to moderate yield in previously reported Ugi reactions.[26] Therefore, it could be expected to provide a range of yields of the desired products and help determine which conditions provide best purity

TABLE 9.2

Solid-Phase Ugi Reaction at Room Temperature and 40°C[a]

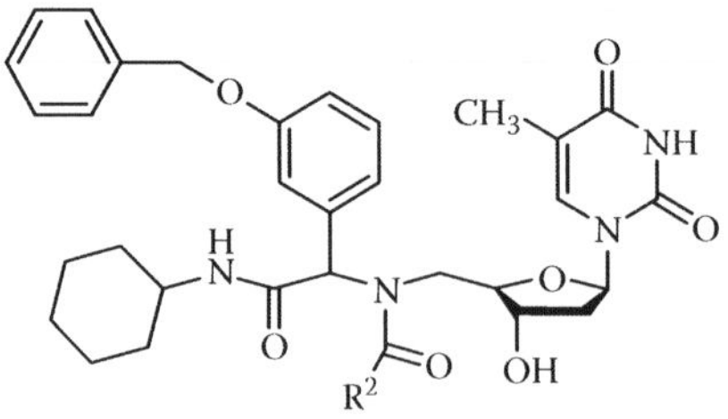

Compound	R^2	HPLC yield[b] (r.t.)	HPLC yield[b] (40°C)	t_R[c] (min)	MS[d][MH+]
1	 F–⟨⟩–	43%	61%	13.4	685.5
2	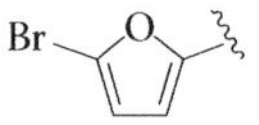 Br–⟨O⟩–	61%	75%	13.4	735.5

[a] Optimization was done on a Quest 210 synthesizer: Solid supported 5′-aminothymidine resin **6** (R = CH₃) was obtained from PS-DES resin (144 mg, 0.107 mmol) after 3 steps (resin activation, loading and azide reduction), then a solution of 3-benzyloxybenzaldehyde (10 eq.) in CH₂Cl₂ (1 mL) was added to the reaction vessel containing resin **6**, followed by the addition of a solution of its corresponding carboxylic acid (10 eq.) in methanol (1 mL). The reactions were agitated for 30 min, then a solution of cyclohexyl isocyanide (10 eq.) in CH₂Cl₂ (1 mL) was added. The reaction mixture was agitated for 24 h at room temperature or 40°C. The resin was filtered, washed with DMF (3 × 4 mL), MeOH (3 × 4 mL), CH₂Cl₂ (3 × 4 mL), and THF (3 × 4 mL). For cleavage: 1.6 mL of HF/pyridine in THF (0.4 M) was added and the reaction mixture was allowed to agitate for 2.5 h, followed by the addition of MeOSiMe₃ and agitated for another 3.5 h. The cleavage solution was collected and the resin was further washed with MeOH (3 × 4 mL). The combined solution was evaporated to afford crude product, which was subject to HPLC-MS analysis. [b] Analytical RP-HPLC was conducted on an Agilent 1100 chemstation, using an Alltech platinum EPS C18 column (100Å, 5 μm, 4.6 × 150 mm) with precolumn 4.6 × 10 mm, flow rate 1.0 mL/min and a gradient of solvent A (water with 1% acetic acid) and solvent B (acetonitrile): 0–2.00 min 100% A; 2.00–17.00 min 0–100% B. UV detection at 254 nm. [c] Retention time. [d] Mass spectra were recorded on a Bruker Esquire LC-MS using ESI.

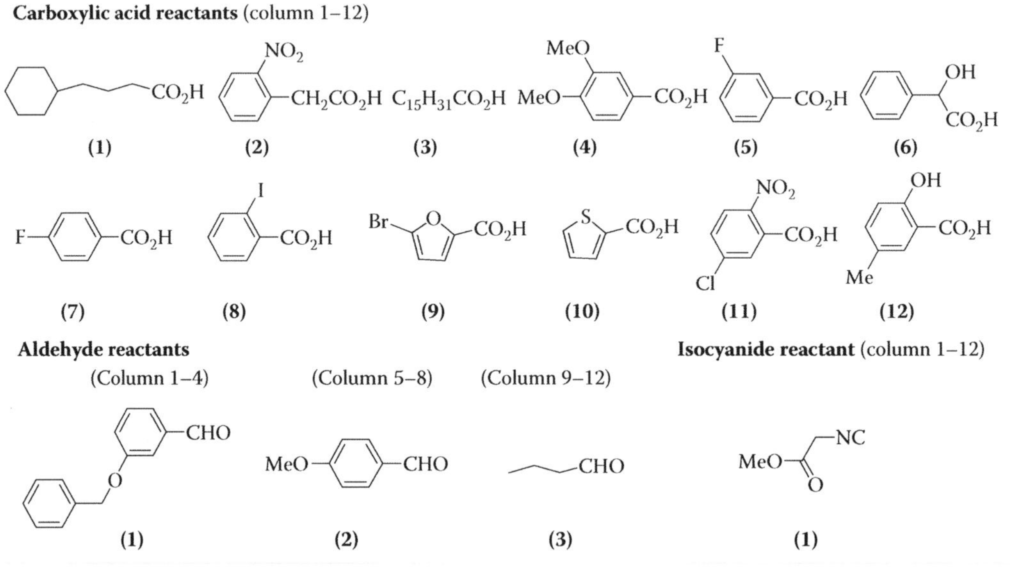

Carboxylic acid reactants (column 1–12)

Aldehyde reactants　　　　　　　　　　　　　　　　　　　　　**Isocyanide reactant** (column 1–12)
(Column 1–4)　　　　　(Column 5–8)　　　(Column 9–12)

FIGURE 9.3　Building blocks for Ugi optimization.

TABLE 9.3
HPLC Purity of Ugi Products under Solid-Phase Optimization

	Solvent (by volume)	$R^1CHO/$ R^2CO_2H/R^3NC (molar ratio)	1	2	3	4	5	6	7	8	9	10	11	12
A	DCM:MeOH (1:1, 0.6 mL)	10:10:10	○	○	■	■	○	■	○	■	■	■	■	■
B	DCM:MeOH (2:1, 0.9 mL)	10:10:10	○	○	■	■	○	■	○	■	○	○	◆	◆
C	DCM:MeOH (2:1, 1.5 mL)	20:20:10	○	■	■	■	■	■	■	△	■	○	■	◆
D	DCM:MeOH (1:1, 1.2 mL)	10:20:10	○	■	■	■	■	■	■	■	■	■	△	◆
E	DCM:MeOH (1:1, 1.2 mL)	10:10:10	○	■	■	■	■	■	■	■	■	○	■	◆
F	CHC$_{13}$:MeOH (3:1, 1.2 mL)	10:10:10	■	△	■	■	■	■	■	■	■	■	■	◆
G	PhMe:MeCN (3:1, 1.2 mL)	10:10:5	○	△	■	■	■	△	■	△	△	△	△	○
H	THF, 0.9 mL	10:10:10	■	△	■	■	○	△	○	△	△	△	△	■

Product yield index: ◆: >80%; ○: 50–80%; ■: 20–50%; △: <20%.

and yields for the widest variety of building blocks. The reaction time was increased to 2.5 days (1.5 days at 40°C and 1 day at room temperature) in an effort to drive the reaction to completion. After the final cleavage step, the purity of the target products was determined by LC-MS analysis and the results are shown in Table 9.3. It was found that using mixtures of CH$_2$Cl$_2$:MeOH (2:1, by volume, 0.9 mL) as solvent and a 10:10:10 molar excess of reactants to resin loading offered the optimal results, and this condition was selected for use in final library synthesis. The use of toluene and acetonitrile (3:1) or THF as a solvent gave the lowest purities (due to the poor solubility of reactants in these solvents).

9.2.6 SELECTION OF BUILDING BLOCKS

To generate a final set of eight aldehydes and twelve carboxylic acids suitable for the final library synthesis in parallel arrays, a variety of carboxylic acids and aldehydes with aliphatic, aromatic, and heterocyclic moieties were evaluated for their compatibility with the Ugi conditions in the 96-well filter plates (Figure 9.4). After surveying these building blocks, we found that the overall purity of the Ugi products was more sensitive to aldehyde than to carboxylic acid reactants. The aliphatic aldehydes gave the best results, followed by aromatic aldehydes, with the heterocyclic aldehydes yielding the worst results. This pattern is in good agreement with the ease of imine formation. The building blocks that led to poor HPLC purity of the desired products or had poor solubility in reaction solvents were eliminated from further library synthesis. After this study, twelve carboxylic acids (R^{2}COOH, columns 1 through 12) and eight aldehydes (R^{1}CHO, rows A through H) were selected for final library production (Figure 9.5).

9.2.7 SYNTHESIS OF UGI LIBRARY 1

Library synthesis was performed in fourteen 96-well filter plates. Solid-supported nucleoside amine **6** was prepared from PS-DES resin in a three-step bulk reaction (resin activation, loading, and azide reduction) in a solid-phase peptide synthesizer. It was then distributed evenly to a

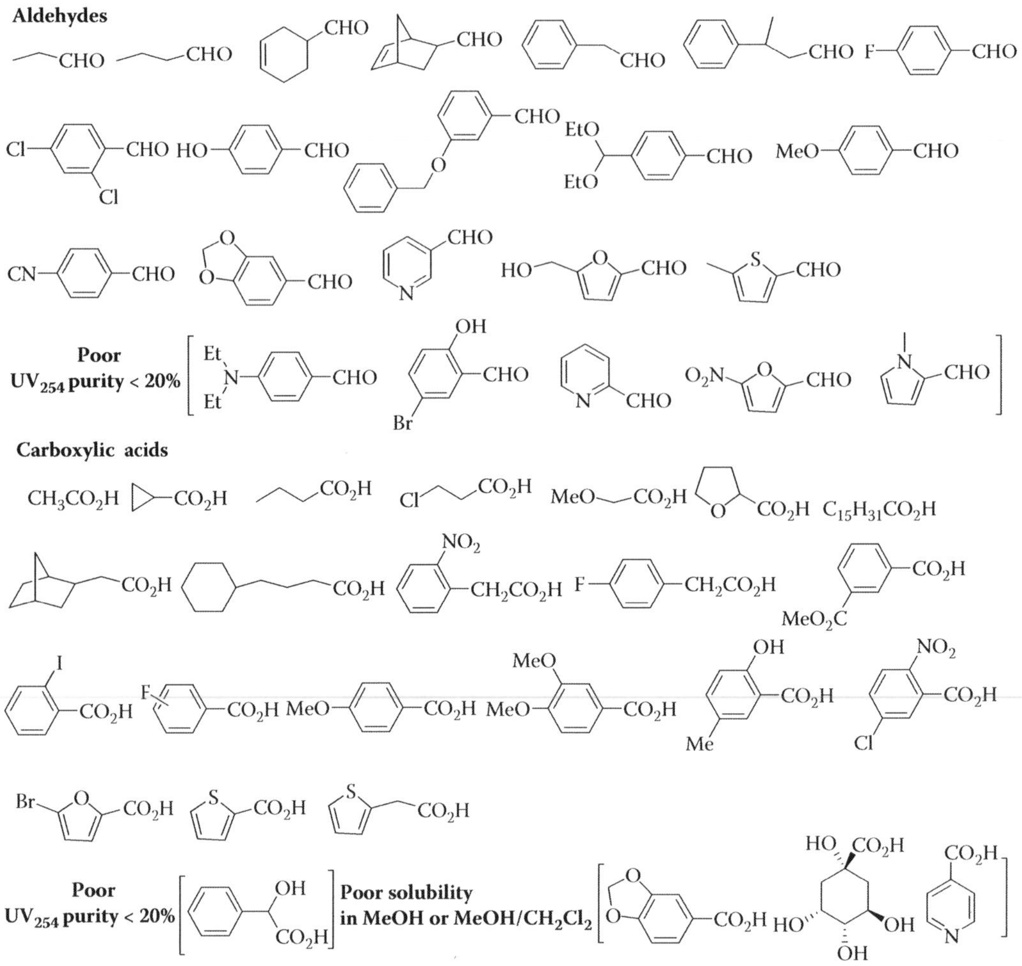

FIGURE 9.4 Building blocks evaluated on solid-phase Ugi reaction.

96-well filter plate using a neutral buoyancy suspension method, which resulted in 0.019 mmol amine resin **6** per well. Resin **6** was reacted with eight aldehydes (R^1CHO) {**A** to **H**}, twelve carboxylic acids (R^2COOH) {**1** to **12**}, and one of seven isocyanides (R^3NC) {**1** to **7**} per plate to afford solid-supported Ugi products **9** in an 8 × 12 array for each of the two nucleosides {**1** and **2**} (Scheme 9.2; Figure 9.5). After cleavage with HF:pyridine in THF, the excess HF was quenched by adding $MeOSiMe_3$ to form volatile by-products Me_3SiF and MeOH, which allows a simple evaporation step to yield a product that is suitable for direct antimicrobial screening.[27] Using this solid-phase synthesis protocol,[14] a 1344-member nucleoside Ugi library **1** (Figure 9.5) was synthesized in parallel arrays. As anticipated, diastereomeric Ugi products (ca. 1:1 ratio) were obtained based on HPLC and ^{1}H NMR analyses.

Quality assessment of the Ugi library **1** was performed by LC-MS using one randomly selected column (eight samples) per plate (112 samples for the library, ca. 8% of library size). Some 47% of analyzed samples had HPLC UV_{254nm} purity greater than 80%, 28% samples gave the expected product with purity ranging from 50% to 80%, 10% samples had purities ranging from 20% to 50%, and 15% had purities less than 20% (Figure 9.6). As expected, the unreacted starting material 5′-aminonucleoside was the major side-product in this library.

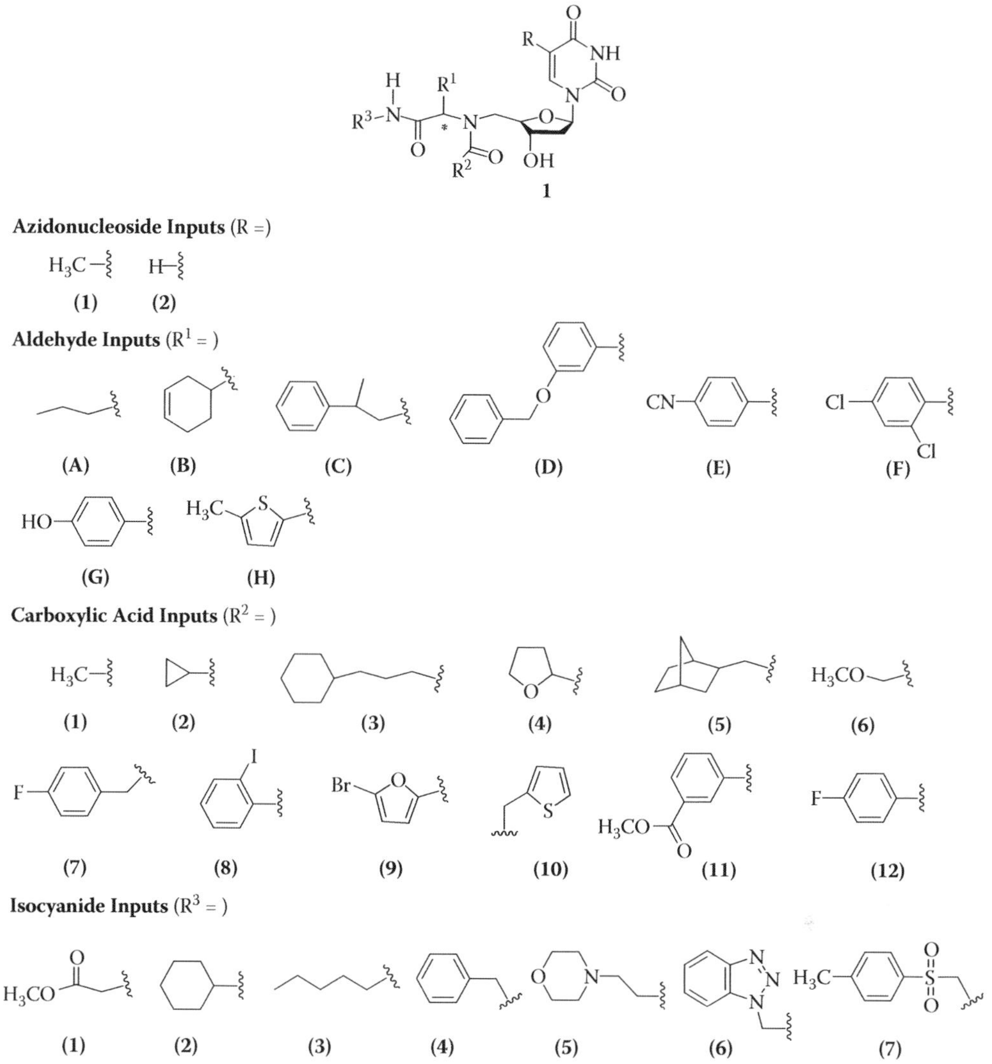

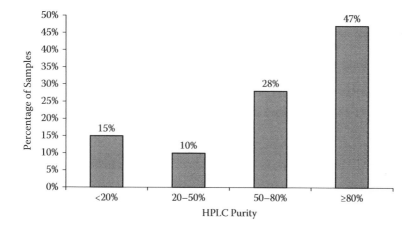

Azidonucleoside Inputs (R =)

H₃C—ξ   H—ξ
  (1)   **(2)**

Aldehyde Inputs (R¹ =)

(A)   **(B)**   **(C)**   **(D)**   **(E)**   **(F)**

HO—ξ   H₃C—S—ξ
  (G)   **(H)**

Carboxylic Acid Inputs (R² =)

H₃C—ξ   ▷—ξ   ...   .   H₃CO—ξ
  (1)   **(2)**   **(3)**   **(4)**   **(5)**   **(6)**

F—...   I ...   Br—O—ξ   S ...   H₃CO—ξ   F—ξ
  (7)   **(8)**   **(9)**   **(10)**   **(11)**   **(12)**

Isocyanide Inputs (R³ =)

H₃CO—ξ   cyclohexyl—ξ   ...   benzyl   O—N—ξ   benzotriazole   H₃C—SO₂—ξ
  (1)   **(2)**   **(3)**   **(4)**   **(5)**   **(6)**   **(7)**

FIGURE 9.5   A 1344-member thymidinyl and 2′-deoxyuridinyl Ugi library **1**.

FIGURE 9.6   HPLC purity distribution of Ugi library **1**.

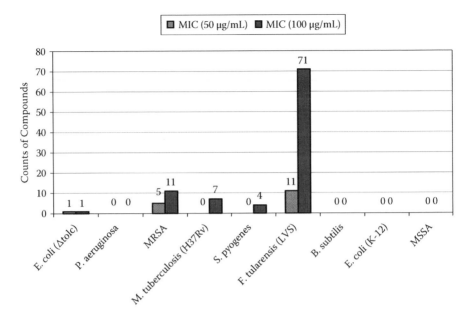

FIGURE 9.7   Counts of library members with MIC values at 50 or 100 µg/mL after screening.

9.2.8 BIOLOGICAL EVALUATION

This library has been screened for the inhibition of growth against nine different bacterial strains, including some of the most clinically important Gram-positive and Gram-negative bacterial pathogens: *Pseudomonas aeruginosa*, methicillin-resistant *Staphylococcus aureus* (MRSA), *Mycobacterium tuberculosis* (H37Rv), *Streptococcus pyogenes*, *Francisella tularensis* (LVS), *Bacillus subtilis*, *Escherichia coli* (K–12), and methicillin-susceptible *Staphylococcus aureus* (MSSA). A number of library members showed promising activity against efflux-deficient *E. coli* (Δtolc), MRSA, *M. tuberculosis* (H37Rv), *S. pyogenes*, and *F. tularensis* (LVS) at minimum inhibitory concentration (MIC) values of 50 and 100 µg/mL. No activity was detected against *P. aeruginosa, B. subtilis, E. coli* (K–12), or MSSA (Figure 9.7). Active compounds were subsequently resynthesized at 0.4 mmol scale on a Quest 210 synthesizer, purified by preparative HPLC, and accurate MICs were determined.

9.2.9 SUMMARY AND CONCLUSION

A solid-phase synthesis strategy was developed to make a thymidinyl and 2′-deoxyuridinyl Ugi library using PS-DES resin support. This required a lengthy developmental process, including maximum loading capacity determinations, Ugi reaction optimization, selection and validation of building blocks, and optimization to 96-well filter plate format. Once the library synthesis conditions had been optimized, a 1344-member library was synthesized with adequate purity for direct antibacterial screening, which produced a number of antimicrobial substances.

9.3 CASE STUDY II: SOLID-PHASE SYNTHESIS OF A 1000-MEMBER THYMIDINYL DIPEPTIDE UREA LIBRARY 2[28]

Solid-phase synthesis has proven to be an efficient and robust approach to develop small molecule, including nucleoside-based, libraries[29] and thus becomes a mainstay in the early stages of modern drug discovery. This technique is an even more powerful tool for generating vast libraries when split and pool techniques are applied.[30] The downside of this approach is that extra steps are often required for the identification of the active compounds. To do so, many elegant tagging or encoding technologies have been developed.[31] One popular solution is to perform library construction on a solid support utilizing IRORI directed sorting technology.[32] This offers an increased ease of simultaneous synthesis using split and pool solid-phase synthesis in small, solvent-permeable, tagged vessels (Kans) that contain a set portion of resin. Using radiofrequency (Rf) or barcode tagging, libraries are produced as discrete compounds in a semi-parallel approach. This automated IRORI platform was selected for the synthesis of thymidinyl dipeptide urea library **2** over 96-well filter plates because it allowed for a more rapid and homogeneous synthesis. The adaptation of our approach to use IRORI MiniKans and the consequent library synthesis are discussed in detail in this section.

9.3.1 PARALLEL CHEMISTRY OPTIMIZATION AND SIDECHAIN DEPROTECTION EVALUATION

To test the chemistry feasibility and validate the designed synthesis strategy prior to library production in the MiniKans, a test library of 22 compounds {**11a–p** and **13a–f** (Scheme 9.3)} was synthesized in filter tubes on a Argonaut Quest 210 synthesizer using solid-supported 5′-aminothymidine **6** prepared as described in Section 9.2.4. Several Fmoc-amino acid coupling methods were explored using activating reagents benzotriazole-1-yl-oxy-tris-pyrrolidino-phosphonium hexafluorophosphate (PyBOP),; 2-(1H-benzotriazole-1-yl)-1,1,3,3-tetramethylaminium hexafluorophosphate (HBTU); and DIC. In all cases, HOBt additive was used to form the activated OBt ester. Unlike PyBOP and HBTU methods, the base is not required in the DIC method. Preliminary experiments showed that DIC-HOBt coupling gave a better result with a cleaner product. Additionally, DIC costs less than PyBOP or HBTU; therefore, the DIC-HOBt method was chosen for the Fmoc-amino acid coupling protocol. Subsequently, the Fmoc protecting group was removed by 20% piperidine in

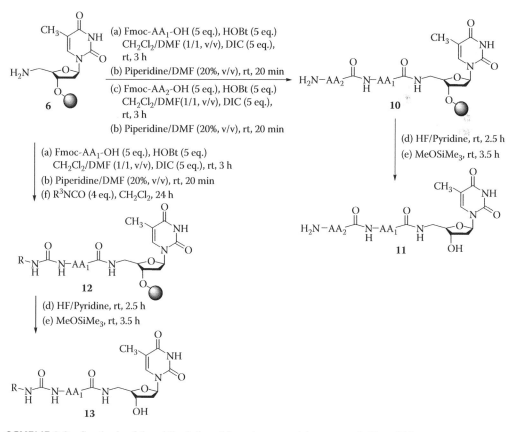

SCHEME 9.3   Synthesis of thymidinyl dipeptide and urea model compounds **11** and **13**.

DMF. This was either followed by another round of Fmoc-amino acid coupling and Fmoc deprotection to give polymer-bound nucleoside dipeptides **10**, or followed by reaction with various isocyanates to give polymer-bound nucleoside ureas **12**. Both schemes were subjected to HF:pyridine in THF to afford thymidinyl dipeptide and urea model compounds **11a–p** and **13a–f**, respectively. All compounds were obtained in excellent purity (≥90%) based on ¹H NMR and HPLC analyses. Their HPLC profiles are summarized in Table 9.4.

Encouraged by these results, we further studied the compatibility of different sidechain protection strategies to determine library stability and to maximize library diversity by the introduction of functionalized sidechains. Various orthogonally protected Fmoc amino acids with sidechain protection schemes were evaluated (Table 9.5), including specific Pd-catalyzed cleavage

TABLE 9.4
HPLC Profiles of Model Compounds 11 and 13ᵃ

Compound	AA₂	AA₁	t_R (min)	HPLC Purity (%)	Compound	AA₂ R for 13a-f	AA₁	t_R (min)	HPLC Purity (%)
11a	Ala	Ala	7.3	95	**11l**	Phe(4-NO₂)	Ala	8.6	100
11b	Phe	Ala	9.3	92	**11m**	ε-Ahx	Ala	7.5	99
11c	Ala	Phe	10.3	98	**11n**	Asn(Trt)	Ala	11.6	96
11d	Phe	Phe	12.0	90	**11o**	Cit	Ala	7.0	96
11e	β-Ala	Ala	7.6	98	**11p**	Cys(Acm)	Ala	7.4	90
11f	Ala	β-Ala	7.3	96	**13a**	n-Pr	Ala	8.2	97
11g	Phe	β-Ala	9.3	92	**13b**	i-Pr	Ala	8.1	100
11h	Ile	Ala	7.7	100	**13c**	C₆H₅	Ala	9.0	91
11i	Nle	Ala	7.9	100	**13d**	n-Pr	Phe	10.0	98
11j	Pro	Ala	7.3	97	**13e**	i-Pr	Phe	10.0	100
11k	Trp	Ala	8.7	97	**13f**	C₆H₅	Phe	10.8	97

ᵃ All compounds were characterized by mass spectrometry and ¹H NMR.

TABLE 9.5
Sidechain Deprotection, Library Stability, and Resin Cleavage Evaluation

Entry	Fmoc-AA-OH	Deprotection Conditions	Results
1	Glu-OAll	Pd(Ph₃P)₄/CHCl₃/AcOH/NMM, 2.5 h	selectively deprotected with no cleavage
2	Lys(Dde)	2% N₂H₄ in DMF, 20 min	selectively deprotected with no cleavage
3	Lys(Boc)	TFA/H₂O/TIS (95/2.5/2.5), 2 h	complete co-cleavage
4	Asn(Trt)	TFA/H₂O/TIS (95/2.5/2.5), 2 h	complete co-cleavage
5	Arg(Pbf)	TFA/H₂O/TIS (95/2.5/2.5), 2 h	complete co-cleavage
6	His(Mtt)	TFA/DCM/TIS (1/94/5), 20 min	partial (ca. 25%) co-cleavage
7	Hse(Trt)	TFA/DCM/TIS (1/94/5), 20 min	partial (ca. 25%) co-cleavage
8	Thr(tBu)	TFA/H₂O/TIS (95/2.5/2.5), 2 h	complete co-cleavage
9	Tyr(tBu)	TFA/H₂O/TIS (95/2.5/2.5), 2 h	complete co-cleavage

of allyl protected glutamic acid (Glu-OAll); selective hydrazine cleavage of Dde-protected lysine (Lys(Dde)); acid labile Boc-protected lysine (Lys(Boc)); trityl-protected asparagine (Asn(Trt)) and homoserine (Hse(Trt)); Pbf-protected arginine (Arg(Pbf)); methyltrityl-protected histidine (His(Mtt)); and *t*-butyl-protected threonine (Thr(*t*Bu)) and tyrosine (Tyr(*t*Bu)). The results in Table 9.5 are based on the HPLC-MS analysis of the samples from the sidechain deprotection solutions (step 2 after **14**), the samples from their corresponding final HF cleavage solutions (step 3 after **14**), and the mass comparison of these collected samples. As anticipated, allyl-protected glutamic acid and Dde-protected lysine were selectively removed by palladium and 2% hydrazine, respectively, and no cleavage was detected (entries 1 and 2). Under 95% trifluoric acid (TFA) conditions, complete sidechain deprotection and simultaneous resin cleavage occurred (entries 3 to 5, 8, and 9). Under diluted 1% TFA deprotection conditions, approximately 25% of products were released from the resin while the Mtt and Trt protecting groups were removed (entries 6 and 7). The representative HPLC profiles are shown in Figure 9.8.

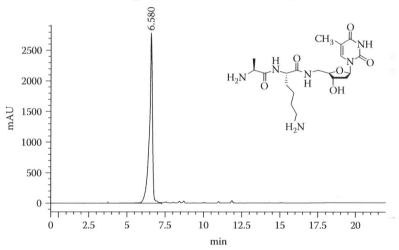

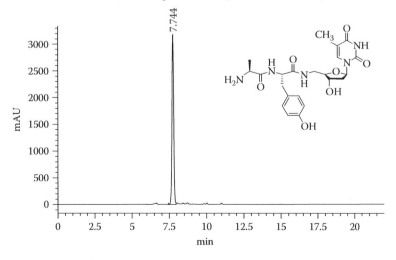

FIGURE 9.8   Representative HPLC profiles.

9.3.2 FURTHER OPTIMIZATION AND DEVELOPMENT OF A CO-CLEAVAGE COCKTAIL IN MINIKANS

In the test synthesis, final resin cleavage was achieved by the standard HF:pyridine condition and no product degradation was observed. From the perspective of target library synthesis using the selected IRORI technology, HF:pyridine cleavage was not compatible with this tagging system due to its potential to damage the glass-coated Rf tags; therefore, HF:pyridine was replaced by acidic TFA cleavage. Using an acidic cleavage step also allows for co-cleavage of sidechain protecting groups, which simplifies the overall number of reaction steps and thus reduces the synthesis time-line and the chemical cost for library production. As a part of the chemistry optimization, reactions and cleavage were moved from the polyethylene Quest filter tubes to the more expensive and less accessible IRORI MiniKan reactors.

TFA was chosen for acidic cleavage and deprotection, which at 95% gave us rapid and complete sidechain deprotection results with good purity. However, TFA has a highly corrosive and toxic nature, it is relatively difficult to remove, and it has the potential to degrade the sensitive nucleoside bond during cleavage and TFA removal processes. We therefore sought to optimize conditions that reduced the amount of TFA used during the cleavage step but still afforded simultaneous sidechain deprotection. Two test experiments were designed that included Fmoc amino acids with various acid-labile trityl, t-butyl, and t-Boc functionalized sidechains (Table 9.6). The first was designed to examine the effects of TFA concentration (5, 10, and 20% TFA for 2 hr in CH_2Cl_2) or different solvent systems (TFA:THF:H_2O for 2 hr or AcOH:THF:H_2O for 4 hr). The second was designed to evaluate the effects of cleavage time (3 hr, 6 hr, or overnight) under 10% and 15% TFA conditions. The HPLC purities of the desired products and sidechain-protected products are summarized in Table 9.6. Based on these results, we chose to use 10% TFA overnight with triisopropylsilane (TIS) added as a cation quench reagent (TFA:TIS:CH_2Cl_2, 10:2.5:90) for library 2 synthesis.

TABLE 9.6
HPLC Purity of Product 17 under Various Co-Cleavage Study

16a	R′ = Asp(OtBu)
b	Lys(Boc)
c	Thr(tBu)
d	Gln(Trt)

17a	R = Asp	a′	R = Asp(OtBu)
b	Lys	b′	Lys(Boc)
c	Thr	c′	Thr(tBu)
d	Gln	d′	Gln(Trt)

		HPLC purity (%) of product 17			
Entry	Co-cleavage conditions	a(a′)	b(b′)	c(c′)	d(d′)
1	20% TFA/CH$_2$Cl$_2$, 2 h	81(19)	100	92	90
2	10% TFA/CH$_2$Cl$_2$, 2 h	51(49)	100	84(13)	85
3	5% TFA/CH$_2$Cl$_2$, 2 h	34(66)	94	41(57)	90
4	TFA/THF/H$_2$O (5/100/10), 2 h	48(52)	98	3(88)	0(97)
5	AcOH/THF/H$_2$O (6/6/1), 4 h	0(79)	0(88)	0(66)	0(83)
6	10% TFA/CH$_2$Cl$_2$, 6 h	75(25)		88(5)	
7	10% TFA/CH$_2$Cl$_2$, o/n	93(4)		88(7)	
8	15% TFA/CH$_2$Cl$_2$, 3 h	84(16)		95(5)	
9	15% TFA/CH$_2$Cl$_2$, 6 h	95(5)		94(6)	
10	15% TFA/CH$_2$Cl$_2$, o/n	100		89(7)	

9.3.3 SYNTHESIS OF A 64-MEMBER TEST LIBRARY

To explore the scope and limitation of the designed synthesis with the automated IRORI system, a three-step ($4 \times 4 \times 4$) rehearsal library **18** was designed and synthesized in IRORI MiniKan reactors (Figure 9.9). A structurally and functionally diverse set of building blocks was used, which included orthogonally protected Fmoc-amino acids. Different types of capping procedures were evaluated, including benzoylation, urea formation, and reductive amination (Table 9.7). Each MiniKan was loaded with an Rf tag and 5′-aminothymidine-attached resin **6** (ca. 80 μmol per MiniKan), and directed sorting was used at each step. Two rounds of Fmoc-amino acid coupling (AA$_1$ and AA$_2$) and subsequent Fmoc removal were performed using the previously developed protocol described in Section 9.3.1, except that the reaction times for the coupling and deprotection processes were extended from 3 hr to overnight and from 20 min to 1 hr, respectively, to ensure reaction completeness.

This synthesis was initiated before the optimal cleavage and sidechain deprotection conditions in TFA were determined. Therefore, acid co-cleavage was achieved using 95% TFA in CH$_2$Cl$_2$, which produced rehearsal library **18** with good overall purity according to LC-MS analysis. Based on this rehearsal library synthesis, three problems were identified and addressed before scaling up to full library synthesis. First, library members subject to a reductive amination step resulted in a mixture of the desired products and unidentified impurities. Second, we found that library members

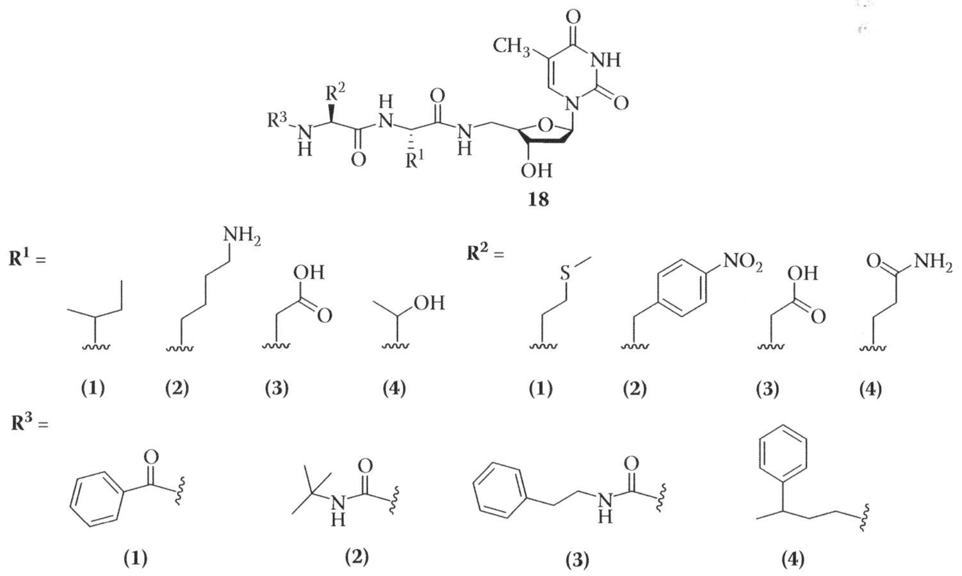

FIGURE 9.9 A 64-member test library **18**.

TABLE 9.7
Building Blocks for Test Library 18 ($4 \times 4 \times 4$) Synthesis

Step	Building Block	{1}	{2}	{3}	{4}
1	Fmoc-AA$_1$-OH	Ile	Lys(Boc)	Asp(OtBu)	Thr(tBu)
2	Fmoc-AA$_2$-OH	Met	Phe(4-NO$_2$)	Asp(OtBu)	Gln(Trt)
3	R^{3}Cl for {1}	Bz	t-Bu	PhCH$_2$CH$_2$	CH$_3$CH(Ph)CH$_2$
	R^{3}NCO for {2–3}				
	R^{3}CHO for {4}				

derived from *t*-butyl isocyanate were de-*t*-butylated under concentrated TFA conditions. Third, for product library members containing a free carboxylic acid functionality, significant quantities of the corresponding methyl ester by-product also formed. This esterification was determined to occur during the resin washing and pooling steps post cleavage. To eliminate these problems, we excluded the building block *t*-butyl isocyanate and the reductive amination capping procedure from the final target library production; 10% H_2O:THF was instead employed for final washes after cleavage to avoid the formation of methyl ester from carboxylic acid.

9.3.4 SYNTHESIS OF LIBRARY 2 AND QUALITY CONTROL ASSESSMENT

A 1000 ($10 \times 10 \times 10$) member library was designed to achieve maximum diversity at the R^1, R^2, and R^3 positions (Scheme 9.4; Table 9.8). Selection of building blocks was based on mimicking the natural product moiety, chemistry feasibility, and producing structural diversity in the library.

Resin activation, loading, and solid-phase azide reduction synthesis were carried out in bulk in a standard glass solid-phase peptide synthesizer vessel (500 mL). Solid-supported 5'-aminothymidine 6 (ca. 75 g) was prepared in a three-step sequence from PS-DES resin (60 g, 100–200 mesh, stated capacity: 1.45 mmol/g). A neutral buoyancy suspension of the solid-supported 5'-aminothymidine resin in CH_2Cl_2:THF (2:1) was evenly distributed into 1000 IRORI MiniKan reactors containing Rf tags to afford resin 6 (ca. 75 mg, 87 µmol per MiniKan).

The 1000 resin-loaded MiniKans with Rf tags were scanned and sorted into ten Erlenmeyer glass reaction flasks (1 L), at 100 MiniKans per flask. The scanning procedure allowed us to record and identify the unique reaction path for each MiniKan. Fmoc-amino acid coupling was done using the same HOBt-DIC coupling protocol as described in the 64-member rehearsal library, except that the corresponding excess molar amount of Fmoc-protected amino acids, HOBt, and DIC was

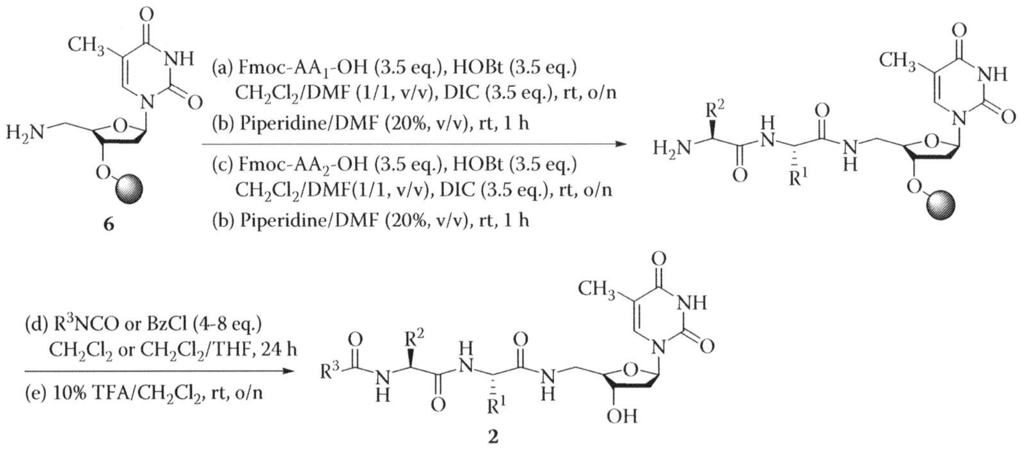

SCHEME 9.4 Synthesis of a 1000-member thymidinyl dipeptide urea library 2 in MiniKans.

TABLE 9.8
Building Blocks for Library 2 ($10 \times 10 \times 10$) Synthesis

Step	Building Block	{1}	{2}	{3}	{4}	{5}	{6}	{7}	{8}	{9}	{10}
1	Fmoc-AA₁-OH										
2	Fmoc-AA₂-OH	Ala	β-Ala	Ile	Met	Asp(OtBu)	Lys(Boc)	Asn(Trt)	Tyr(tBu)	Phe(4-NO₂)	Trp(Boc)
3	R³NCO for {1–9} R³Cl for {10}	*n*-Hex	*n*-Pr	*i-Pr*	Bn	Ph	4-CNPh	4-MeOPh	2,4-Dichlorophenyl	PhCH₂CH₂	Bz

changed from 5 equiv to 3.5 equiv to reduce cost. For the first AA$_1$ {1–10} attachment, the 100 MiniKans in each reaction flask were reacted with one Fmoc-amino acid, HOBt, and DIC (each 3.5 equiv, 0.12 M) in anhydrous CH$_2$Cl$_2$:DMF (1:1). After the reaction mixture was shaken at room temperature overnight, the flask was decanted and the MiniKans were washed twice with DMF and THF before all MiniKans were pooled together for common washes. After multiple washes with DMF, MeOH, THF, and CH$_2$Cl$_2$, the Fmoc protecting group was removed under 20% piperidine in DMF for 1 hr to provide free amines with AA$_1$ {1–10} input attached. All 1000 MiniKans were then directed sorted into ten reaction flasks (1 L) for the second Fmoc-amino acid coupling and subsequent Fmoc removal following the same acylation, deprotection, and washing protocol. All Mini-Kans were scanned once more and directed sorted into ten reaction flasks for the final capping step. Flasks 1 through 9 were treated with nine corresponding isocyanates in CH$_2$Cl$_2$ and flask 10 was reacted with benzoyl chloride in CH$_2$Cl$_2$ in the presence of diisopropylethylamine (DIPEA) base. After the capping reaction was complete, the MiniKans in each flask were then washed sequentially with DMF, THF, MeOH, CH$_2$Cl$_2$, and dried. The MiniKans were finally scanned and sorted into individual vessels and subjected to final cleavage and simultaneous sidechain deprotection with TFA:TIS:CH$_2$Cl$_2$ (10:2.5:90, by volume) (3 mL). The cleavage solution was decanted to a labeled glass vial, each MiniKan was washed twice with 3 mL of 10% H$_2$O:THF, and the combined solution was evaporated to dryness *in vacuo* to give thymidinyl dipeptide urea library 2 (Figure 9.10), which was archived in the same 8 × 12 array format.

Library characterization and analysis were performed using the following sampling protocol. Nineteen samples from two randomly selected rows (one horizontal and one vertical) per array were used for HPLC and MS analysis (190 samples for the library, 19% of library size), and five samples per array (50 samples for the library, 5% of library size) were selected for ^{1}H NMR analysis. All

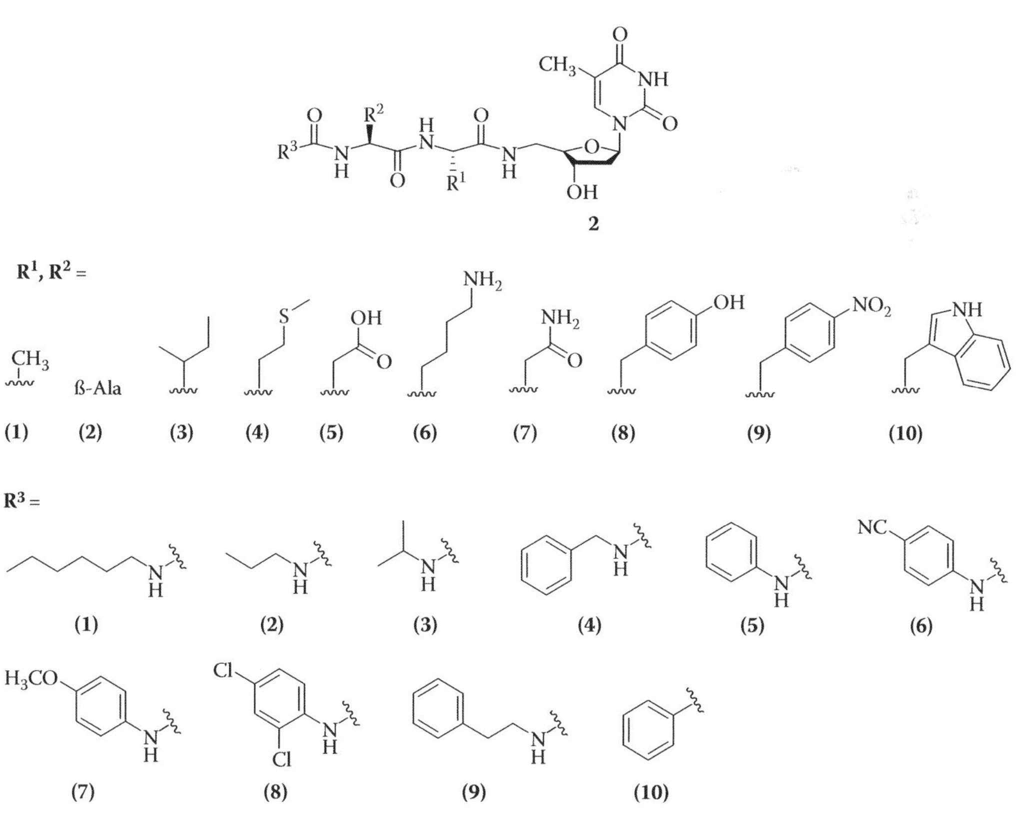

FIGURE 9.10   A 1000-member thymidinyl dipeptide urea library 2.

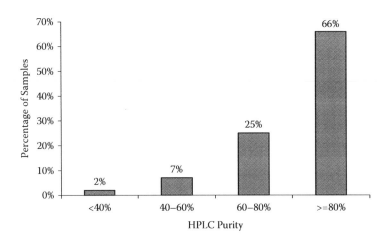

FIGURE 9.11   HPLC purity distribution of target library **2**.

analyzed samples gave the expected molecular weight and had a good ^{1}H NMR spectra. From HPLC analysis with UV_{254nm} detection, 66% of the analyzed samples contained the desired compounds with over 80% purity, 25% samples with a purity of 60% to 80%, 7% samples gave purity ranging from 40% to 60%, and 2% samples had purity less than 40% (Figure 9.11). The overall yield (71%) of this library was calculated based on the average weight of library members in the first two archived 8 × 12 arrays. This data correlates well with the predicted values based on loads determined in the loading experiment.

Sixteen of the 190 compounds analyzed had an HPLC purity of less than 60%. The structural analysis of these compounds is shown in Figure 9.12. Fourteen of them had a tryptophan residue at either R^1 or R^2 position. The lower purity was attributed to steric effects of this bulky residue and incomplete sidechain deprotection, resulting in lower yields in the coupling and cleavage reactions. Half of these lower purity samples possessed a phenethyl or benzoyl substituent at the R^3 position. The three compounds with an HPLC purity of less than 40% (2{*4,10,9*}, 2{*2,10,10*}, 2{*1,10,10*}) all have a tryptophan residue at the R^2 position.

9.3.5 BIOLOGICAL EVALUATION

This library has been screened for enzyme inhibition against the enzymes involved in mycobacterial L-rhamnose biosynthesis pathway and for whole cell MIC activity against *M. tuberculosis* (H37Rv). Initial hits from library screening were further purified by preparative HPLC and retested. The characterization and anti-tuberculosis activity of purified library members from primary screening hits are summarized in Table 9.9. The most active compounds identified were 2{*10,3,1*} and 2{*6,9,1*}, with MIC values of 50 µg/mL.

9.3.6 SUMMARY AND CONCLUSION

We developed a solid-phase synthesis approach to make a thymidinyl dipeptide urea library on PS-DES support using IRORI technology. Following optimized protocols, a 1000-member thymidinyl dipeptide urea library was synthesized with good purity and yield. After the biological evaluation of this library for anti-tuberculosis activity and inhibition of enzymes in the dTDP-L-rhamnose biosynthesis pathway, the most active library members identified were 2{*10,3,1*} and 2{*6,9,1*} with MIC values of 50 µg/mL. Overall, the use of IRORI MiniKans offered a number of advantages over filter plate synthesis, including more rapid synthesis, higher purity, and a larger yield of each

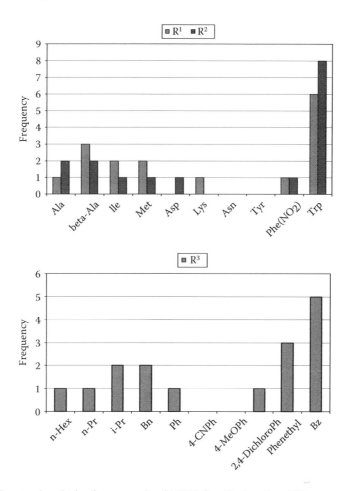

FIGURE 9.12 Structural analysis of compounds with HPLC purity less than 60%.

TABLE 9.9

Characterization and Anti-tuberculosis Activity of Purified Library Members

Library Members	MS [M+Na][a] (m/z)	HPLC Purity[b] (%)	t_R[c] (min)	MIC$_{90}$ (μg/mL)
2{10,1,1}	648.4	100	5.44	>200
2{10,9,2}	727.5	90	5.43	200
2{2,10,1}	648.4	100	5.38	100
2{10,3,1}	690.4	100	5.78	50
2{10,8,1}	740.5	100	5.53	100
2{6,9,1}	711.5	100	5.04	50
2{10,7,1}	691.4	100	5.26	200

[a] Mass spectra were recorded on a Bruker Esquire LC-MS using ESI.

[b] Analytical RP-HPLC was conducted on a Shimadzu HPLC system with a Phenomenex C18 column (100 Å, 3 μm, 4.6 × 50 mm), flow rate 1.0 mL/min, and a gradient of solvent A (water with 0.1% TFA) and solvent B (acetonitrile): 0–2.00 min 100% A; 2.00–8.00 min 0–100% B (linear gradient). UV detection at 254 nm.

[c] Retention time.

compound. The major drawbacks of this technique were the cost of IRORI MiniKans for an academic laboratory and increased difficulty in monitoring reaction progress. Although these libraries produced a small number of active compounds, these compounds were interesting and warranted subsequent follow-up. We believe that these product libraries have high intrinsic value for testing against other nucleoside binding enzymes for drug leads and chemical probes. Using the approaches described here, further more complex nucleoside libraries are currently in production and testing.

ACKNOWLEDGMENTS

This work was supported by grant AI057836 from the National Institutes of Health. We thank Victoria Jones and Dr. Michael McNeil at Colorado State University and Elizabeth Carson and Robin Lee at the University of Tennessee Health Science Center for assistance in antimicrobial screening of these libraries. We also thank Robin Lee for assistance in the preparation of this manuscript.

REFERENCES

1. World Health Organization. Antimicrobial Resistance, WHO Fact Sheet No 194. Geneva: Health Communications, WHO, 2002, http://www.who.int/inf-fs/en/fact194.html.
2. (a) Drews, J. *Science,* 2000, 287, 1960–1964. (b) Payne, D.J.; Gwynn, M.N.; Holmes, D.J.; and Pompliano, D.L. *Nature Rev. Drug Discov.,* 2007, 6, 29–40.
3. For a recent review, see: Butler, M.S. *Nat. Prod. Rep.,* 2005, 22, 162–195.
4. (a) Lee, V.J.; and Hecker, S.J. *Med. Res. Rev.,* 1999, 19, 521–542. (b) Kimura, K.; and Bugg, T.D.H. *Nat. Prod. Rep.,* 2003, 20, 252–273.
5. Rachakonda, S.; and Cartee, L. *Curr. Med. Chem.,* 2004, 11, 775–793.
6. For articles related to mureidomycin analogs, see (a) Gentle, C.A.; and Bugg, T.D.H. *J. Chem. Soc., Perkin Trans. 1,* 1999, 1279–1285. (b) Gentle, C.A.; Harrison, S.A.; Inukai, M.; and Bugg, T.D.H. *J. Chem. Soc., Perkin Trans. 1,* 1999, 1287–1294. (c) Bozzoli, A.; Kazmierski, W.; Kennedy, G.; Pasquarello, A.; and Pecunioso, A. *Bioorg. Med. Chem. Lett.,* 2000, 10, 2759–2763. (d) Howard, N.I.; Bugg, T.D.H. *Bioorg. Med. Chem.,* 2003, 11, 3083–3099. For articles related to pacidamycin analogues, see (e) Boojamra, C.G.; Lemoine, R.C.; Lee, J.C.; Léger, R.; Stein, K.A.; Vernier, N.G.; Magon, A.; Lomovskaya, O.; Martin, P.K.; Chamberland, S.; Lee, M.D.; Hecker, S.J.; and Lee, V.J. *J. Am. Chem. Soc.,* 2001, 123, 870–874. (f) Boojamra, C.G.; Lemoine, R.C.; Blais, J.; Vernier, N.G.; Stein, K.A.; Magon, A.; Chamberland, S.; Hecker, S.J.; and Lee, V.J. *Bioorg. Med. Chem. Lett.,* 2003, 13, 3305–3309.
7. For reviews, see (a) Lee, A.; and Breitenbucher, J.G. *Curr. Opin. Drug Disc. Dev.,* 2003, 6, 494–508. (b) Sanchez-Martin, R. M.; Mittoo, S.; and Bradley, M. *Curr. Top. Med. Chem.,* 2004, 4, 653–669.
8. (a) Merrifield, R.B. *J. Am. Chem. Soc.,* 1963, 85, 2149–2154. (b) Merrifield, B. *Science,* 1986, 232, 341–347.
9. For review, see Thompson, L.A.; and Ellman, J.A. *Chem. Rev.,* 1996, 96, 555–600.
10. *Solid phase Organic Syntheses*; Czarnik, A.W., Ed.; John Wiley & Sons, New York, 2001.
11. For a review regarding *Mycobacterium tuberculosis* cell envelope, see Lee, R.E.; Brennan, P.J.; and Besra, G.S. *Curr. Top. Microbiol. Immunol.,* 1996, 215, 1–27.
12. Ma, Y.; Pan, F.; and McNeil, M. *J. Bacteriol.,* 2002, 184, 3392–3395.
13. Giraud, M.-F.; and Naismith, J.H. *Curr. Opin. Struct. Biol.,* 2000, 10, 687–696.
14. A short communication regarding this library synthesis has been recently disclosed, see Sun, D.; and Lee, R.E. *Tetrahedron Lett.,* 2005, 46, 8497–8501.
15. For reviews, see (a) Domling, A.; and Ugi, I. *Angew. Chem., Int. Ed.,* 2000, 39, 3168–3210. (b) Ugi, I.; and Heck, S. *Comb. Chem. High Throughput Screening,* 2001, 4, 1–34. (c) Ugi, I. *Pure Appl. Chem.,* 2001, 73, 187–191. (d) Nair, V.; Rajesh, C.; Vinod, A.U.; Bindu, S.; Sreekanth, A.R.; Mathen, J.S.; and Balagopal, L. *Acc. Chem. Res.,* 2003, 36, 899–907. (e) Zhu, J. *Eur. J. Org. Chem.,* 2003, 1133–1144.
16. Ugi, I.; Meyr, R.; Fetzer, U.; and Steinbrückner, C. *Angew. Chem.,* 1959, 71, 386–388.
17. (a) Tempest, P.A.; Brown, S.D.; and Armstrong, R.W. *Angew. Chem., Int. Ed.,* 1996, 35, 640–642. (b) Kim, S.W.; Shin, Y.S.; and Ro, S. *Bioorg. Med. Chem. Lett.,* 1998, 8, 1665–1668. (c) Hoel, A. M.L.; and Nielsen, J. *Tetrahedron Lett.,* 1999, 40, 3941–3944. (d) Golebiowski, A.; Jozwik, J.; Klopfenstein,

S.R.; Colson, A.-O.; Grieb, A.L.; Russell, A.F.; Rastogi, V.L.; Diven, C.F.; Portlock, D.E.; and Chen, J.J. *J. Comb. Chem.,* 2002, 4, 584–590. (e) Gedey, S.; Van der Eycken, J.; and Fülöp, F. *Org. Lett.,* 2002, 4, 1967–1969. (f) Liu, L.; Li, C.P.; Cochran, S.; and Ferro, V. *Bioorg. Med. Chem. Lett.,* 2004, 14, 2221–2226. (g) Chapman, T.M.; Davies, I.G.; Gu, B.; Block, T.M.; Scopes, D.I.C.; Hay, P.A.; Courtney, S.M.; McNeill, L.A.; Schofield, C.J.; and Davis, B.G. *J. Am. Chem. Soc.,* 2005, 127, 506–507. (h) Lin, Q.; O'Neill, J.C.; and Blackwell, H.E. *Org. Lett.,* 2005, 7, 4455–4458. (i) Lin, Q.; and Blackwell, H.E. *Chem. Commun.,* 2006, 2884–2886.

18. For examples of solid-phase Ugi reactions, see (a) Lee, D.; Sello, J.K.; and Schreiber, S.L. *Org. Lett.,* 2000, 2, 709–712. (b) Cheng, J.-F.; Chen, M.; Arrhenius, T.; and Nadzan, A. *Tetrahedron Lett.,* 2002, 43, 6293–6295. (c) Campian, E.; Lou, B.; and Saneii, H. *Tetrahedron Lett.,* 2002, 43, 8467–8470. (d) Portlock, D.E.; Naskar, D.; West, L.; Ostaszewski, R.; and Chen, J.J., *Tetrahedron Lett.,* 2003, 44, 5121–5124. (e) Cristau, P.; Vors, J.-P.; and Zhu, J. *Tetrahedron Lett.,* 2003, 44, 5575–5578. (f) Henkel, B.; Sax, M.; and Dömling, A. *Tetrahedron Lett.,* 2003, 44, 7015–7018. (g) Hanyu, M.; Murashima, T.; Miyazawa, T.; and Yamada, T. *Tetrahedron Lett.,* 2004, 45, 8871–8874. (h) Gedey, S.; Van der Eycken, J.V.; and Fülöp, F. *Lett. Org. Chem.,* 2004, 1, 215–220.

19. Bannwarth, W. *Helv. Chim. Acta,* 1988, 71, 1517–1527.

20. (a) Bartra, M.; Romea, P.; Urpí, F.; and Vilarrasa, J. *Tetrahedron,* 1990, 46, 587–594. (b) Kick, E.K.; and Ellman, J.A. *J. Med. Chem.,* 1995, 38, 1427–1430. (c) Maltais, R.; Tremblay, M.R.; and Poirier, D. *J. Comb. Chem.,* 2000, 2, 604–614.

21. For a review regarding linker chemistries and cleavage strategies on solid support, see Guillier, F.; Orain, D.; and Bradley, M. *Chem. Rev.,* 2000, 100, 2091–2157.

22. PS-DES resin (100–200 mesh, loading: 0.74–1.58 mmol/g) is commercially available from Argonaut Technologies, now a Biotage company. For resin activation and loading of alcohols, see (a) Hu, Y.; Porco, J.A. Jr.; Labadie, J.W.; Gooding, O.W.; and Trost, B.M. *J. Org. Chem.,* 1998, 63, 4518–4521. (b) Dragoli, D.R.; Thompson, L.A.; O'Brien, J.; and Ellman, J.A. *J. Comb. Chem.,* 1999, 1, 534–539.

23. Kaiser, E.; Colescott, R.L.; Bossinger, C.D.; and Cook, P.I. *Anal. Biochem.,* 1970, 34, 595–598.

24. Chan, W.C.; and White, P.D. In *Fmoc Solid-Phase Peptide Synthesis: A Practical Approach*; Chan, W.C.; and White, P.D., Eds.; Oxford University Press, Oxford, 2000, p. 63.

25. König, W.; and Geiger, R. *Chem. Ber.,* 1970, 103, 788–798.

26. Krelaus, R.; and Westermann, B. *Tetrahedron Lett.,* 2004, 45, 5987–5990.

27. Hu, Y.; and Porco, J.A. Jr. *Tetrahedron Lett.,* 1998, 39, 2711–2714.

28. A full article describing this library synthesis has been published; see Sun, D.; and Lee, R.E. *J. Comb. Chem.,* 2007, 9, 370–385.

29. For examples of nucleoside library synthesis on solid support, see (a) Khan, S.I.; and Grinstaff, M.W. *Tetrahedron Lett.,* 1998, 39, 8031–8034. (b) Crimmins, M.T.; and Zuercher, W.J. *Org. Lett.,* 2000, 2, 1065–1067. (c) Bozzoli, A.; Kazmierski, W.; Kennedy, G.; Pasquarello, A.; and Pecunioso, A. *Bioorg. Med. Chem. Lett.,* 2000, 10, 2759–2763. (d) Choo, H.; Chong, Y.; and Chu, C.K. *Org. Lett.,* 2001, 3, 1471–1473. (e) De Champdoré, M.; De Napoli, L.; Di Fabio, G.; Messere, A.; Montesarchio, D.; and Piccialli, G. *Chem. Commun.,* 2001, 2598–2599. (f) Rodenko, B.; Wanner, M.J.; and Koomen, G.-J. *J. Chem. Soc., Perkin Trans. 1,* 2002, 1247–1252. (g) Epple, R.; Kudirka, R.; and Greenberg, W.A. *J. Comb. Chem.,* 2003, 5, 292–310. (h) Ding, Y.; Habib, Q.; Shaw, S.Z.; Li, D.Y.; Abt, J.W.; Hong, Z.; and An, H. *J. Comb. Chem.,* 2003, 5, 851–859. (i) Varaprasad, C.V.; Habib, Q.; Li, D.Y.; Huang, J.; Abt, J.W.; Rong, F.; Hong, Z.; and An, H. *Tetrahedron,* 2003, 59, 2297–2307. (j) Aucagne, V.; Berteina-Raboin, S.; Guenot, P.; and Agrofoglio, L.A. *J. Comb. Chem.,* 2004, 6, 717–723. (k) Ramasamy, K.S.; Amador, R.B.; Habib, Q., Rong, F.; Han, X.; Li, D.Y.; Huang, J.; Hong, Z.; and An, H. *Nucleosides Nucleotides Nucleic Acids,* 2005, 24, 1947–1970. (l) Rodenko, B.; Detz, R.J.; Pinas, V.A.; Lambertucci, C.; Brun, R.; Wanner, M.J.; and Koomen, G.-J. *Bioorg. Med. Chem.,* 2006, 14, 1618–1629. (m) Oliviero, G.; Amato, J.; Borbone, N.; D'Errico, S.; Piccialli, G.; and Mayol, L. *Tetrahedron Lett.,* 2007, 48, 397–400.

30. Furka, Á.; Hamaker, L.K.; and Peterson, M.L. In *Combinatorial Chemistry: A Practical Approach*; Fenniri, H., Ed.; Oxford University Press, Oxford, 2000, p. 1–31.

31. For general reviews regarding tagging or encoding technologies, see (a) Czarnik, A.W. *Curr. Opin. Chem. Biol.,* 1997, 1, 60–66. (b) Barnes, C.; Scott, R.H.; and Balasubramanian, S. *Rec. Res. Dev. Org. Chem.,* 1998, 2, 367–379. (c) Barnes, C.; and Balasubramanian, S. *Curr. Opin. Chem. Biol.,* 2000, 4, 346–350.

32. (a) Nicolaou, K.C.; Xiao, X.-Y.; Parandoosh, Z.; Senyei, A.; and Nova, M.P. *Angew. Chem., Int. Ed. Engl.*, 1995, 34, 2289–2291. (b) Xiao, X.–Y.; Zhao, C.; Potash, H.; and Nova, M.P. *Angew. Chem., Int. Ed. Engl.*, 1997, 36, 780–782. (c) Xiao, X.-Y.; Li, R.; Zhuang, H.; Ewing, B.; Karunaratne, K.; Lillig, J.; Brown, R.; and Nicolaou, K.C. *Biotechnol. Bioeng.*, 2000, 71, 44–50. For application examples of a 10,000-member IRORI library, see (d) Nicolaou, K.C.; Pfefferkorn, J.A.; Mitchell, H.J.; Roecker, A.J.; Barluenga, S.; Cao, G.-Q.; Affleck, R.L.; and Lillig, J.E. *J. Am. Chem. Soc.*, 2000, 122, 9954–9967. (e) Herpin, T.F.; Van Kirk, K. G.; Salvino, J.M.; Yu, S.T.; and Labaudinière, R.F. *J. Comb. Chem.*, 2000, 2, 513–521. For an example of a 25,000-member nucleoside library, see (f) Epple, R.; Kudirka, R.; and Greenberg, W.A. *J. Comb. Chem.*, 2003, 5, 292–310.

Index